Fracture mechanics of concrete: Structural application and numerical calculation

ENGINEERING APPLICATION OF FRACTURE MECHANICS
Editor-in-Chief: George C. Sih

G.C. Sih and L. Faria (eds.), Fracture mechanics methodology: Evaluation of structure components integrity. 1984. ISBN 90-247-2941-6.

E.E. Gdoutos, Problems of mixed mode crack propagation. 1984. ISBN 90-247-3055-4.

A. Carpinteri and A.R. Ingraffea (eds.), Fracture mechanics of concrete: Material characterization and testing. 1984. ISBN 90-247-2959-9.

G.C. Sih and A. DiTommaso (eds.), Fracture mechanics of concrete: Structural application and numerical calculation. 1984. ISBN 90-247-2960-2.

Fracture mechanics of concrete: Structural application and numerical calculation

Edited by

George C. Sih

Lehigh University,
Institute of Fracture and Solid Mechanics,
Bethlehem, PA 18015, USA

A. DiTommaso

University of Bologna
Bologna, Italy

1985 **MARTINUS NIJHOFF PUBLISHERS**
a member of the KLUWER ACADEMIC PUBLISHERS GROUP
DORDRECHT / BOSTON / LANCASTER

Distributors

for the United States and Canada: Kluwer Academic Publishers, 190 Old Derby Street, Hingham, MA 02043, USA
for the UK and Ireland: Kluwer Academic Publishers, MTP Press Limited, Falcon House, Queen Square, Lancaster LA1 1RN, England
for all other countries: Kluwer Academic Publishers Group, Distribution Center, P.O. Box 322, 3300 AH Dordrecht, The Netherlands

Library of Congress Cataloging in Publication Data

Main entry under title:

Fracture mechanics of concrete.

(Engineering application of fracture mechanics ; v. 4)
Includes bibliographical references and indexes.
1. Concrete--Fracture. 2. Fracture mechanics.
3. Structures, Theory of. 4. Numerical calculations.
I. Sih, G. C. (George C.) II. DiTommaso, A. III. Series.
TA440.F735 1984 620.1'366 84-3996

ISBN-13: 978-94-009-6154-8 e-ISBN-13: 978-94-009-6152-4
DOI: 10.1007/978-94-009-6152-4

Contents

Series on engineering application of fracture mechanics

Fracture mechanics technology has received considerable attention in recent years and has advanced to the stage where it can be employed in engineering design to prevent against the brittle fracture of high-strength materials and highly constrained structures. While research continued in an attempt to extend the basic concept to the lower strength and higher toughness materials, the technology advanced rapidly to establish material specifications, design rules, quality control and inspection standards, code requirements, and regulations for safe operation. Among these are the fracture toughness testing procedures of the American Society of Testing Materials (ASTM), the American Society of Mechanical Engineers (ASME) Boiler and Pressure Vessel Codes for the design of nuclear reactor components, etc. Step-by-step fracture detection and prevention procedures are also being developed by the industry, government and university to guide and regulate the design of engineering products. This involves the interaction of individuals from the different sectors of the society that often presents a problem in communication. The transfer of new research findings to the users is now becoming a slow, tedious and costly process.

One of the practical objectives of this series on *Engineering Application of Fracture Mechanics* is to provide a vehicle for presenting the experience of real situations by those who have been involved in applying the basic knowledge of fracture mechanics in practice. It is time that the subject should be presented in a systematic way to the practicing engineers as well as to the students in universities at least to all those who are likely to bear a responsibility for safe and economic design. Even though the current theory of linear elastic fracture mechanics (LEFM) is limited to brittle fracture behavior, it has already provided a remarkable improvement over the conventional methods not accounting for initial defects that are inevitably present in all materials and structures. The potential of the fracture mechanics technology, however, has not been fully recognized. There remains much to be done in constructing a quantitative theory of material damage that can reliably translate small specimen data to the design of large size structural components. The work of the physical metallurgists and the fracture mechanicians should also be brought together by reconciling the details of the material microstructure with the assumed continua of the computational methods. It is with the aim of developing a wider appreciation of the fracture mechanics technology applied to the design of engineering structures such as aircrafts, ships, bridges, pavements, pressure vessels, off-shore structures, pipelines, etc. that this series is being developed.

Undoubtedly, the successful application of any technology must rely on the soundness of the underlying basic concepts and mathematical models and how they reconcile with each other. This goal has been accomplished to a large extent by the book series on *Mechanics of Fracture* started in 1972. The seven published volumes offer a wealth of information on the effects of defects or cracks in cylindrical bars, thin and thick plates, shells, composites and solids in three dimensions. Both static and dynamic loads are considered. Each volume contains an introductory chapter that illustrates how the strain energy criterion can be used to analyze the combined influence of defect size, component geometry and size, loading, material properties, etc. The criterion is particularly effective for treating mixed mode fracture where the crack propagates in a non-self similar fashion. One of the major difficulties that continuously perplex the practitioners in fracture mechanics is the selection of an appropriate fracture criterion without which no reliable prediction of failure could be made. This requires much discernment, judgement and experience. General conclusion based on the agreement of theory and experiment for a limited number of physical phenomena should be avoided.

Looking into the future the rapid advancement of modern technology will require more sophisticated concepts in design. The micro-chips used widely in electronics and advanced composites developed for aerospace applications are just some of the more well-known examples. The more efficient use of materials in previously unexperienced environments is no doubt needed. Fracture mechanics should be extended beyond the range of LEFM. To be better understood is the entire process of material damage that includes crack initiation, slow growth and eventual termination by fast crack propagation. Material behavior characterized from the uniaxial tensile tests must be related to more complicated stress states. These difficulties could be overcome by unifying metallurgical and fracture mechanics studies, particularly in assessing the results with consistency.

This series is therefore offered to emphasize the applications of fracture mechanics technology that could be employed to assure the safe behavior of engineering products and structures. Unexpected failures may or may not be critical in themselves but they can often be annoying, time-wasting and discrediting of the technical community.

Bethlehem, Pennsylvania G.C. Sih
1984 Editor-in-Chief

Editors' preface

Concrete has traditionally been known as a material used widely in the construction of roads, bridges and buildings. Since cost effectiveness has always been one of the more important aspects of design, concrete, when reinforced and/or prestressed, is finding more use in other areas of application such as floating marine structures, storage tanks, nuclear vessel containments and a host of other structures. Because of the demand for concrete to operate under different loading and environmental conditions, increasing attention has been paid to study concrete specimens and structure behavior. A subject of major concern is how the localized segregation of the constituents in concrete would affect its global behavior. The degree of nonhomogeneity due to material property and damage by yielding and/or cracking depends on the size scale and loading rate under consideration. Segregation or clustering of aggregates at the macroscopic level will affect specimen behavior to a larger degree than it would to a large structure such as a dam. Hence, a knowledge of concrete behavior over a wide range of scale is desired. The parameters governing micro- and macro-cracking and the techniques for evaluating and observing the damage in concrete need to be better understood. This volume is intended to be an attempt in this direction.

The application of Linear Elastic Fracture Mechanics to concrete is discussed in several of the chapters. Depending on the specimen size, loading rate and type, concrete can behave linearly up to fracture or nonlinearly to failure by plastic collapse and/or fracture. Such behavior can be generally observed in other materials as well. This indicates the importance of identifying damage in concrete with load history as a path *dependent* process.

Chapter 1 considers modeling crack extension in concrete by the finite element method. An approach assuming that fracture is simulated by a smeared crack band in concrete is presented. The model aims to reflect the densely distributed cracks and to offer computational convenience. Energy consumed in the crack band and the remaining portion of the specimen is calculated. Strain-softening is included in the analysis such that the results exhibit the gradual transition from failure predicted by strength at one extreme and by linear fracture mechanics at the other. In most situations, failure occurs in this transition range depending on the structural size. The stability aspect of fracture is also analyzed by focusing attention on strain localization instability and crack spacing. Furthermore a study is proposed of triaxial strain-softening and triaxial constitutive relations for the fracture process zone. An exhaustive amount of references is given at the end of this chapter.

Chapter 2 applies the principle of dimensional analysis to investigate crack behavior in concrete. Fracture or damage patterns in the large and small structural components are recognized to be different. The large structure can fail by brittle fracture, while the smaller structure can fail by plastic collapse. A brittleness number is defined which is directly dependent on the critical stress intensity factor and inversely on the beam height and yield strength. Bounds on this number are presented for failure by plastic collapse and brittle fracture. The idea is also extended to treat fracture stability in plain and reinforced concrete structural members.

Chapter 3 emphasizes the need to include softening due to the damage in the fracture zone for describing concrete behavior. Such data can be collected from a simple tension test. Application of the single crack model in Linear Elastic Fracture Mechanics may not lead to realistic results because of the inability to identify a real crack in practice. The concept of fictitious crack model is discussed. Data on material properties are still lacking for making reliable theoretical predictions.

The application of finite element to treat crack extension in concrete is presented in Chapter 4. A revival of interest considering the discrete character of cracking is discussed in contrast to the smeared crack approach. This is mainly due to the advent of interactive graphics providing capability in remeshing of elements that were not available previously. The tensile strength criterion is compared with the critical stress intensity factor approach in Linear Elastic Fracture Mechanics. Crack growth direction can be predicted from the stationary values of the strain energy density criterion. The results are compared with those obtained from the maximum circumferential stress and maximum energy release rate criteria. Discussed is an example on the cracking of a full scale dam. This involved stress and failure analysis in three dimensions.

Chapter 5 deals with the fracture of steels used for reinforcing and prestressing concrete. Such information is essential for understanding the load transfer character between the steel and concrete which can significantly alter the structure behavior. The break down of the interfacial bond between steel and concrete is emphasized. Analysis of defects can be involved as it may require the three-dimensional stress analysis with elastoplastic constitutive relations. Initiation and growth of fatigue cracks are also discussed in connection with aggressive environments.

The valuable time spent by the authors to complete this work is acknowledged. The contribution belongs solely to those who have made the publication of this volume possible.

Lehigh University G.C. Sih
University of Bologna A. DiTommaso
April 1983

Contributing authors

Z.P. Bažant
Northwestern University, Evanston, Illinois

A. Carpinteri
University of Bologna, Bologna, Italy

M. Elices
Universidad Politécnica de Madrid, Madrid-3, Spain

A. Hillerborg
Lund Institute of Technology, Lund 7, Sweden

A.R. Ingraffea
Cornell University, Ithaca, New York

V. Saouma
University of Pittsburgh, Pittsburgh, Pennsylvania

Z.P. BAŽANT

1

Mechanics of fracture and progressive cracking in concrete structures

1.1 Introduction

Cracking is an essential feature of the behavior of concrete structures. Even under service loads, concrete structures are normally full of cracks. Clearly, cracking should be taken into account in predicting ultimate load capacity as well as behavior in service.

To fracture mechanics specialists, it appears natural that concrete structures should be designed according to fracture mechanics. Yet, none of the existing code provisions are based on fracture mechanics. The reason is not ignorance on the part of concrete engineers. Fracture mechanics analysis was tried, and was found to yield predictions that deviate from measurements, on the average, at least as much as those based on the tensile strength approach. These were, however, predictions of the classical, linear fracture mechanics.

In various recent studies, especially those at the Technical University of Lund, Northwestern University and Politecnico di Milano, it became apparent that fracture mechanics does work for concrete, provided that one uses a proper, nonlinear form of fracture mechanics in which a finite nonlinear zone at the fracture front is taken into account. This may be done in various ways. In the first part of the present work, an exposition of one particularly efficient approach will be given. In this approach [8], which is based on the work recently pursued with success by a group of researchers at Northwestern University and Politecnico di Milano, cracking is modeled in a continuous, or smeared manner, and fracture is treated as a propagation of a smeared crack band through concrete. Continuous modeling of cracks in concrete, introduced by Rashid [119], has become popular in finite element analysis, not just because it reflects the reality of densely distributed cracks, but mainly because it is computationally convenient.

In the present engineering practice, tensile strength is used as the cracking criterion. This criterion, however, does not give objective results and does not agree with fracture tests. Remedy can be obtained by introducing an energy criterion. This approach will be described in detail, along with the finite element implementation, comparisons with fracture tests, and some examples of application. Considered will be the consequence for the structural size effect, and how this effect should be manifested in code formulas for brittle failures, such as the diagonal shear failure of beams. Furthermore, the stability aspects of fracture will be analyzed focusing attention on the strain localization instability as well as crack spacing. Finally, the conclusion will center on a more fundamental study of strain-softening triaxial constitutive relations for the fracture process zone.

The principal intent of this work* is to highlight various new research directions, rather than present a systematic review and description of all the existing knowledge.

1.2 Blunt crack band theory

Basic hypothesis. The analysis which follows is based on the hypothesis that fracture in a heterogeneous material such as concrete can be modeled as a band of parallel, densely distributed microcracks having a blunt front [1–3]. This hypothesis may be justified as follows.

For the purpose of analysis, a heterogeneous material is approximated by an equivalent homogeneous continuum. One must then distinguish the continuum stresses and strains (macrostresses and macrostrains) from the actual stresses and strains in the microstructure, called the microstresses and microstrains. In the theory of randomly inhomogeneous materials, the homogenized continuum stresses and strains are defined as the averages of the microstresses and microstrains over a certain representative volume (Figure 1.1). Its size must be sufficiently large compared to the size of the inhomogeneities. Even for a crude description, this size must be considered to be at least several times the size of inhomogeneities, i.e., several times the maximum aggregate size.

In the usual analysis, only the average elastic (or inelastic) material properties are considered and the geometry of the microstructure with the differences in the elastic constants between the aggregate and the cement paste is not taken into account. The detailed distribution of stress or strain over distances less than several times the aggregate size (Figure 1.1). is then meaningless, and only the stress resultants and the accumulated strain over the cross section of the representative volume have physical meaning. In finite element analysis, it makes, therefore, no sense to use finite elements smaller than several aggregate sizes. In case of fracture, this further means that if an equivalent homogeneous con-

* This work was partially supported under AFOSR grant 83-0009 to Northwestern University.

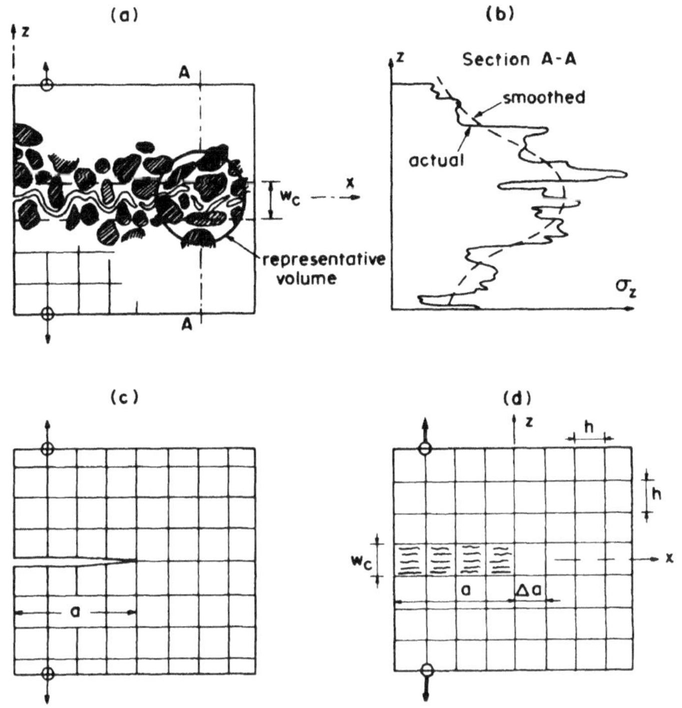

Fig. 1.1. Random microstructure, scatter of microstresses, and crack band or sharp crack model.

tinuum is assumed, it makes no sense to consider concentrations of stress (or of microcrack density) within volumes less than several aggregate sizes (Figure 1.1).

Similarly, a straight-line crack is an approximation. The actual crack path in concrete is not smooth but highly tortuous, since the crack tends to pass around the hard aggregate pieces and randomly sways to the side of a straight path by distances roughly equal to the aggregate size (Figure 1.1). Therefore, the actual stress (microstress) variation over such distances cannot be relevant for the macroscopic continuum model.

In view of the foregoing arguments, one should not subdivide the width of the crack band front into several finite elements. There is, however, also another reason. A strain-softening continuum is unstable and exhibits a strain-localization instability [4,5], in which the deformation localizes into one of the elements across the width of the crack band front. This instability will be illustrated later.

For an elastic material in which the stress drops suddenly to zero at the fracture front (Figure 1.2), it is found (regardless of the aggregate size) [6,7] that a sharp interelement crack and a smeared crack band in a square mesh (without any singularity elements) give essentially the same

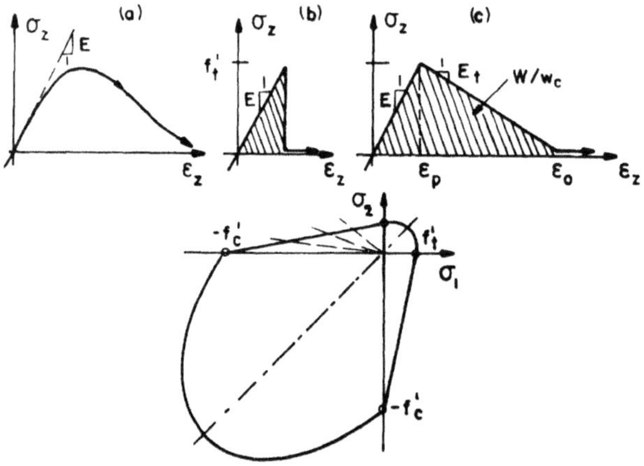

Fig. 1.2. Gradual strain-softening and sudden stress drop (a–c), and biaxial failure envelope (d).

results for the energy release rate and agree closely (within a few percent) with the exact elasticity solution, provided that the finite element is not larger than about 1/15 of the cross section dimension. To demonstrate it here, Figure 1.3 shows some of the numerical results for a line crack (left) and crack band (right) [6]. The finite element mesh covers a cut-out of an infinite elastic medium loaded at infinity by uniform normal stress $\bar{\sigma}$ perpendicular to a line crack of length 2a. The nodal loads applied at the boundary are calculated as the resultants (over the element width) of the exact stresses in the infinite medium at that location. Westergaard's exact solution is shown as the solid curve. The data points show the calculated results for the square mesh shown (mesh A), as well as for meshes B and C (not shown) in which the element size is reduced to 1/2 and 3/8, respectively. Each element consisted of two constant-strain triangles (and calculations were made for $\bar{\sigma} = 0.981\ \alpha$ (MPa), $E_c = 2256$ MPa, $\nu = 0.2$, and stress intensity factor 0.6937 $\mathrm{MNm}^{-3/2}$).

A similar equivalence of line crack and blunt crack band may be expected when a gradual stress drop is considered (Figure 1.2). This is confirmed in Figure 1.3c by the fact that a reduction of mesh size does not affect the results. The reason for this equivalence is the fact that fracture propagation depends essentially on the flux of energy into the fracture process zone at the crack front, which represents a global characteristic of the entire structure and depends little on the detailed distributions of stress and strain near the fracture front.

It may be also noted that the results for the stress intensity factor [6] obtained with nonsingular finite elements agree with the exact elasticity solution quite closely, usually within 1% for typical meshes. There is no need to use singularity elements in fracture analysis. Moreover, one

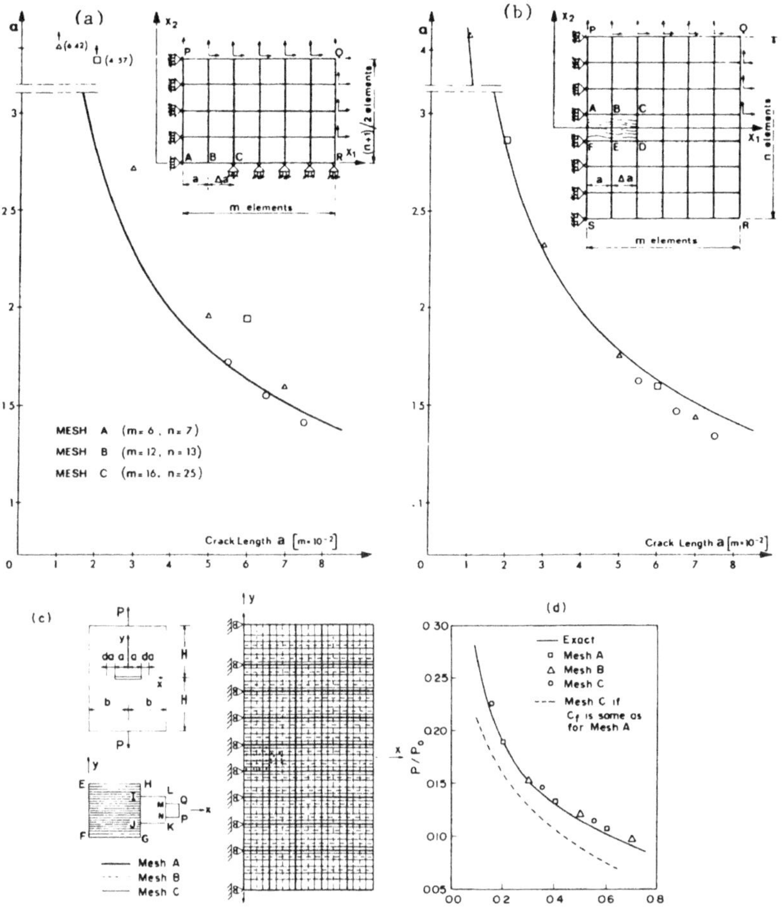

Fig. 1.3. Finite element results of Bažant and Cedolin (1979) for sudden stress drop (a, b), showing equivalence of blunt crack band and sharp crack modeling, and results of Bažant and Oh (c, d), showing the results for gradual stress decrease and meshes of different size.

should realize that the nonuniform stress distribution implied in a singularity element is meaningful only if this element is many times larger than the representative volume, i.e., is at least 20 aggregate pieces in size, which is too large for most applications.

Because the line crack and the crack band models are essentially equivalent, the choice of one or the other is principally a question of computational convenience. The line crack model appears to be less convenient. When a line crack extends through a certain node, the node must be split into two nodes. This increases the total number of nodes and changes the topological connectivity of the mesh. Unless one renumbers the nodes, the band structure of the structural stiffness matrix is destroyed. All this complicates programming. Furthermore, when the direction in which a Mode I crack should extend is not known in advance, one must make trial calculations for various possible locations of the node ahead of the crack front through which the crack should pass, in order to determine the location which gives the maximum energy release rate.

The smeared cracking approach, introduced by Rashid [8] avoids these difficulties. The cracking is modeled simply by changing the isotropic elastic moduli matrix to an orthotropic one, reducing the material stiffness in the direction normal to the cracks in the band. This is easily implemented by finite elements. Moreover, a crack propagation in an arbitrary direction with respect to the mesh lines, or a crack following a curved path, may be easily modeled as a zig-zag crack band (Section 1.3) whose overall path through the mesh approximates the actual crack path. Another advantage of the crack band model is that the known properties of stress-strain relations and failure envelopes can be applied to fracture; this includes, e.g., the effect of the compressive normal stress parallel to the crack, or the effect of creep. Still another advantage of the crack band model is the fact that, as it will be shown in the sequel, one can treat the case when principal stress directions in the fracture process zone rotate during the progressive fracture formation, i.e., during the strain softening,

Ahead of the tip of a propagating major crack in concrete, there is always a relatively large zone of discontinuous microcracks. Formation of microcracks at the fracture front has recently been observed experimentally [9–11]. From measurements of tensile strain fields by Moiré interferometry [10,11], it appears that there is at the fracture front a zone of microcracks whose width is about the aggregate size. From microscopic observations, it seems that the larger, easily discernible microcracks are not spread over a band of a large width but are concentrated essentially on a line. However, the line along which the microcracks are scattered is not straight but is highly tortuous (Figure 1.1), deviating to each side of the overall fracture axis by a distance equal roughly to the aggregate size, as the crack is trying to pass around the harder aggregate pieces. In the equivalent, smoothed macroscopic continuum which is implied in structural analysis, the scatter in the locations of visible microcracks is characterized by a microcrack band better than by a straight row of microcracks. Further, it should be realized that the boundary of the fracture process zone should not be defined as the boundary of visible microcracks but as the boundary of the strain-softening region, i.e., the region in which the maximum stress decreases with increasing maximum strain. Since the strain-softening is caused not just by microcracking but also by unobservable bond ruptures and submicroscopic flaws, the fracture process zone is probably much wider (as well as longer) than the region of

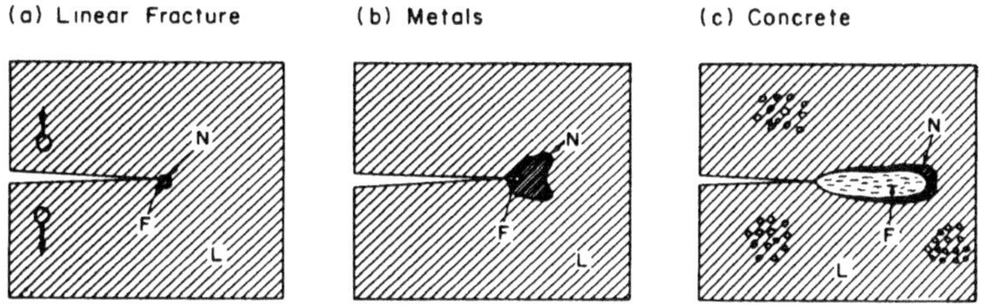

Fig. 1.4. Shape of nonlinear zone (N) and fracture process zone (F).

visible microcracks. Measurements of Cedolin et al. [10,11] appear to support this view. However, because of the foregoing justifications, the question of the actual shape of the microcrack zone is unimportant for the macroscopic continuum modeling.

In ductile fracture of metals, there is a large plastic zone in which the metal is yielding but does not undergo strain-softening, and the strain-softening zone itself (i.e., the fracture process zone) is only a small part of the yielding zone (Figure 1.4). On the other hand, in concrete, the nonlinear zone is not much larger than the strain-softening zone (Figure 1.4) since concrete in tension is not capable of plastic deformation. Thus, one may consider that the concrete outside the fracture process zone behaves as nearly elastic.

If the relation of the normal stress σ_z and the relative displacement δ_f across a line crack is identical to the relation of σ_z to the displacement δ_f obtained by accumulating the strains ε_f due to microcracking over the width w_c of the crack band [1], then, according to the above reasoning, the line crack model and the crack band model are essentially equivalent. Line crack models with softening stress-displacement relations were proposed in many previous works [12–18]. These include especially the works of Knauss [16], Wnuk [17], and Kfouri, Miller and Rice [14,15] on polymers and metals, which considered a gradual release of the forces between the opposite surfaces as the opening displacements grow, and modeled it by a gradual decrease of the internodal force as one node in a finite element grid is being separated by fracture into two nodes. For concrete, the concept of a gradually decreasing stress-displacement relation was first applied in the outstanding original work of Hillerborg, Modéer and Petersson [18,19] in their model of fictitious sharp interelement crack. Their pioneering work provided inspiration for developing the present blunt crack model with gradual strain-softening.

Stress-strain relation for fracture process zone. Cartesian coordinates $x_1 = x$, $x_2 = y$, and $x_3 = z$ will be introduced with the cracks being assumed to be normal to the axis z. The normal stress and strain components may be grouped into the column matrices $\boldsymbol{\sigma} = (\sigma_x, \sigma_y, \sigma_z)^{\mathrm{T}}$, $\boldsymbol{\varepsilon} = (\varepsilon_x, \varepsilon_y, \varepsilon_z)^{\mathrm{T}}$, where T denotes the transpose. The strains are assumed to be linearized or small. The elastic stress-strain relation for the normal components may then be written as $\boldsymbol{\sigma} = \boldsymbol{D}\boldsymbol{\varepsilon}$, in which $\boldsymbol{D}$ is the stiffness matrix of the uncracked material,

$$\boldsymbol{D} = \begin{bmatrix} D_{11} & D_{12} & D_{13} \\ & D_{22} & D_{23} \\ \text{sym.} & & D_{33} \end{bmatrix}. \tag{1.1}$$

If the elastic material is intersected by continuously distributed parallel cracks normal to z, the stress-strain relation has the form [8,20] $\boldsymbol{\sigma} = \boldsymbol{D}^{\mathrm{fr}}\,\boldsymbol{\varepsilon}$, in which

$$\boldsymbol{D}^{\mathrm{fr}} = \begin{bmatrix} D_{11} - D_{13}^2 D_{33}^{-1}, & D_{12} - D_{13} D_{23} D_{11}^{-1}, & 0 \\ & D_{22} - D_{23} D_{33}^{-1}, & 0 \\ \text{sym.} & & 0 \end{bmatrix}. \tag{1.2}$$

This matrix, representing the stiffness matrix of a fully cracked material, may be derived from the condition that strain $\varepsilon_z^{\mathrm{m}}$ of the material between the crack is unrelated to ε_z (except for $\varepsilon_z^{\mathrm{m}} < \varepsilon_z$) and that the stress normal to the cracks must be zero, assuming that the material between the cracks behaves as an uncracked elastic material in a plane stress state. This, of course, is a simplification, because the material between the cracks becomes damaged by discontinuous microcracks.

To describe progressive development of microcracks in the fracture process zone, it is necessary to formulate a stiffness matrix which continuously changes from the form given in Eq. 1.1 to that in Eq. 1.2. This objective is hard to achieve by direct reasoning, since every element of the 6×6 stiffness matrix changes. It was found [1,3], however, that the task becomes easier if the compliance matrix, $\boldsymbol{C}$, is used. For a crack-free material,

$$\boldsymbol{\varepsilon} = \boldsymbol{C}\boldsymbol{\sigma}, \tag{1.3}$$

where

$$\boldsymbol{C} = \boldsymbol{D}^{-1} = \begin{bmatrix} C_{11} & C_{12} & C_{13} \\ & C_{22} & C_{23} \\ \text{sym.} & & C_{33} \end{bmatrix} \tag{1.4}$$

Now, if only cracks normal to x are permitted, the appearance of cracks at constant stresses increases only the overall strain ε_z normal to the cracks and has no effect on the lateral strains ε_x and ε_y. Therefore, the compliance matrix after appearance of partial discontinuous cracks should have the form [1,3]:

$$\boldsymbol{C}(\mu)=\begin{bmatrix} C_{11} & C_{12} & C_{13} \\ & C_{22} & C_{23} \\ \text{sym.} & & C_{33}\mu^{-1} \end{bmatrix} \tag{1.5}$$

where μ is a certain parameter, called cracking parameter, which increases C_{33}. This formulation must, in the limit, be equivalent to the well-known generally accepted stiffness matrix $\boldsymbol{D}^{\mathrm{fr}}$ for a fully cracked material (Eq. 1.2). Indeed, as generally proved in [1], the matrix in Eq. 1.2 results as

$$\boldsymbol{D}^{\mathrm{fr}}=\lim_{\mu\to 0}\boldsymbol{C}^{-1}(\mu). \tag{1.6}$$

In writing a computer program, it is convenient for the programmer to note that instead of setting $\mu=0$, one may assign in the program μ^{-1} as a very large number (e.g., 10^{40}) and let the computer carry out the inversion of the matrix numerically; the result is a stiffness matrix like that in Eq. 1.2 except that extremely small numbers (10^{-40}) are obtained instead of 0.

A characteristic feature of the compliance matrix for progressive microcracking (Eq. 1.4) is the fact that cracking has no effect on lateral strains. This can be true only if all microcracks are normal to axis z, which is certainly a simplification. In reality, a certain distribution of the orientations of the microcracks is expected, the orientation normal to axis x being just the prevalent one, not the only one. If inclined microcracks were considered, than it would be necessary to also introduce a gradual change of the off-diagonal terms in Eq. 1.4 as the formation of microcracks advances.

Comparing now the compliance matrices in Eqs. 1.5 and 1.6, it is seen that a continuous transition from a crack-free state to a fully cracked state may be very simply obtained by continuously varying the cracking parameter μ between the limits

$$\text{uncracked state;}\quad \mu=1 \quad \text{and} \quad \text{fully cracked state:}\quad \mu=0 \tag{1.7}$$

The law governing the variation of the cracking parameter, μ, may be determined on the basis of the uniaxial tensile test. It has been proven independently by several investigators [21–25] that concrete exhibits tensile strain-softening, i.e., a gradual decrease of stress at increasing strain. Tests of strain-softening are stable only if the loading frame is much stiffer than the specimen and if the specimen is not too long. The observed stress-strain relation appears [21,22,24,26] to be smoothly curved. Although a curved strain-softening can be easily implemented with parameter μ, we assume, for the sake of simplicity, a bilinear stress-strain relation (Fig. 1.2), the declining (strain-softening) branch of which is characterized by negative compliance C_{33}^{t}. For uniaxial tension σ_z it follows that

$$\varepsilon_z = C_{33}\mu^{-1}\sigma_z \quad \text{or} \quad \sigma_z = C_{33}\mu^{-1}\varepsilon_z \tag{1.8}$$

which must be equivalent to the relation $\sigma_z = (\varepsilon_z - \varepsilon_0)C_{33}^{t}$ for the straight-line softening in which C_{33}^{t} is negative and ε_0 represents the terminal point of the strain-softening branch at which the tensile stress vanishes (Fig. 1.2). This point is related to the strain ε_p at the peak stress point, $\varepsilon_0 = \varepsilon_p + (-C_{33}^{t})f_c'$. Comparing the foregoing two expressions for σ_z, the following result is obtained [1]:

$$\mu^{-1} = \frac{-C_{33}^{t}}{C_{33}} \frac{\varepsilon_z}{\varepsilon_0 - \varepsilon_z} \tag{1.9}$$

as the law governing the variation of cracking parameter μ. Substituting Eq. 1.9 and Eq. 1.5 and inverting the matrix, the stiffness matrix $\boldsymbol{D}$ to be used in the finite element program is then obtained.

Concrete may be considered to be isotropic. It follows that

$$C_{33} = 1/E, \quad C_{33}^{t} = 1/E_t \tag{1.10}$$

where E = Young's modulus, and the compliance matrix for partially cracked concrete takes the following special form:

$$\boldsymbol{C}(\mu) = \frac{1}{E}\begin{bmatrix} 1 & -\nu & -\nu \\ -\nu & 1 & -\nu \\ -\nu & -\nu & \mu^{-1} \end{bmatrix} \tag{1.11}$$

and the limit of its inverse at $\mu \to 0$ is

$$\boldsymbol{D}^{\mathrm{fr}} = \frac{E}{1-\nu^2}\begin{bmatrix} 1 & \nu & 0 \\ \nu & 1 & 0 \\ 0 & 0 & 0 \end{bmatrix} \tag{1.12}$$

in which E = Young's modulus and ν = Poisson's ratio.

In computer finite element analysis, it is most convenient to use the incremental loading technique. For this purpose, the incremental stress-strain relations may be obtained differentiating Eq. 1.3 in which μ from

Eq. 1.9 is substituted. In a finite element program, it is also necessary to enlarge the compliance and stiffness matrices to a 6×6 form, including the rows and columns for shear strains and stresses. Most simply, these may be considered to be the same as for a crack-free material, except that the shear stiffness in the diagonal term needs to be reduced by an empirical shear stiffness reduction factor [27,20]. More accurately, the columns and rows for the shear behavior should reflect the frictional-dilatant properties of cracks; see, e.g., [28,29]. The question of shear terms is, however, often unimportant since fracture is geomaterials usually occurs in principal stress directions.

During the passage of the fracture process zone through a fixed station, the principal stress directions usually do not rotate significantly. This justifies another simplifying assumption which has been implied in the preceding formulation. It is the fact that we use total stress-strain relations (Eqs. 1.3 and 1.5) which are path-independent. In reality, all inelastic behavior is path-dependent. Nevertheless, the assumption of path-independence of the stress-strain relation in the vicinity of the fracture front may be adequate for many situations. Note that the microplane model outlined in the sequel provides the possibility to take into account the path-dependence if one is willing to accept a more complicated method of analysis (Eq. 1.93 in the sequel).

The difference of the total strain ε_z at the strain-softening branch from the strain predicted for an uncracked material, i.e., $\varepsilon_f = \varepsilon_z - \sigma_z/E$, represents the strain which is caused by microcracking. If this strain is integrated over the width of the crack band, i.e., $\delta_f = \varepsilon_f w_c$, one may obtain from our stress-strain relation a stress-displacement relation. For models in which the fracture is treated as a sharp interelement crack, this displacement is analogous to the opening displacement, δ_f, of such a crack. In this sense, the present theory is equivalent to previous models based on stress-displacement relations, especially the model of Hillerborg et al. [18,19].

The blunt crack band approach lends itself logically to describing the effect on fracture of the triaxial stress state in the vicinity of the crack front. From extensive testing, it is known that in the presence of transverse normal compression stresses, the tensile strength is diminished [27,30,31,1,32]. The measured biaxial failure envelope (Figure 1.2d) seems to consist approximately of a straight line which connects the failure point for uniaxial tensile failure to that for uniaxial compression failure in the (σ_x, σ_y) plane. Accordingly, we may suppose that transverse compressive stresses reduce the peak stress f_t' to the value f_{tc}' given as:

$$\text{for } \Delta f_t' \leqslant 0: \quad f_{tc}' = f_t' + \mathrm{k}(\sigma_x + \sigma_y)$$

$$\text{for } \Delta f_t' > 0: \quad f_{tc}' = f_t' \qquad (1.13)$$

where $f_t' =$ uniaxial tensile strength, $f_c' =$ uniaxial compression strength, and $k = f_t'/f_c'$.

It is worth noting that if μ is replaced by $1 - \omega$, then ω resembles the damage parameter used in the so-called continuous damage mechanics, which has recently been applied to concrete [33–37]. There is, however, a fundamental difference in that the damage due to microcracking is considered to be inseparable from a zone of a certain characteristic width, w_c, which is a fixed parameter, a material property.

Fracture characteristics. The fracture energy is defined as the energy consumed by crack formation per unit area of the crack plane. It may be calculated as

$$G_f = W_f w_c, \quad W_f = \int_0^{\varepsilon_0} \sigma_z \mathrm{d}\varepsilon_z \tag{1.14}$$

in which w_c is the width of the crack band (fracture process zone) and W_f the work of tensile stress per unit volume which is equal to the area under the tensile stress-strain curve (Figure 1.2).

In theory, it should be possible to determine the crack band width w_c by analyzing the strain-localization instability that leads to fracture. It should be possible to do this by extending the previous simple analysis of this instability [4,5]. The practical calculation would be, however, quite complicated in case of a large fracture process zone with a nonhomogeneously stressed specimen. Aside from that, if both W_f and G_f are considered as constants, w_c should also be a constant. This constant may be determined empirically. For the bilinear tensile stress-strain relation (Figure 1.2), one has

$$W_f = \tfrac{1}{2}(C_{33} - C_{33}^t) f_t'^2 w_c = \tfrac{1}{2} f_t' \varepsilon_0, \quad \text{or} \quad w_c = \frac{2G_f}{f_t'^2} \frac{1}{C_{33} - C_{33}^t} \tag{1.15}$$

in which C_{33}^t is negative. For isotropic material, $C_{33} = 1/E$, $C_{33}^t = 1/E_t$. This equation indicates that the width of the fracture process zone, precisely, the effective width corresponding to a uniform transverse distribution of tensile strain over the crack band, may be determined by measuring the softening compliance, the tensile strength, and the fracture energy. To ensure that C_{33}^t be negative, Eq. 1.15 requires that

$$w_c < w_0, \qquad \text{where } w_0 = \frac{2C_f}{f_t'^2 C_{33}} \quad \text{or} \quad w_0 = \frac{2G_f E}{f_t'^2}. \tag{1.16}$$

Note that the expression for w_0 is similar to the well-known Irwin's expression for the size of the yielding zone [39–41], in which the yield stress appears instead of f_t'.

Because of the aforementioned approximate equivalence of the fracture models utilizing stress-displacement relations for sharp cracks [18], it seems that the precise width w_c of the fracture process zone should not matter, provided that correct energy dissipation due to crack formation is assured. In other words, we should get essentially the same results utilizing different widths of the crack band, provided we adjust the softening compliance C_{33}^t so as to assure that the energy consumed in the fracture process zone equals the given value G_f. Thus, we may choose the value w_c, and then we may calculate C_{33}^t from Eq. 1.15, thereby assuring the energy consumption to be correct. It has been numerically demonstrated [1] (Figure 1.3) that indeed the analyses with different w_c yield essentially the same numerical results. If insistance is made on using the correct experimentally observed softening compliance C_{33}^t, then, of course, only one value of the crack band width w_c is correct. It has been from this condition that the value of w_c has been determined (Eq. 1.17 below).

Although the stress intensity factor, K_I, cannot be defined here as a limiting property of the stress field, one may introduce an "effective" K_I employing the relation known from linear fracture mechanics: $K_I = \sqrt{G_f E'}$ where $E' = E$ for plane stress, and $E' = E/(1 - \nu^2)$ for plane strain. All the subsequent expressions involving G_f could be stated in terms of K_I, but there is no need for this.

Comparison with fracture test data. Most of the important test data from the literature [25,43–57] have been successfully fitted in report [1] with the present nonlinear fracture model. Some of the fits, obtained in [1] by finite element analysis using square meshes, are shown in Figures 1.5 and 1.6 by the solid lines. The best possible fits obtainable with linear fracture mechanics are shown for comparison in these figures as the dashed lines (these fits were calculated also by the finite element method using square meshes). In computations, the loading point was displaced in small steps. Reaction, representing the load P, was evaluated at each loading step. The same stress-strain relation was assumed to hold for all finite elements. However, only some elements entered nonlinear behavior. A plane stress state was assumed for all calculations.

In optimizing the data fits, it was discovered that the optimum width on the crack band front was for all cases between $2d_a$ and $5d_a$ and that the crack band front width

$$w_c = 3d_a, \tag{1.17}$$

where d_a = maximum aggregate size, was nearly optimum for all calculations. It was for this width w_c that the area under the stress-strain curve

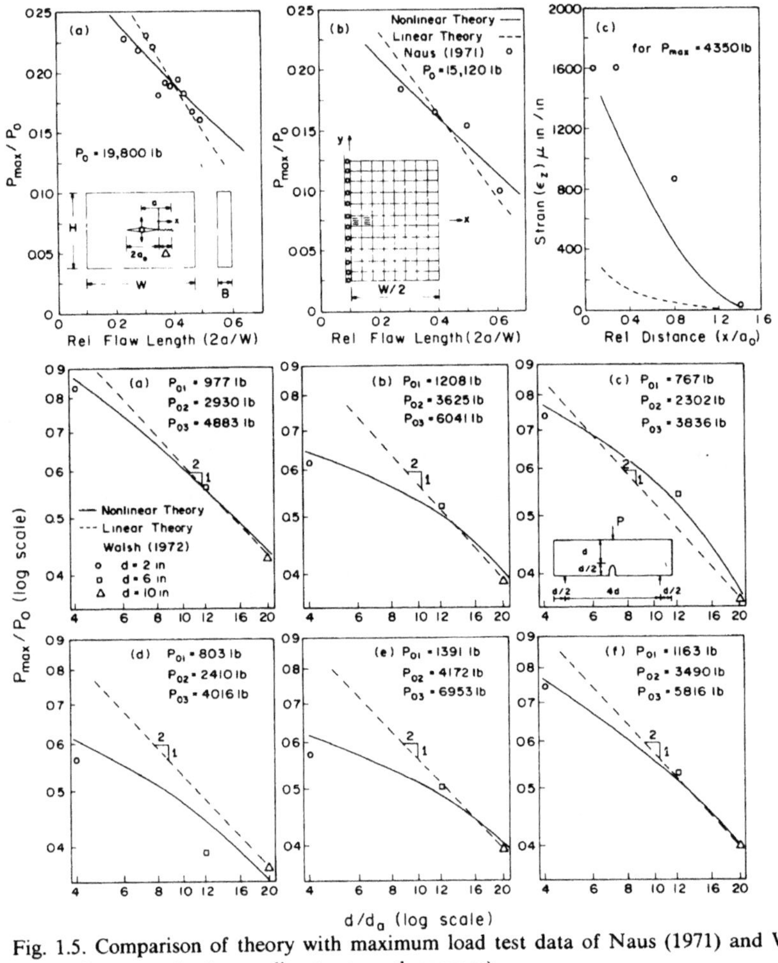

Fig. 1.5. Comparison of theory with maximum load test data of Naus (1971) and Walsh (1972) (P_0 = max, load according to strength concept).

yielded the correct value of the fracture energy needed to obtain good fits of the test data. It thus appears that, at least for plain concretes, the width of the crack band front may be predicted from the maximum aggregate size. However, we must caution that the foregoing simple relation might not hold for high strength concretes, in which the crack band is, no doubt, more concentrated, since the difference between the strengths of aggregate and matrix is less.

In view of Eq. 1.17, the present fracture theory is essentially a two parameter theory. The two material parameters to be determined by experiment are G_f and f_c'.

As for the length of the fracture zone (the strain-softening zone), its value is not constant. A typical value is roughly 12 d_a, but it can be as

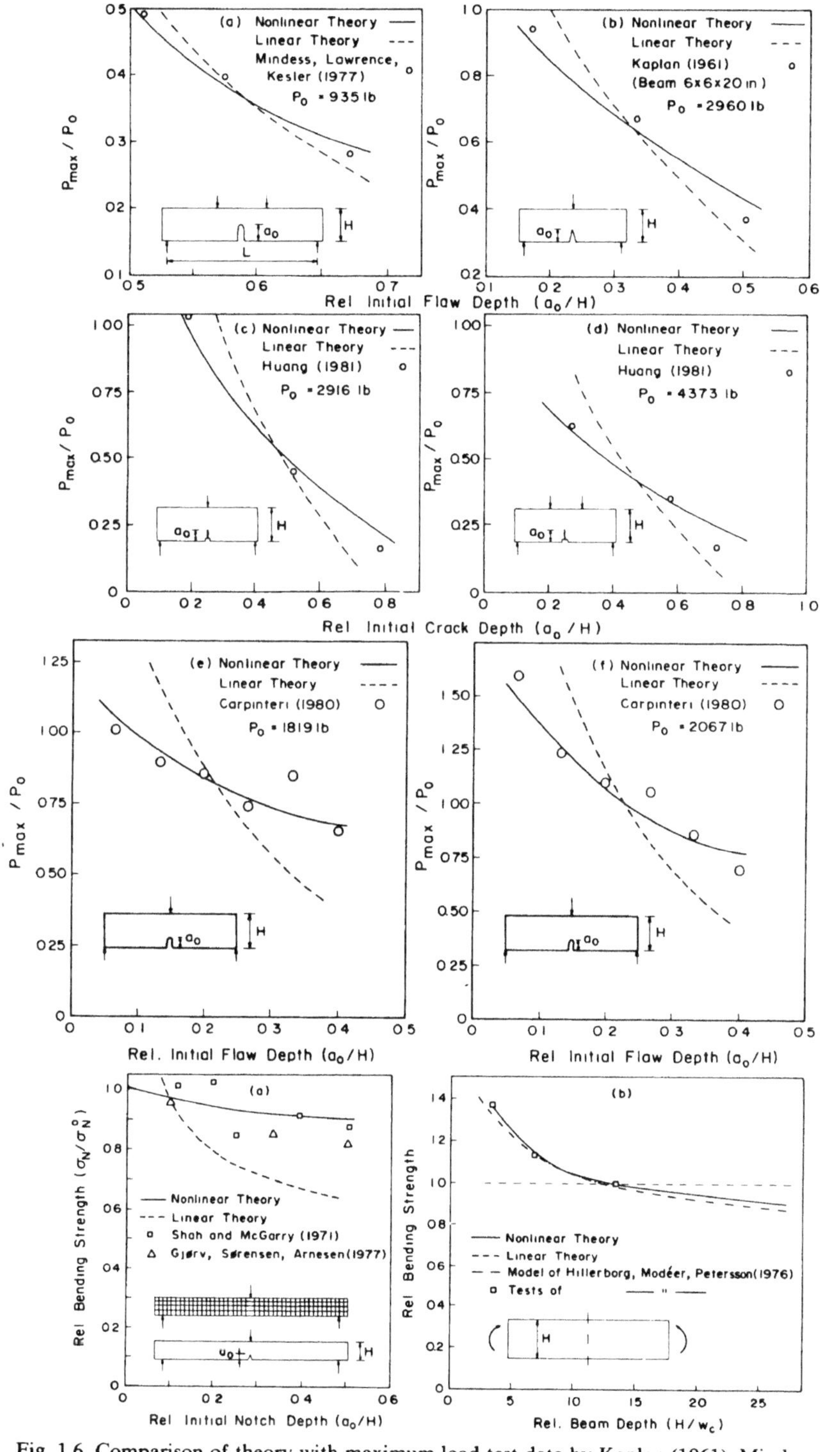

Fig. 1.6. Comparison of theory with maximum load test data by Kaplan (1961), Mindess et al. (1977), Huang (1981), Carpinteri (1980), Shah and McGarry (1971), Gjørv et al. (1971), and Hillerborg et al. (1976).

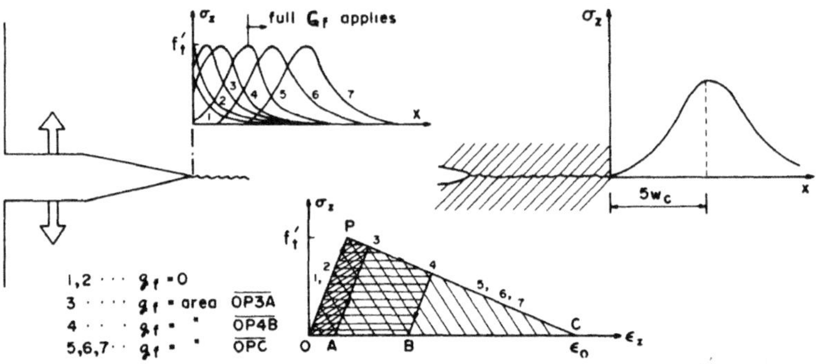

Fig. 1.7. Finite element results of Bažant and Oh (1983) on stress distribution (1,2,...7) ahead of fracture front at successive loading stages, and corresponding states on stress-strain curves.

small as d_a. This is illustrated by the calculated stress profiles at various stages of loading, shown in Figure 1.7 along with the associated states on the stress-strain diagram.

Figures 1.5 and 1.6 represent maximum load data, for which the P_{max}-values are normalized with regard to the engineering strength theory predictions. The data are plotted as a function of either crack length or structure size. Note that the theory also yields the size effect in bending failure of unnotched beams (data of Hillerborg et al. in Figure 1.6). This confirms that this effect is principally a fracture size effect, and not an effect of statistical inhomogeneity of cross section. It is due to the fact that, in a small beam, the fracture process zone cannot develop to full length, causing the energy consumed by fracture to be less than in a deep beam.

Nonlinear fracture properties are sometimes characterized by means of R-curves (resistance curves), which represent the variations of apparent fracture energy as a function of the crack extension from a notch. In some recent works, the R-curves were considered as a basic material property [41]. This is not so, however, according to the present theory, since the R-curves can be unique only in the asymptotic sense, for infinitely small crack extension from a notch. In the present calculations, the resulting R-curves are somewhat different for different specimen geometries, loading arrangements, etc. At the beginning of crack extension from a notch, the theory gives a smaller value of fracture energy because the fracture process zone is not yet fully developed.

The apparent fracture energies needed to generate the curves in Figure 1.8 were obtained by evaluating for each crack length increment the total loss of strain energy from all finite elements outside the crack band, which behave elastically. The specimens for Sok et al.'s data in Figure 1.8 were prestressed in the direction parallel to crack, and this had to be taken into account; Eq. 1.13 was used.

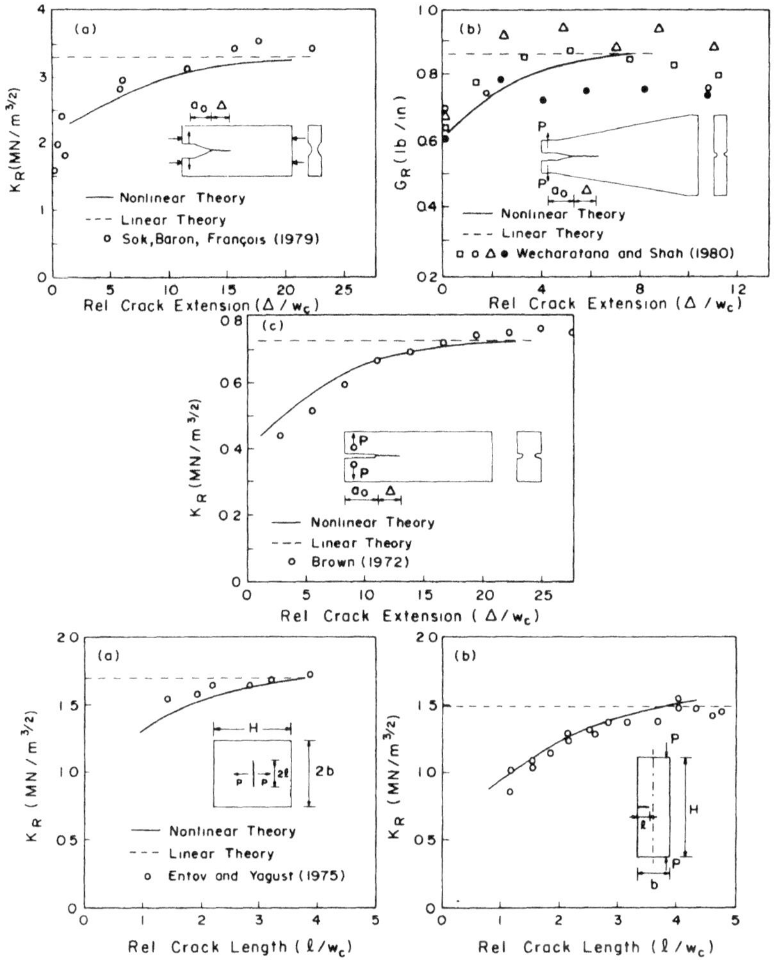

Fig. 1.8. Comparison of theoretical R-curves with test data by Sok et al. (1979), Wecharatana and Shah (1980), Brown (1972), and Entov and Yagust (1975) (K_R is the stress intensity factor corresponding to fracture energy $G_R = G_F$).

For details of data fitting and interpretation of test data, see [1]. Table 1.1 [1] summarizes the values of material parameters for all the data fits shown. Note that the values of G_f obtained with the present formulation are considerably different from the G_f values obtained by the experimentalists in a different manner. Table 1.1 also lists the values G_f^{lin} which give the best fits for linear fracture mechanics (dashed lines).

The present theory has also been used to analyze the most important test data on the fracture of rock available in the literature [58]. For rocks of very different types, involving Indiana limestone, Carrara marble,

TABLE 1
Parameters for fracture test data (after Bažant and Oh, 1983)

Test series	f_t' (psi)	E (ksi)	G_f (lb./in.)	d_a (in.)	w_c (in.)	G_f^{lin} (lb./in.)	c_f	$\tilde{G}_f$ (lb./in.)
1. Naus – no. 1	460 *	4,450 *	0.205 *	0.375	1.125 *	0.430 *	7.664 *	0.224 *
2. Naus – no. 2	360 *	4,500 *	0.099 *	0.375	1.125 *	0.249 *	6.111 *	0.113 *
3. Walsh – no. 1	347 *	3,299 *	0.174 *	0.50	1.50 *	0.188 *	6.356 *	0.185 *
4. Walsh – no. 2	430 *	4,083 *	0.188 *	0.50	1.50 *	0.173 *	5.535 *	0.270 *
5. Walsh – no. 3	273 *	2,593 *	0.126 *	0.50	1.50 *	0.158 *	5.845 *	0.123 *
6. Walsh – no. 4	286 *	2,716 *	0.133 *	0.50	1.50 *	0.162 *	5.888 *	0.133 *
7. Walsh – no. 5	495 *	4,697 *	0.224 *	0.50	1.50 *	0.173 *	5.725 *	0.348 *
8. Walsh – no. 6	414 *	3,928 *	0.193 *	0.50	1.50 *	0.176 *	5.897 *	0.253 *
9. Mindess, Lawrence, Kesler	370 *	6,260	0.088 *	0.375	1.125 *	0.170 *	7.154 *	0.087 *
10. Shah, McGarry	300 *	3,000 *	0.108 *	0.375	1.125 *	0.047 *	6.400 *	0.103 *
11. Gjørv, Sørensen, Arnesen	300 °	3,000 *	0.108 *	0.375	1.125 *	0.047 *	6.400 *	0.103 *
12. Kaplan	300 *	4,190	0.101	0.50	1.50 *	0.177 *	6.269 *	0.098 *
13. Huang – no. 1	360 *	3,122 *	0.225 *	0.50	1.50 *	0.337 *	7.227 *	0.217 *
14. Huang – no. 2	360 *	3,122 *	0.225 *	0.50	1.50 *	0.245 *	7.227 *	0.217 *
15. Carpinteri – no. 1	313 *	2,700 *	0.207 *	0.375	1.125 *	0.147 *	10.14 *	0.128 *
16. Carpinteri – no. 2	356 *	3,130 *	0.280 *	0.752	2.256 *	0.201 *	6.130 *	0.315 *
17. Hillerborg, Modéer, Petersson	400 *	3,300 *	0.100 *	0.157	0.471 *	0.118 *	8.758 *	0.086 *
18. Sok, Baron, François	740 *	3,000 *	2.800 *	0.472	1.416 *	2.910 *	21.66 *	1.600 *
19. Brown	690 *	2,200 *	0.182 *	0.047	0.141 *	0.185 *	11.93 *	0.178 *
20. Wecharatana, Shah	740 *	3,000 *	0.855 *	0.250 *	0.750 *	0.860 *	12.49 *	0.848 *
21. Entov, Yagust – no. 1	450 *	3,000 *	0.746 *	0.787	2.360 *	0.755 *	9.366 *	0.657 *
22. Entov, Yagust – no. 2	440 *	3,000 *	0.640 *	0.787	2.360 *	0.630 *	8.405 *	0.617 *

Note: psi = 6895 N/m^2, lb./in. = 175.1 N/m, in. = 25.4 mm, ksi = 1000 psi; $c_f \simeq 1 - E/E_t$.
* Asterisk indicates numbers estimated by calculations; without asterisk as reported.

Colorado oil shale, and Westerly granite, it was found that nearly optimum fits are achieved for all these rocks with $w_c = 5d_g$ where d_g = grain size of rock. These studies involved the measurements of maximum load, as well as of the resistance curves (R-curves). Statistical regression analysis for the test data for various rocks also indicated a significant improvement. The coefficient of variation of the deviation from the regression line for the plot of relative maximum load values was found to be 10.6%. This is to be compared with the value 15.2% for linear fracture theory, and the value 79.6% for strength-based predictions.

The test data on concrete fracture available in the literature are sufficiently numerous for a statistical regression analysis of the errors. Figure 1.9 shows a regression analysis of the maximum load data for twenty-two different concretes [1]. In this plot, the abscissa is $X = P_t/P_0$ and the ordinate is $Y = P_m/P_0$, in which P_m = measured maximum load P_{max}, P_t = theoretical value of P_{max}, and P_0 = failure loads calculated according to the strength theory. If the theory were perfect, then the plot of Y vs. X would have to be a straight line of slope 1.0, passing through the origin. Thus, the vertical deviations of the data points from the regression line characterize the errors of the theory. The coefficient of variation, ω, of the vertical deviation from the regression line in Figure 1.9 is [1]:

$$\begin{array}{ll} \text{For the present fracture theory} & \omega = 0.066 \\ \text{For linear fracture theory} & \omega = 0.267 \\ \text{For strength criterion} & \omega = 0.650 \end{array} \tag{1.18}$$

These results confirm that the improvement achieved with the present non-linear fracture theory is quite significant.

The test data available in the literature on the R-curves may be analyzed similarly [1]. In this regression analysis, the fracture energy values were normalized with regard to the product $f_t' \ d_a$, and the theoretical values of $G_f/f_t'd_a$ were plotted against the measured values of this ratio. Again, if everything worked perfectly, this plot would have to be a straight line of slope 1.0 and intercept 0.0. The standard errors for the vertical deviations from this regression line have been calculated for the sets of various test data available in the literature,

$$\begin{array}{ll} \text{For the present fracture theory:} & s = 0.083 \\ \text{For linear fracture theory:} & s - 0.317 \end{array} \tag{1.19}$$

The values of the fracture energy obtained for the optimum fits of various fracture data on concrete were further examined to see whether the fracture energy could be approximately predicted from the elemen-

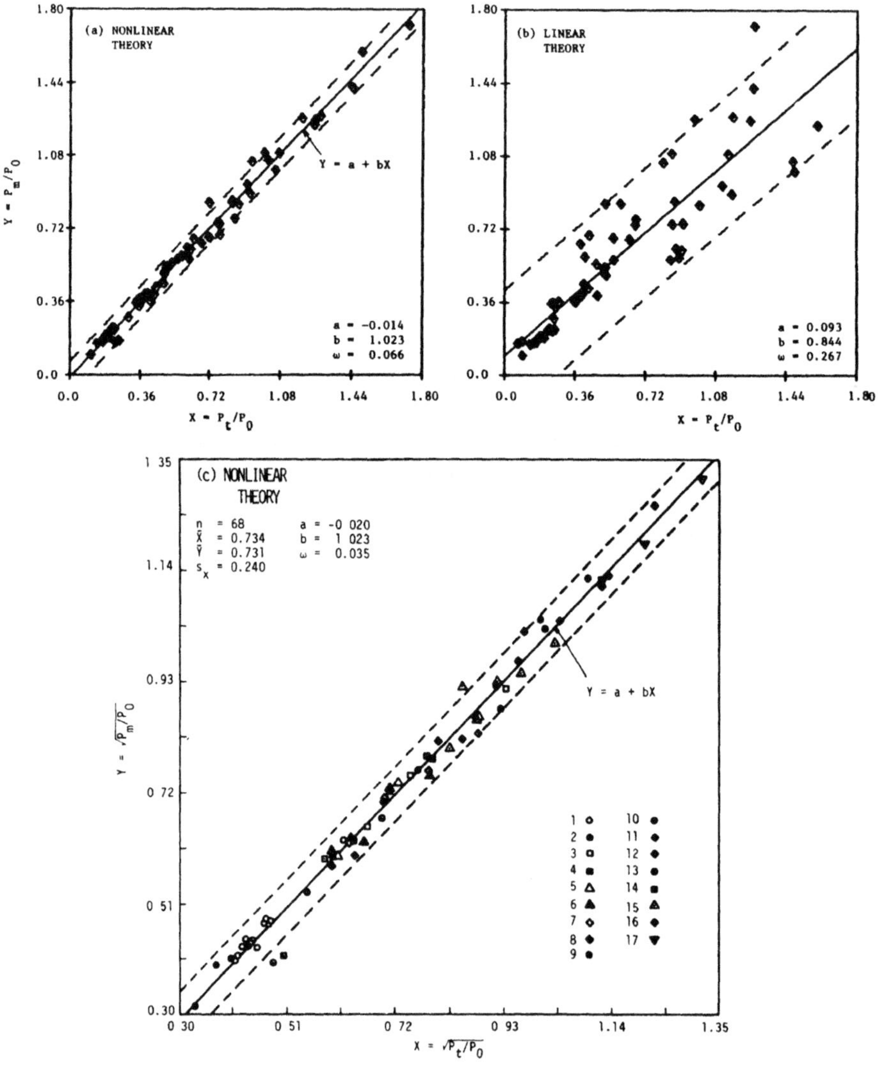

Fig. 1.9. (a) Statistical regression analysis of maximum load data from Figs. 5–6 based on present theory (P_m = measured maximum load, P_f = theoretical maximum load); (b) same, but for linear fracture mechanics; (c) same as (a) but in a different scale (data set numbers – see Table 1).

tary characteristics of concrete. The following approximate formula was found [1]

$$\tilde{G}_f \simeq 0.0214(f_t' + 127)f_t'^2 d_a/E \tag{1.20}$$

in which f_t' must be in psi (psi = 6895 Pa), d_a = maximum aggregate size,

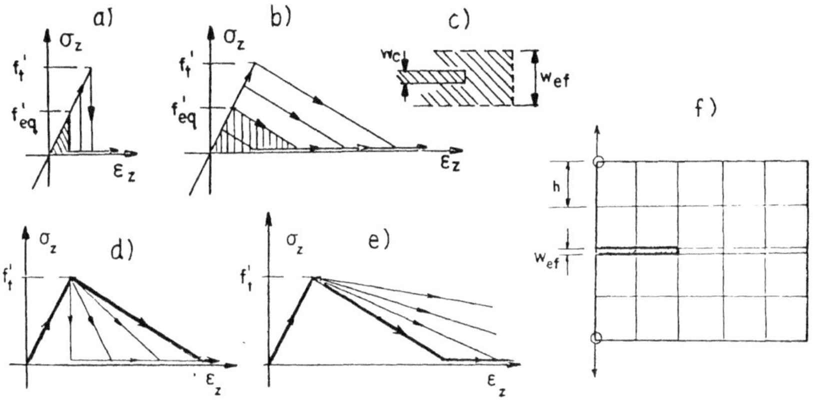

Fig. 1.10. Change of stress-strain relation for fracture process zone needed to ensure correct energy dissipation.

E = elastic modulus, and $\tilde{G}_f$ is in lb/in. The values $\tilde{G}_f$ predicted from Eq. 1.20 are listed in Table 1.1. The coefficient of variation of the errors $G_f - \tilde{G}_f$ is about 16% [1].

It must be emphasized, however, that Eq. 1.20 yields only the fracture energy values for the present nonlinear theory, and not the apparent fracture energy values determined according to linear fracture mechanics or those determined from other theories.

Noting that $G_f \simeq 3d_a f'^2_t(E^{-1} - E_t^{-1})/2$, the following prediction formula for E_t further ensues:

$$E_t \simeq \frac{-69.9\,E}{f'_t + 56.7} \tag{1.21}$$

1.3 Finite element implementation

Effect of element size. Eq. 1.16 gives the upper bound on the finite element size for the present formulation. However, for large structures such as dams or reactor vessels, much larger finite elements need to be used and should be sufficient. This is indeed possible if the correct value of fracture energy G_f is preserved. To preserve it, the strength limit f'_t needs to be reduced to a lower value f'_{eq}, called the equivalent strength [6,59,60]. We may either consider a vertical stress drop (Figure 1.10a) or keep the correct declining slope E_t (Figure 1.10b).

Consider a vertical stress drop and assume a uniform stress distribution across the crack band. The condition of preserving the correct fracture energy for crack band advance Δa is $\Delta U = \Delta U_2 - \Delta U_1 = G_1 \Delta a$

where $\Delta U_1 = (c_f^2 w_{ef} \Delta a / 2E')[\sigma_1(\sigma_1 - \nu'\sigma_3) + \sigma_3(\sigma_3 - \nu'\sigma_1)]$ = strain energy in the frontal element before its cracking, $\Delta U_2 = (k_w w_{ef} \Delta a / 2E')\sigma_1'^2$ = strain energy in this element after cracking; σ_3, σ_2 = initial transverse and longitudinal normal stresses in the element, σ_1' = longitudinal normal stress after cracking; $w_{ef} = h$ = element side if the crack band runs in a square mesh parallel to a mesh line; c_f = empirical coefficient (close to 1) taking into account the actual nonuniform stress distribution in the finite element; and for plane stress $E' = E$, $\nu' = \nu$, while for plane strain $E' = E/(1 - \nu^2)$, $\nu' = \nu/(1 - \nu)$ where E = Young's modulus of concrete and ν = its Poisson ratio. The formation of cracks has no effect on stress σ_1 parallel to cracks, and assuming that no loading change occurs which would change σ_1 during the crack advance, it follows that $\sigma_1' = \sigma_1$. Also, set $\sigma_3 = f_{eq}'$ = equivalent strength, and substitute $G_f = w_c f_t'^2 (E^{-1} - E_t^{-1})/2$ as determined from a uniaxial test. From the condition $\Delta U_2 - \Delta U_1 = G_f \Delta a$, it is found that

$$f_{eq}' = c_f \left(\frac{2G_f E'}{w_{ef} r_f} \right)^{1/2} = c_f f_t' \left[\left(1 + \frac{E}{-E_t} \right) \frac{w_c}{w_{ef} r_f} \right]^{1/2},$$

$$r_f = 1 - 2\nu' \frac{\sigma_3}{\sigma_1} > 0 \text{ (vertical stress drop)} \tag{1.22}$$

The strength value f_{eq}' used in analysis must be increased as the element size, w_{ef}, decreases. Furthermore, a compressive normal stress parallel to the crack plane causes a reduction of f_{eq}' or the effective fracture energy. This result agrees with experience. However, the decrease of f_{eq}' due to r might be too strong since, at $\nu \simeq 1/6$, a compression $\sigma_1 = -6\sigma_3$ would reduce the tensile strength to zero, while biaxial failure test data indicate that the tensile strength is nonzero for any ratio σ_3/σ_1. This means that the effective fracture energy is probably also variable, as a function of σ_1/σ_3, which is neglected in Eq. 1.22 (cf. Eq. 1.13). Nevertheless, for $\sigma_1/\sigma_3 \leqslant -6$, the tensile strength is, no doubt, greatly reduced and so Eq. 1.22 might be acceptable for practical purposes. In previous works [6,59,60], coefficient r_f was not used ($r_f = 1$); the effect, however, was almost nil since σ_1/σ_3 was negligible in the examples solved.

Coefficient c_f, which takes into account the nonuniformity of strain distribution enforced in the frontal finite element by its shape function, may be calibrated empirically, so that the results would agree with those obtained from the energy criterion (Eq. 1.33 in the sequel). For a square element consisting of two identical constant strain triangles, $c_f\sqrt{2} = 0.921$ [6], while for that consisting of four identical constant strain triangles (with a central node condensed out), $c_f\sqrt{2} = 0.826$ [59]. For a four-node square with a single-point numerical integration and Flanagan-Belytschko's optimal orthogonal (elastic) hour-glass control [61], $c_f = 0.74$, as determined by P. Pfeiffer at Northwestern University.

If the finite element size w_{ef} is up to a few time w_c then it is best to modify the stress-strain diagram in such a manner that both G_f and f_t' are preserved. This may be achieved by replacing the actual downward slope E_t by an effective one $\tilde{E}_t$ (Figure 1.10d). The condition of equal energy dissipation is $w_c f_t'^2(E^{-1} - E_t^{-1})/2 = w_{ef} f_t'^2(E^{-1} - \tilde{E}_t^{-1})/2$, which yields the rule

$$-\tilde{E}_t = \frac{E}{\dfrac{w_c}{w_{ef}}\left(1 - \dfrac{E}{E_t}\right) - 1} \tag{1.22a}$$

Thus, the downward slope $\tilde{E}_t$ must be made steeper as the finite element is made larger. There is a limit for this; a vertical stress drop, for which $1/\tilde{E}_t = 0$, and Eq. 1.22a indicates that this happens when $w_{ef} = w_c(1 - E/E_t)$. So, a change of downward slope (Eq. 1.22) can be used to achieve correct energy dissipation only if

$$w_{ef} \leqslant w_0 = w_c\left(1 + \frac{E}{-E_t}\right) \tag{1.22b}$$

If the finite element needs to be made larger, then one must keep a vertical stress drop and adjust the strength limit according to Eq. 1.22; see Figure 1.10a.

For relatively small structures it may be sometimes desirable to use finite elements smaller than $w_c = 3d_a$ (Figure 1.10f). Leaving aside (for lack of data) the question of how important it is to keep a blunt fracture front, the correct fracture energy may then be preserved by using a downward slope $\tilde{E}_t$ that is milder (less steep) than the actual one, E_t. This slope may again be calculated from Eq. 1.22a, in which $w_c/w_{ef} > 1$. Obviously, there is now no mathematical limit on how small w_{ef} can get. Note that if one reduces only the width of the finite elements in the crack band, permitting these elements to become elongated rectangles (Figure 1.10f), then a reduction of w_{ef} to a very small volume makes this approach equivalent to that of Hillerborg et al. [18], in which one uses a stress displacement relation, the displacement being $\delta = w_{ef}\varepsilon_z$.

Another possibility of preserving the correct fracture energy is to keep the actual downward slope E_t and change the peak stress value from f_t' to f_{eq} (Figure 1.10b). This may be less realistic than the previous method (Eq. 1.22a) for $w_{ef} \leqslant w_0$, but is simpler and avoids the limitation in Eq. 1.22b. In this approach the stress-strain diagram remains geometrically similar, and so this approach is easier to implement if one uses a curved stress-strain diagram. The energy balance condition remains the same as before except that $1/E'$ must now be replaced with $(1 - E/E_t)/E'$. Making this replacement in Eq. 1.22, one obtains

$$f'_{eq} = c_f \left(1 + \frac{E}{-E_t}\right)^{-1/2} \left(\frac{2G_f E'}{w_{ef} r_f}\right)^{1/2} = c_f f'_t \left(\frac{w_c}{w_{ef} r_f}\right)^{1/2} \quad \text{(correct slope } E_t\text{)} \tag{1.23}$$

where again $r_f = 1 - 2\nu'\sigma_3/\sigma_1$. In contrast to Eq. 22, this equation should apply even for small structures (nonlinear fracture range). One should also realize that the sudden stress drop is an inappropriate assumption for dynamic finite element programs since it generates spurious shock waves [62,63,64].

For larger structures, it is found that the present method with an abrupt stress drop gives results which are in excellent agreement with the exact solutions for sharp cracks, and approximate these solutions just as well as the method of sharp interelement cracks. This has been demonstrated by Bažant and Cedolin [59,6,60,7] and one of these demonstrations is shown in Fig. 1.12 in which a nondimensionalized load parameter is plotted versus the crack length, a. The specimen is a rectangular panel with a center crack, loaded by a uniform normal stress at top and bottom. The calculation has been carried out for three different meshes shown in Fig. 1.11, with finite element sizes in the ratios 4 : 2 : 1. (Note that the exact solutions are slightly different for each mesh because the boundary of each mesh was not exactly the same.) We see that, with finer meshes, the present method can be used to obtain linear fracture mechanics solutions.

The energy actually dissipated per unit extension of the crack band in the finite element mesh is $W' = wf'^2_t(E^{-1} - E_t^{-1})/2$, which is proportional to the width w of the frontal finite element. So, by reducing the element size to zero ($w \to 0$), the energy that needs be supplied to produce the fracture becomes vanishingly small if f'_t, E and E_t are constant. This conspicuously demonstrates the irrationality of using the same complete stress-strain diagram regardless of the element size.

The foregoing deductions regarding the effect of element size are based on the premise that the cracking front is single-element wide. This premise is justified by two reasons:

1) If the fracture front is considered to be two or more elements wide, then one finds that a deformation increment of localization type (see Sec. 4.2 in the sequel) consumes negative energy, i.e. the multi-element width of the cracking front is unstable.

2) If the loading step is so small that the strain in only one element goes over the peak stress point within this step, it is impossible to obtain

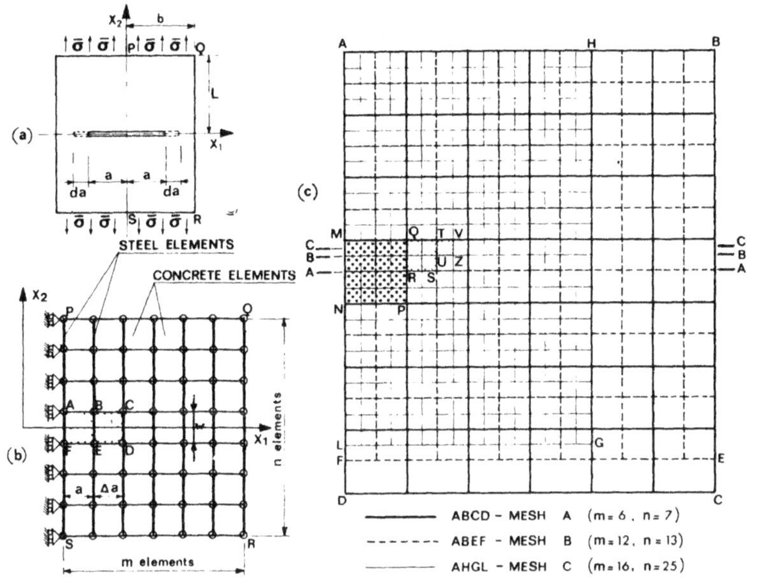

Fig. 1.11. Numerical example of center-cracked rectangular panel and finite element meshes used (after Bažant and Cedolin, 1980).

a multiple-element cracking front. Cracking of one finite element relieves the stress from its neighbor element on the side, and for this reason this neighbor element can never be made to crack in subsequent loading steps. One can get both finite elements to crack only if the loading step is sufficiently large.

In the currently existing large finite element codes, propagation of distributed (smeared) cracking from one element to another is being determined on the basis of the tensile strength criterion. It is well known that such a calculation cannot converge to correct results, since refinement of the element size to zero leads to infinite stress concentrations just ahead of the front cracked element, causing that the load needed for further extension of the crack band tends always to zero. It has not been however generally recognized that the use of the strength criterion can lead to very large errors. According to the numerical results of Bažant and Cedolin [6,59,60], the differences in the results can be as large as 100% when the finite element sizes differ as 4 : 2 : 1. To demonstrate it, some of these results are reproduced in Fig. 1.11, in which the failure load needed to cause further extension of the crack band is plotted for the same panel as in Fig. 1.11 against the length of the crack band. The curves obtained for meshes A, B, C of finite element sizes 4 : 2 : 1 are seen to be very far apart, whereas the curves for the finite element results on the basis of the equivalent strength approach for the abrupt stress drop

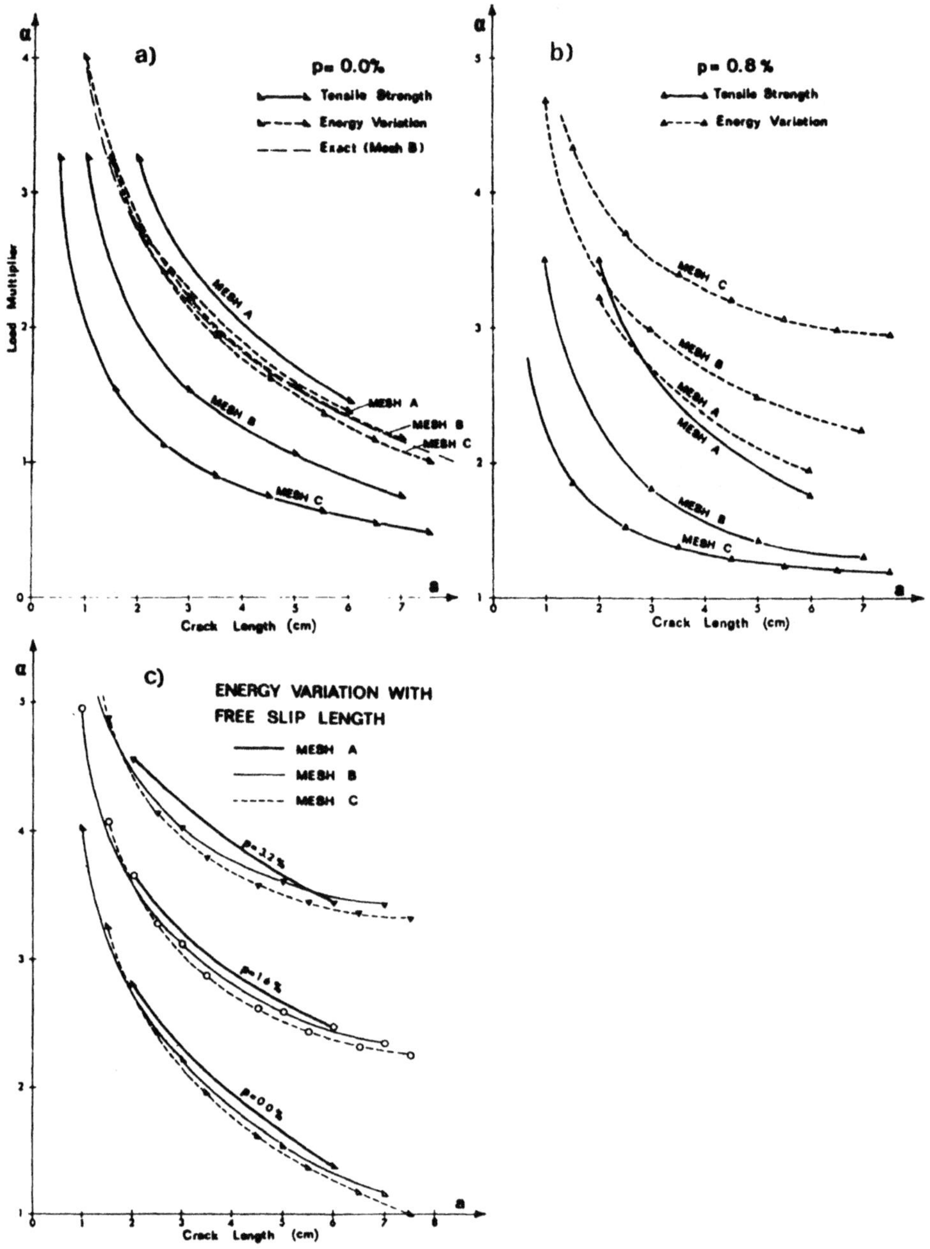

Fig. 1.12. Results for different mesh sizes for the panel from Fig. 1.11 (a) unreinforced, (b) reinforced, no bond slip, (c) various reinforcements, bond slip (after Bažant and Cedolin, 1980), (p = reinforcement ratio).

(as well as those obtained on the basis of energy criterion in Eq. 1.33 below) agree with each other well; see Figure 1.12. The difference between the curves is negligible and tends to zero as the mesh is refined. The curves for the equivalent strength criterion closely agree with those for the energy criterion.

The calculation results in Figure 1.2c further demonstrate that also in the case of gradual strain softening, the finite element results strongly depend on the chosen mesh size, and thus are unobjective. When, however, the downward slope is varied to preserve the same energy, the results for various meshes are about the same; see Figure 1.12c.

The foregoing examples demonstrate that strength criteria are fundamentallly at fault, except for plastic failures, characterized by constant rather than decreasing stress during failure. This is true not only of tensile failure of plain concrete but also of reinforced concrete, and of shear and compression failures whenever they exhibit strain softening. Eventually, it will be necessary to develop fracture mechanics for compression and shear failures of concrete if consistent results, independent of the chosen mesh size, should be achieved.

On the other hand, the use of strength criterion in the literature often yielded results that agreed well with measurements. This must have been due to one of the following two reasons:

(1) The measurements were made on laboratory specimens the size of which was the minimum possible with regard to the aggregate size. In this case, $f'_{eq} \simeq f'_t$ and the strength analysis is correct (see the discussion below Eq. 1.46). However, engineers need to extrapolate from such laboratory structures to much larger real structures, and this is in question.

(2) Many concrete structures are fracture-insensitive, i.e., the tensile strength of concrete has very little effect on the failure load. These include the bending failure of beams or plates (which must be designed, according to ACI Standard 318 [65], so that steel fails plastically before concrete fails in compression), or failure of spiral or tied columns, in which the confinement makes concrete relatively ductile. To determine whether the structure if fracture-insensitive, the analyst needs to analyze his structure twice – once for the actual strength value, and once for a zero strength value. If the results of both analyses are not approximately the same, fracture mechanics analysis is required.

As the size of the structure, and thus the size of the finite element, becomes very large, the value of the equivalent strength (Eq. 1.22 or 1.23) obviously tends to zero. In the limit, the no-tension material is obtained. This approach was pioneered in the mid-1960's by Zienkiewicz et al. as a method for the cracking analysis of large rock masses.

Effect of mesh inclination. In a general situation, the fracture direction need not be parallel to the mesh lines. A smoothly curved or inclined

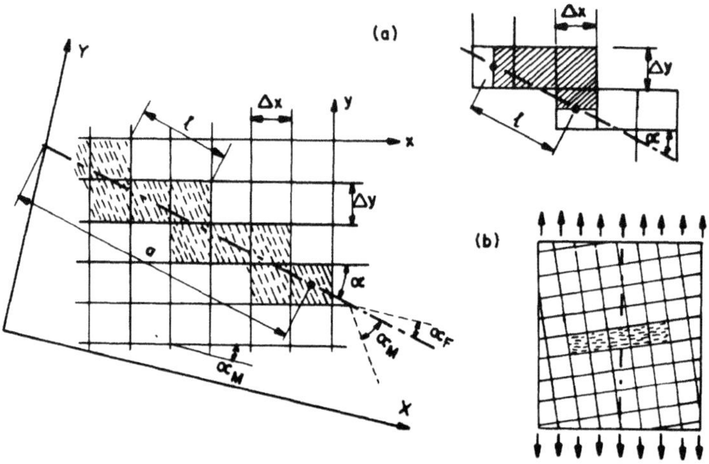

Fig. 1.13. (a) Zig-zag crack band propagation in finite element mesh in skew direction; (b) False bias in crack direction caused by a slightly slanted mesh.

crack or crack band may then be conveniently represented as a zig-zag crack band through the finite element mesh; see Fig. 1.13. Numerical studies indicated that the equivalent strength and energy criteria may still be used but must be modified.

Consider a rectangular mesh of mesh sizes Δx and Δy (Figure 1.13). Let α_F be the orientation angle of the zig-zag crack band (overall fracture direction), α_M be the orientation angle of mesh lines x, and α_C be the direction of the cracks (microcracks) within the finite element (Figure 1.13). To be determined is the effective width w_{ef} of a smooth crack band which is equivalent to the zig-zag band. Consider one cycle, of length l, on the line connecting the centroids of the elements in the zig-zag band. The number of elements per cycle l in the x-direction is $N_x = l \cos \alpha/\Delta x$, and the number of those in the y-direction is $N_y = l \sin \alpha/\Delta y$ where $\alpha = |\alpha_F - \alpha_M|$ ($0° \leqslant \alpha \leqslant 90°$) (Figure 1.13). The area of the zig-zag band per cycle l is $(N_x \Delta y)\Delta x + (N_y \Delta x)\Delta y$. This area must equal the area lw_{ef} of the equivalent smooth crack band, in order to assure the same energy content (assuming same stresses). This condition yields

$$w_{ef} = \Delta x \sin \alpha + \Delta y \cos \alpha \quad \text{or} \quad w_{ef} = c_\alpha h, \quad c_\alpha = \sqrt{2}\cos(45° - \alpha) \qquad (1.24)$$

where the second equation applies to a square mesh ($\Delta x = \Delta y = h$). Thus, we see that the value of w_{ef} to be substituted into Eqs. 1.22 or 1.23 for the equivalent strength depends on the inclination α of the mesh with regard to the fracture direction. Note also that the correction factor due to α is always between 1.0 and 1.41.

By a similar argument, for a three-dimensional orthogonal mesh of steps Δx, Δy, Δz, the equivalent width of a three-dimensional zig-zag band may be shown to be

$$w_{ef} = |\nu_1|\Delta x + |\nu_2|\Delta y + |\nu_3|\Delta z \tag{1.25}$$

where ν_1, ν_2, ν_3 are the direction cosines of the normal of the fracture plane (overall fracture direction) with regard to the mesh coordinates.

Instead of Eq. 1.24 for effective width, a somewhat different equation, namely, $w_{ef} = h/\cos\alpha$, was used in previous work based on a different argument. For $\alpha = 0$ and $\alpha = 45°$, this equation gives the same values of w_{ef} as Eq. 1.24, and between 0 and 45° it gives slightly smaller values of w_{ef} (not smaller by more than 17%). However for α close to 90°, the equation $w_{ef} = h/\cos\alpha$ is inapplicable; it cannot be correct when $\alpha \to 90°$ since it would give $w_{ef} \to \infty$, which in turn, would yield $f'_{eq} \to 0$, causing the equivalent strength criterion to always incorrectly indicate that a crack band parallel to the mesh would always jump to the side, perpendicular to the crack direction [66]. Eq. 24 or 25 avoids this problem.

Although the foregoing equations give correct overall energy dissipation by a zig-zag crack band, they do not completely avoid a directional bias due to the mesh as far as determining the direction of individual jumps of the crack band front is concerned. For example, if a square mesh in the center-cracked rectangular panel is slanted, but only moderately so (Figure 1.13b), then the criterion in Eq. 1.24, used in comparison with the maximum principal stress, indicates the crack band to run straight along the mesh line, i.e. in the inclined direction, while correctly it should zig-zag so as to conform to an overall horizontal direction. It appears rather difficult to avoid this type of bias. On the other hand, for a 45° slant of a square mesh, this problem does not occur and the crack band propagates zig-zag in an overall horizontal direction. Various methods to avoid the bias due to the slant of the mesh are being studied [66–68].

The calculation results must be also objective not only with regard to the choice of the element size but also with regard to the choice of mesh inclination. To demonstrate it, the example shown in Figure 1.11 has been calculated [60] for a square mesh whose sides are inclined at 45° with regard to the side of the rectangular panel. The results of this calculation are shown in Figure 1.15, in which case 1b corresponds to this inclined mesh and case 1a to a square mesh whose sides are parallel to the sides of the panel. An excellent agreement of these two calculations is seen. Similar agreement has been found for the inclined meshes when the element size is varied [60]. A 26.6° inclination of the square mesh has also been considered, and the results were again satisfactory, although the scatter was larger than for the 45° inclination [60].

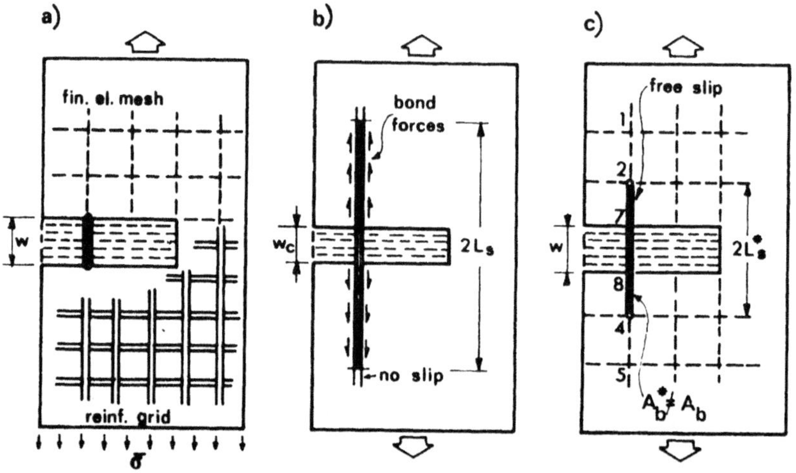

Fig. 1.14. Illustration of bond slip and equivalent free bond slip length L_s^* (after Bažant and Cedolin, 1980).

Effect of reinforcement and bond slip. It has been customary in finite element analysis of reinforced concrete to assume that the steel bars are rigidly attached to concrete in the nodes of the mesh. This treatment is, however, not only physically unjustified but also unobjective with regard to the choice of mesh, and causes incorrect convergence. The bars connecting the nodes on the opposite sides of the crack band represent an elastic connection, the stiffness of which varies inversely as the distance between the nodes, i.e., the width w of the crack band. Thus, as the mesh size is refined to zero (and the crack band width tends to zero), the stiffness of the connection across the crack band increases to infinity, which prevents opening of the crack band. So it is clear that no cracking can be obtained in the limit of a zero element size.

The effect of the mesh size on the results is demonstrated in Figure 1.12b where the load parameter is plotted vs. crack length a for a rectangular panel [59]. The panel is the same as before, but is reinforced by regularly and densely spaced vertical bars of various reinforcement ratios p. We see that the results for the three meshes of sizes 4 : 2 : 1 differ greatly, not only for the constant strength criterion but also for the energy criterion.

To obtain an objective and properly convergent formulation, one must take into account the bond slip. The bond slip occurs over a certain length, L_s (Figure 1.14). The most realistic treatment of bond slip would call for using separate nodes for concrete and steel connected by some nonlinear linkage elements representing forces transmitted by bond. However, this approach would be too cumbersome. In the spirit of the approximations involved in the smeared crack band model, it should be sufficient to introduce the bond slip in such a way that the stiffness of the connection provided by the steel bars crossing the crack band would be roughly correct and independent of the finite element mesh.

Thus, to simplify the formulation, the actual curvilinear variation of the bond forces and the axial forces in the bars may be replaced by an idealized piece-wise constant variation of the bond force and the corresponding piece-wise linear variation of the actual axial force in the bars (Figure 1.14). The latter may further be replaced by a piece-wise constant variation of the axial force, such that the overall extension of the bar over the distance of the bond slip would be roughly the same.

An estimate of L_s will be made. Let A_b = cross section area of bar. Since the bar force $A_b\sigma_s$ must equilibrate the remote bar force $A_b\sigma_s'$, plus the bond force $U_b'L_s$, the bond slip length is $L_s = (\sigma_s - \sigma_s')A_b/U_b'$ (Figure 1.14) where U_b' = ultimate bond force per unit length of bar, as determined by pull-out tests; σ_s, σ_s' = tensile stress in the steel bar at the point where it crosses the crack band, and at the end of the slipping segment, respectively. Furthermore, σ_s' may be approximately related to σ_s; $A_b\sigma_s$ must equal the force per bar carried jointly by steel and concrete at the end of the slipping segment where the strain, $\varepsilon_s = \sigma_s'/E_s$, is the same for concrete and steel. Thus $[E_s p + E_c(1-p)]\sigma_s'/E_s = p\sigma_s$ or $\sigma_s' = \sigma_s n'p/(1-p+n'p)$ in which $n' = E_s/E_c$ = ratio of elastic moduli of steel and concrete, and $p = A_s/(A_c + A_s)$ = steel ratio. The following result is obtained [59]:

$$L_s = \frac{A_s}{U_b'}(\sigma_s - \sigma_s') \simeq \frac{A_b}{U_b'}\frac{1-p}{1-p+n'p}\sigma_s. \tag{1.26}$$

This equation gives the bond-slip length as a fixed property characteristic of the steel-concrete composite.

For the purpose of finite element analysis, the actual bond-slip length L_s may be replaced by some modified length L_s^* such that the steel stress over this length is uniform and the slip of steel bar within concrete may be considered as free. The length L_s^* is determined from the condition that the extension of the steel bar over the length L_s would remain the same. In this manner, the following expression for the equivalent free bond-slip length can be derived [59]:

$$L_s^* = \frac{A_b^*(1-p)}{2[A_b(1-p+pn) - pnA_b^*]}\left[L_s + w_0\left(1 - \frac{s_c}{4L_s}\right)\right] \tag{1.27}$$

where w = width of the element-wide crack band, s_c = spacing of cracks within the crack band ($s_c \simeq d_a$) and A_b^* = the cross-section area of bar chosen for computations. One may conveniently choose such A_b^* that Eq. 1.27 give a length which coincides with a distance between two nodes of the mesh. As a crude approximation

$$L_s^* = \tfrac{1}{2} L_s \qquad \text{if} \quad A_b^* = A_b. \tag{1.28}$$

Using Eq. 1.27, the finite element analysis of the center-cracked rectangular panel, the same one as before (Figure 1.11), yields consistent results when finite elements of different sizes are used (Figure 1.12c) [60]. Further, it has been demonstrated [60] that the use of different mesh sizes for a reinforced panel yields consistent results even when the mesh is inclined; see Figure 1.15 for a 45° mesh inclination and Figure 1.16 for a 26.6° inclination.

The formula for the equivalent strength of concrete needs to be generalized to reflect the bond slip effect. The stiffness of the concrete-steel composite over band width w_c for loading normal to the crack band may be written as $C_1 = (1 - p^*)E_c/w_c + c_p p' E_s/L_s'$ where E_s, E_c = Young's moduli of steel and concrete, $L_s' = L_s^* \cos \alpha_s$ = actual free bond-slip length projected on the normal to the crack band, α_s = angle of the reinforcing bars with this normal, $p' = p^* \cos^2\alpha_s$ where $\cos^2\alpha_s$ represents a correction of stiffness of the steel bars due to their inclination (satisfying the condition that the stiffness be zero when $\alpha_s = 0$), and c_p = empirical correction factor introducing the effect of deformation of concrete outside the element that cracks but lies within length L_s^*. The deformation f_{eq}'/C_1 should equal the deformation f_{eq}^0/C_0 where f_{eq}^0 = equivalent strength in absence of reinforcement, as given before, and $C_0 = E_c(1 - p')/w_c$ = stiffness of concrete across the band width w_c. From the condition $f_{eq}'/C_1 = f_{eq}^0/C_0$, the following expression for f_{eq}' for sudden stress drop may be obtained after algebraic rearrangements:

$$f_{eq}' = c_f \left(\frac{2G_f E_c}{w_{ef} r_f} \right)^{1/2} \left(1 + c_p \frac{E_s}{E_c} \frac{p}{L^*} \cos \alpha_s \right) \tag{1.29}$$

if $1 - p$ is replaced by 1 (normally $p \ll 1$). This formula is the same as that derived in [60]. by using the expression for the asymptotic displacement field near the tip of an equivalent sharp crack [40], except for one difference: The dependence on α_s. By solving a number of examples for reinforced panels on the basis of the energy criterion, and requiring that the use of f_{eq}' would yield approximately the same results, a table of optimum values of c_p for various α_p was set up [60]; approximately, $c_p \approx 0.7$ for all α_s. Note that Eq. 1.29 satisfies the obvious condition that f_{eq}' must become the same as given by Eq. 1.22 when $\alpha_s = 90°$, or $p = 0$, or $E_s = 0$, or $L^* \to \infty$.

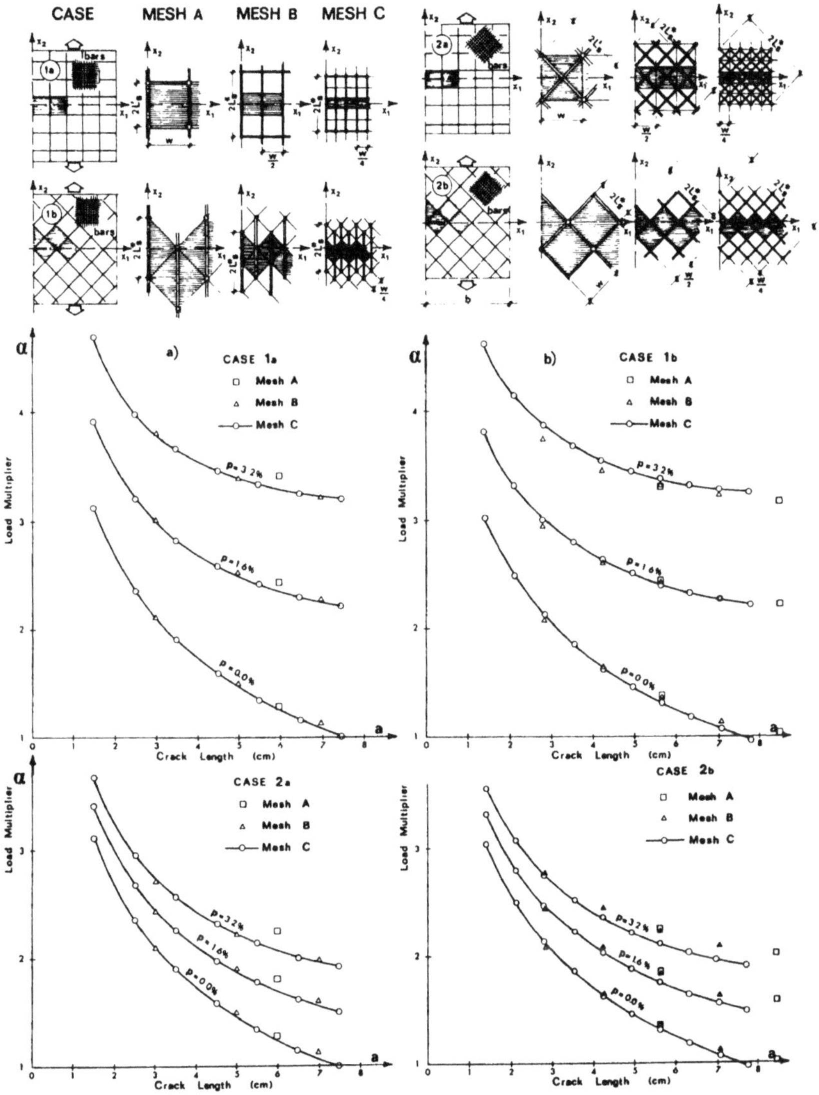

Fig. 1.15. Results for various finite element sizes for zig-zag crack band in a 45° inclined mesh and parallel mesh, both for energy criterion and equivalent strength (after Bažant and Cedolin, 1983).

A more realistic continuum treatment of reinforcement and bond slip would be to approximate the reinforcing net by a continuum that is allowed to slip against the continuum representing concrete, and consider that the distributed (volume) forces transmitted between the two continua depend on the relative slip displacement. This would be, however, more complicated.

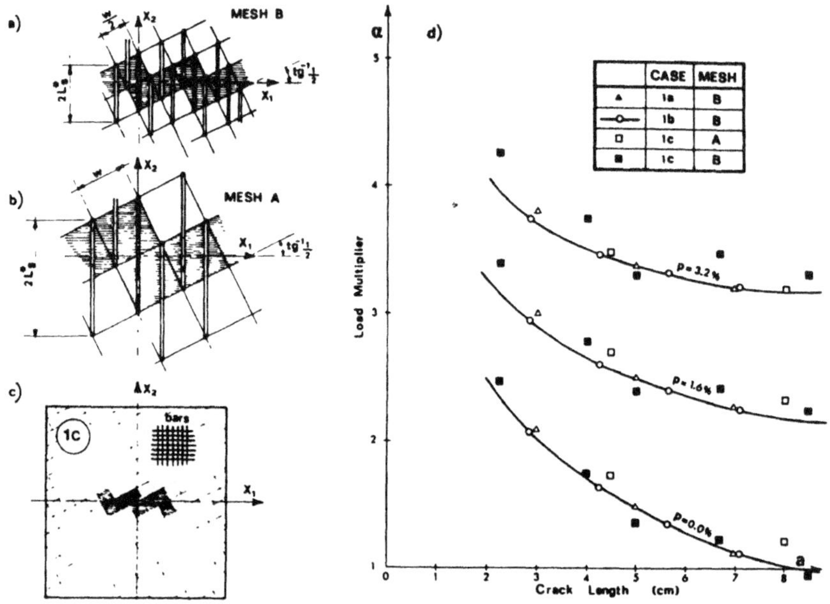

Fig. 1.16. Results for various finite element sizes for zig-zag crack band in a 26.6° inclined mesh, both for energy criterion and equivalence strength (after Bažant and Cedolin, 1983).

1.4 Energy considerations

Energy criterion for crack bands. When a structure is so large that a sudden stress drop may be considered, and fracture energy is the only important fracture property, the energy criterion may be directly implemented in a finite element program [6,59]. One needs to evaluate in the program the amount of energy, ΔW, that is available for fracture as the crack bands extends by length Δa of one finite element.

A similar problem is the extension of a notch, and energy analysis of this case was made by Rice [69]. The case of a crack band differs from that of a notch by the fact that, as the element of volume ΔV ahead of the crack front gets cracked (Figure 1.17), it loses merely the capability of transmitting stresses across the crack plane, but remains capable of carrying normal stresses parallel to the crack planes. Moreover, one must take into account the fact that the volume ΔV may contain reinforcing bars which, in the uncracked state, transmit to concrete interface forces. We consider now the steel-concrete composite, and distinguish steel and concrete by subscripts s and c. Assuming the material to be elastic (and the applied forces to be conservative), the variation of the potential energy of the structure due to the extension of the crack band into volume ΔV is independent of the path in which this extension happens.

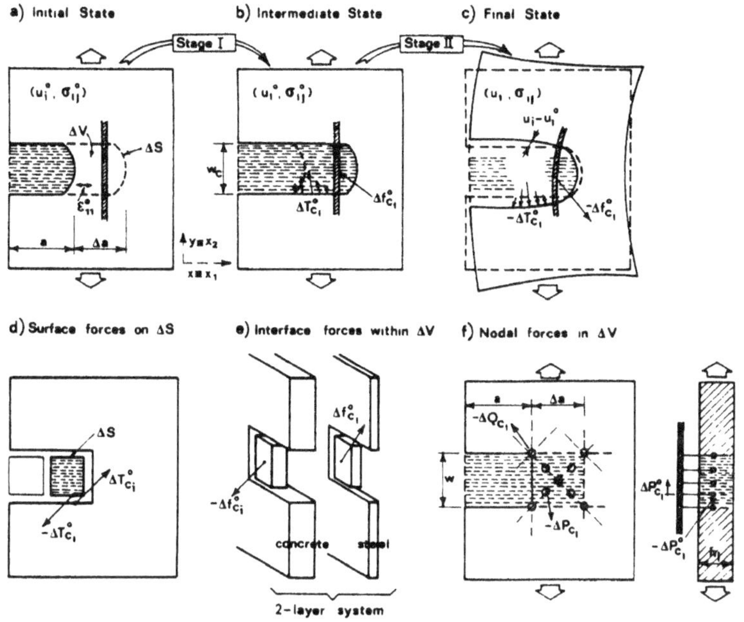

Fig. 1.17. Assumed stages of extension of crack band to explain energy criterion (after Bažant and Cedolin, 1980).

Consequently, as in [6], the crack extension may be decomposed in two stages.

Stage I. Cracks are created in concrete inside volume ΔV of the element ahead of the crack in the direction of principal tensile stress (Figure 1.17b), while, at the same time, the deformations and stresses in the rest of the body are imagined to remain fixed (frozen). This means that one must introduce surface tractions $\Delta T^0_{c_i}$ applied on the boundary ΔS of volume ΔV, and distributed forces $\Delta f^0_{c_i}$ applied at the concrete-steel interface, such that they replace the previous action of concrete that cracked upon the remaining volume $V - \Delta V$ and upon the reinforcement within ΔV.

Stage II. Next, forces $\Delta T^0_{c_i}$ and $\Delta f^0_{c_i}$ (Figure 1.17c) are released (unfrozen) by gradually applying the opposite forces $-\Delta T^0_{c_i}$ and $-\Delta f^0_{c_i}$, reaching in this way the final state.

Let u^0_i and ε^0_{ij} be the displacements and strains before the crack band advance, and let u_i and ε_{ij} be the same quantities after the crack band advance. For the purpose of analysis, the reinforcement may be imagined to be smeared in a separate parallel layer undergoing the same strains as concrete. The interface forces between steel and concrete, $\Delta f^0_{c_i}$, then appear as volume forces applied on the concrete layer.

Upon passing from the initial to the intermediate state (Stage I), the strains are kept unchanged, while the mechanical properties of concrete inside ΔV are varied. Thus, the corresponding stress changes in concrete in ΔV are given by $\Delta\sigma_{11}^{c} = \sigma_{11}^{c^0} - E_c'\varepsilon_{11}^0 = (\varepsilon_{11}^0 + \nu_c'\varepsilon_{22}^0)E_c'/(1-\nu_c'^2) - E_c'\varepsilon_{11}^0$; $\Delta\sigma_{22}^{c} = \sigma_{22}^{c^0}$; $\Delta\sigma_{12}^{c} = \sigma_{12}^{c^0}$. Here, $\sigma_{ij}^{c^0}$ denotes the stress fraction carried before cracking by concrete alone, defined as force in concrete per unit area of the steel-concrete composite; E_c and ν_c are the Young's modulus and Poisson's ratio of concrete. The conditions $E_c' = E_c$ and $\nu_c' = \nu_c$ apply to plane strain and $E_c' = E_c/(1-\nu_c^2)$ and $\nu_c' = \nu_c/(1-\nu_c)$ to plane strain. Assuming that cracks in concrete propagate in the direction of the principal stress just ahead of the crack, one has $\Delta\sigma_{12}^{c} = \sigma_{12}^{c^0} = 0$ in the above expressions. The change in potential energy of the system during Stage I in Figure 1.17b is given by the elastic energy initially stored in ΔV and released by cracking, i.e.,

$$\Delta W = -\int_{\Delta V} \tfrac{1}{2}\left(\sigma_{ij}^{c^0}\varepsilon_{ij}^0 - E_c'\varepsilon_{11}^{0^2}\right)\mathrm{d}V. \tag{1.30}$$

The change in potential energy during Stage II in Figure 1.17 is given by the work done by the forces $\Delta T_{c_i}^0$ and $\Delta f_{c_i}^0$ while they are being released, i.e.,

$$\Delta L = \int_{\Delta S} \tfrac{1}{2}\Delta T_{c_i}^0 (u_i - u_i^0)\mathrm{d}S + \int_{\Delta V} \tfrac{1}{2}\Delta f_{c_i}^0 (u_i - u_i^0)\mathrm{d}V. \tag{1.31}$$

Coefficients 1/2 must be used because, for a sufficiently small Δa, the forces T_{c_i} and f_{c_i} vary almost linearly during Stage II and reduce to zero at the end of Stage II.

Not all of the energy that is supplied to the element that cracks from the rest of the structure and from the unloading of concrete between the cracks is available for producing new crack surfaces. Part of this energy is consumed by the bond slip of reinforcing bars during cracking within volume ΔV. This part may be expressed [60] as:

$$\Delta W_b = \int_s U_b'\delta_b \mathrm{d}s \tag{1.32}$$

where δ_b represents the relative tangential displacement between the bars and the concrete, U_b' is the average bond force during displacement δ_b per unit length of the bar (force during the slip) and s is the length of the bar segment within the fracture process zone w_c (and not within volume ΔV since the energy consumed by bond slip would then depend on the chosen element size and would thus spoil the objectivity and proper convergence of the fracture criterion). Approximately, $U_b' \simeq$ ultimate bond force.

The energy criterion for the crack band extension may now be expressed as

$$\Delta U = G_f \Delta a - \Delta W - \Delta L - \Delta W_b > 0 \text{ stable}$$

$$= 0 \text{ critical}$$

$$> 0 \text{ unstable} \qquad (1.33)$$

where ΔU = energy that must be externally supplied to the structure to extend the crack band of width h by length Δa. (ΔU = total energy in the case of rapid, or adiabatic, fracture, and ΔU = Helmholtz's free energy in the case of slow, or isothermal, fracture.) If $\Delta U > 0$, then no crack extension can occur without supplying energy to the structure, and so the crack band is stable, does not propagate. If $\Delta U < 0$, crack band extension provokes a spontaneous energy release by the structure, which is an unstable situation, and so the crack extension must happen; the crack then extends in a dynamic manner, and the excess energy $-\Delta U$ is transformed into kinetic energy. If $\Delta U = 0$, no energy needs to be supplied and none is released, and so the crack band may extend in a static manner; in this case $G_f + \Delta W_b/\Delta a = (\Delta W + \Delta L)/\Delta a$ = finite difference approximation to the energy release rate of the structure. For this approximation to be second-order accurate, the corresponding crack band length a should be considered to reach up to the centroid of the frontal element that undergoes cracking.

For practical calculation, the volume integral in Eq. 1.30 needs to be expressed in terms of nodal displacements using the shape functions of the finite element. The boundary integral in Eq. 1.31 is evaluated from the change of nodal forces acting on volume ΔV from the outside [59]. Among the terms in Eq. 1.33, ΔW and ΔW_b normally are relatively small and often may be neglected, yielding $\Delta L/\Delta a = G_f$ as the approximate energy criterion [70].

The energy ΔL released from the surrounding body into ΔV may be, alternatively, also calculated as the difference between the total strain energy contained in all finite elements of the structure before and after the crack advance. According to the principle of virtual work, the result is exactly the same as that from Eq. 1.31 [6,59]. This calculation is possible, however, only if the structure is perfectly elastic whereas Eq. 1.31 is correct even for inelastic behavior (assuming Δa to be so small that T_{c_i} and f_{c_i} vary almost linearly during Stage I). It should also be mentioned that Y.T. Pan, A. Marchertas and coworkers at Argonne National Laboratory [66,71] calculate ΔL in their finite element analyses (using the crack band approach) by means of the J-integral. They keep the integration contour the same for various crack lengths. Their calculation yields the same ΔL because their integration contour passes only through the elastic part of the structure (except for crossing the crack band behind the front where, however, the stresses are almost zero).

In the case of a zig-zag, inclined crack band, the value of Δa in Eq. 1.33 must be replaced by the effective extension Δa_{ef} in the direction of the equivalent smoothed crack band. The notation from Figure 1.13a will be adopted. Similarly to the derivation of Eq. 1.24, assume that Δa_{ef} is the same for each crack band advance within the cycle l, whether the advance is in the x- or y-direction. Then $\Delta a_{ef} = l/N$ where $N = N_x + N_y$ = number of elements per cycle l (Figure 1.13a). This condition yields

$$\Delta a_{ef} = \left(\frac{\cos\alpha}{\Delta x} + \frac{\sin\alpha}{\Delta y} \right)^{-1} \quad \text{or}$$

$$\Delta a_{ef} = \frac{h}{\sqrt{2}\,\cos(45^\circ - \alpha)}, \quad (0 \leqslant \alpha \leqslant 90^\circ) \tag{1.34}$$

where the first equation applies to any rectangular mesh, and the second one to a square mesh ($\Delta x = \Delta y = h$).

Various numerical examples confirm the use of Eq. 1.34 (or some similar equation) is objective in that it gives results that are essentially independent of the choice of the mesh [66–68].

Strain localization instability and interpretation of tensile test. The formation of fracture through a gradual deformation of a finite fracture process zone may be treated as an instability of a nonlinear continuum, in which a uniformly distributed strain localizes into a band of finite width, w_c, at the boundary of which there is a jump in the value of strain while the stress is continuous. With regard the shear failures in an infinite medium, the concept of strain-localization instability was analyzed in detail by Rice [72] and others, with particular attention to the effect of geometric nonlinearities. A stability analysis of strain localization in tensile failures, with particular attention to finite size bodies and to a combination of strain-softening and unloading was presented in [4,5].

Following previous work [5], it is instructive to analyze the failure of a uniformly stressed specimen subjected to uniaxial tension (Figure 1.18). Such a specimen may serve as an approximate model for the fracture process zone. The specimen is loaded through a spring of spring constant C which represents either the spring constant of a testing machine per unit cross section area of the specimen, or the stiffness (per unit area) of the elastic support provided to the fracture process zone by the surrounding structure (the dimension of C is N/m per m^2, i.e., N/m^3). Let the cross section of the specimen be $A = 1$. The appearance of the crack band in the specimen may be considered as a sudden finite jump by distance

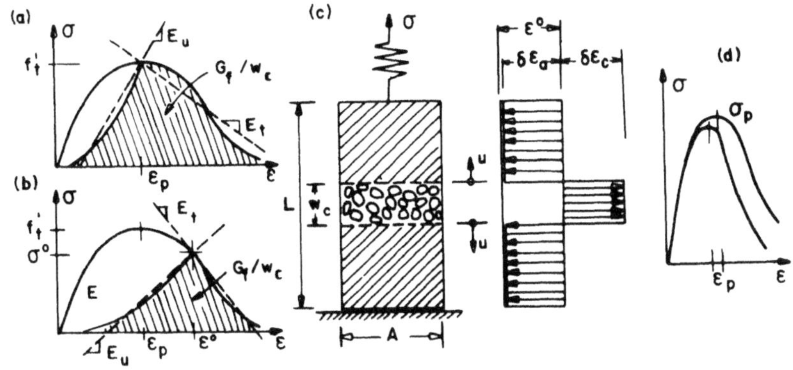

Fig. 1.18. (a–c) Strain-localization in a tensile specimen serving as model for fracture process zone; (d) possible effect of size on stress-strain diagram.

$\Delta a = 1$ in which the front of the crack band moves from the left face to the right face of the specimen (Fig. 1.18). For a uniaxial stress state of an unreinforced specimen of length L, Eqs. 1.30 and 1.31 take the form

$$\Delta W = \frac{w_c}{2}\sigma^0\varepsilon^0 = \frac{w_c}{2}\sigma^0\frac{\sigma^0}{\bar{E}_u} \tag{1.35}$$

$$\Delta L = \tfrac{1}{2}\sigma^0(\Delta u - \Delta u_0) = \tfrac{1}{2}\sigma^0\left[(L - w_c)\frac{\sigma^0}{\bar{E}_u} + \frac{\sigma^0}{C}\right] \tag{1.36}$$

in which σ and ε are the axial stress and strain; $\Delta u - \Delta u_0$ is the change in the relative displacement between the opposite face of the crack band of width w_c; and $\bar{E}_u$ is the average unloading modulus (Figure 1.18c).

Alternatively, ΔW can be calculated from the changes of strains ε_c in the crack band and c_a outside the crack band. To satisfy equilibrium in the tensile specimen in Figure 1.18c, the stress change $\delta\sigma$ must be the same inside and outside the crack band, and thus

$$\delta\varepsilon_c = \delta\sigma/E_t, \quad \delta\varepsilon_a = \delta\sigma/E_u. \tag{1.37}$$

The energy consumed in the crack band and the energy released from the rest of the specimen may now be calculated as

$$\Delta U_c = \frac{w_c}{2}\delta\sigma\delta\varepsilon_c = w_c\frac{(\delta\sigma)^2}{-2E_t} \tag{1.38}$$

$$\Delta U_a = \tfrac{1}{2}\delta\sigma\left[(L - w_c)\delta\varepsilon_a + \frac{\delta\sigma}{C}\right] = (L - w_c)\frac{(\delta\sigma)^2}{2E_u} + \frac{(\delta\sigma)^2}{2C}. \tag{1.39}$$

Expressing now the stability conditions similarly to Eq. 1.33, one has

$$\Delta U = \Delta U_c - \Delta U_a > 0 \text{ stable}$$
$$= 0 \text{ critical}$$
$$< 0 \text{ unstable} \tag{1.40}$$

Considering a finite stress change from σ^0 to 0, i.e., $\delta\sigma = -\sigma^0$, for which E_u and E_t in Eqs. 1.38 and 1.39 must be replaced with the average unloading modulus $\bar{E}_u$ and with the average tangent modulus $\bar{E}_t$ ($\bar{E}_t < 0$), respectively, one may substitute Eqs. 1.33 and 1.35–39 into the criticality condition $\Delta U = \Delta U_c - \Delta U_a = 0$. The resulting equation involves G_f, and solving it for G_f one gets

$$\bar{G}_f = \frac{w_c}{2}\left(\frac{1}{\bar{E}_u} - \frac{1}{-\bar{E}_t}\right)\sigma^{0^2} \tag{1.41}$$

in which a bar is attached to G_f to indicate that this is an apparent fracture energy value. It is not a constant since it depends on the stress σ^0 at which the fracture begins, which is governed by an incremental stability condition and is not necessarily equal to the peak stress σ_p.

In case that the instability which produces fracture happens right at the peak stress point, Eq. 1.41 provides

$$G_f = \frac{w_c}{2}\left(\frac{1}{\bar{E}_u} + \frac{1}{-\bar{E}_t}\right)\sigma_p^2 \tag{1.42}$$

which is a constant and represents the fracture energy value corresponding to the value used in the fracture model to fit test data. The value of G_f is characterized by the cross-hatched area in Figure 1.18a limited by the unloading and the softening branches emanating from the peak stress point.

When a curved stress-strain diagram is considered, one should in general, distinguish two types of instability: incremental instability (in the small, tangential) and instability in the large. The former concerns infinitely small displacements, the latter concerns complete failure. The value of σ^0 at which the instability occurs may be determined from Eq. 1.40 by substituting Eqs. 1.38 and 1.39 in which, for the incremental instability, one uses the incremental moduli E_t and E_u, and for the instability in the large, one uses the $\bar{E}_t$ and $\bar{E}_u$ instead. This yields the critical states [4,3]:

$$\frac{-E_t}{E_u} = \left(\frac{L}{w_c} - 1 + \frac{E_u}{Cw_c}\right)^{-1} \text{ (incremental)} \tag{1.43}$$

$$\frac{-\bar{E}_t}{\bar{E}_u} = \left(\frac{L}{w_c} - 1 + \frac{\bar{E}_u}{Cw_c}\right)^{-1} \text{ (in the large).} \tag{1.44}$$

As the strain in a tensile specimen is increased, E_t, E_u, $\bar{E}_t$ and $\bar{E}_u$ all vary as a function of strain ε. Two types of failure may occur: (1) either the incremental critical state (Eq. 1.43) is reached first, or (2) the critical state in the large (Eq. 1.44) is reached first. In the first case, failure is static and occurs when Eq. 1.44 becomes satisfied. In the second case, failure cannot happen statically, however when Eq. 1.43 is satisfied later, there is an excess energy ($-\Delta U$) for the instability in the large, and then the failure occurs dynamically as a snap-through instability, $-\Delta U$ being converted into kinetic energy.

Considering the stress-strain diagram to be bilinear and E_u to be constant (Fig. 1.2) has the advantage that failure instability occurs always at the peak stress point. The fracture energy G_f is then constant and equals the entire area under the tensile stress-strain diagram. Whether this simplification is adequate for practical analysis of concrete structures should be examined more closely.

From Eqs. 1.43 and 1.44 it is noted that when the loading frame is very soft ($C \to 0$), or when the specimen is very long ($L/w_c \to 0$), the strain localization starts at $E_t = 0$, i.e., at the peak stress point. When the loading frame is very stiff ($C \to \infty$), and when the specimen is very short ($L = w_c$), Eq. 1.43 indicates a large magnitude of $|E_t|$, and so instability never occurs. These are the requirements for being able to measure the complete stress-strain curve. Eqs. 1.43 and 1.44 may be used also to calculate the stiffness C of the loading frame needed to carry out tensile tests with stable strain softening.

The analysis of strain localization sheds light on the interpretation of the direct tensile test. The question is whether the strain can be evaluated from the measured displacement assuming the strain distribution to be uniform. The conditions in Eqs. 1.43 and 1.44 answer this question in the affirmative. Indeed, if the stress can be measured, it means that specimen is not unstable, since otherwise measurement would be impossible. And if it is not unstable, it means that the strain localization has not occurred, i.e., the strain must be uniform (in the macroscopic sense, of course; we do not consider microstresses here). So, in a stable direct tensile test, the strain-softening zone can be wider than w_c, which is a different situation than in fracture process zones. There, since fracture is being formed, the

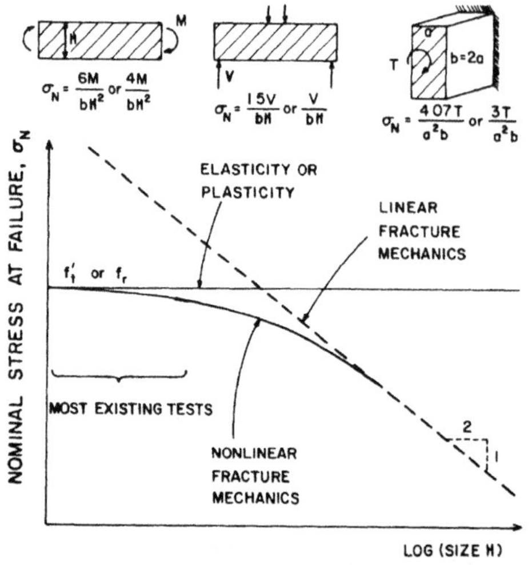

Fig. 1.19. Structural size effect.

strain must localize, and a constant width w_c of the strain-localization zone may be a good approximation.

A related question, raised by some experimentalists, is whether not only the failure point, but also the shape of the stress-strain diagram (Figure 1.18d) is affected by the stiffness of the loading frame. In a deterministic analysis, the stress-strain diagram should be considered as unique, unaffected by the loading frame. The analysis in [5] shows that observed variations of peak stress σ_p (tensile strength) as a function of machine stiffness can be explained by the effect of statistical inhomogeneity of the material on strain localization. These effects are significant only when the material is very inhomogeneous (poor quality concrete).

Structural size effect. The main purpose of fracture mechanics is to correctly capture the size effect in the ultimate load capacity of a structure. The size effect may be illustrated by considering structures of different sizes but the same shape (e.g., beams of the same crack length-to-depth ratio, and the same span-to-depth ratio), and plotting the logarithm of nominal stress at failure, log σ_N, versus log λ where $\lambda = d/d_a$, d = structure size (characteristic dimension), d_a = maximum aggregate size; see Figure 1.19. σ_N may be defined as P/bd (P = failure load, b = thickness), possibly times some constant characterizing the shape of the structure. According to all strength criteria (i.e., stress-based failure criteria), such as those used in elastic, plastic or elastoplastic design (as well as viscoelastic or viscoplastic design), σ_N is independent of d (see the examples of beam bending, shear or torsion in Figure 1.19). Thus, the plot of log σ_N versus log λ is a horizontal line (Figure 1.19), and the only difference between elasticity and plasticity is the level at which this line is drawn.

For linear fracture mechanics, this plot is completely different. It is known that σ_N varies inversely as $\sqrt{d}$ for all linear fracture mechanics solutions, and so the slope of the plot of log σ_N vs. log λ is a straight line of slope $-1/2$; see Figure 1.19.

The finite element solutions for the crack band theory with gradual strain-softening represent a gradual transition from the horizontal line for the strength criterion to the downward sloping straight line for the linear fracture mechanics; see Figure 1.19. With the exception of very large and massive concrete structures, such as dams, most concrete structures fall into this transition range, in which neither the linear fracture mechanics nor the strength criterion is applicable. Failures in this transition range are obviously more difficult to analyze than those for the two limiting cases, and this is the main challenge in failure analysis of concrete structures.

In laboratory testing, the model structures have normally been made the smallest size possible with regard to the aggregate size (cross sections of 5 to 15 aggregate diameters). Thus, most of the laboratory tests of beams, plates, panels, slabs, shells, etc. carried out thus far around the world fall into the initial, nearly horizontal range of the diagram in Figure. 1.19. Obviously, such tests miss the size effect. Present methods of design embodied in the codes are all based on strength criteria, elastic or plastic, and therefore they give an incorrect, unsafe extrapolation to larger sizes characteristic of actual structures. This fact is certainly a matter of concern, and calls for reexamination of existing design procedures for those failures that are of brittle nature; e.g. the diagonal shear failure and torsion failure of beams, punching failure of slabs or shells, shear failure of deep beams and panels, cryptodome failure of top plate in a reactor vessel, etc. Recently it has become popular to apply to these failures plastic analysis, even though the failure is caused by concrete cracking. This trend is, in the writer's opinion, dubious and has led to successful comparisons with test data only because a wide range of sizes has not been tested in the laboratory.

The case of punching shear failure of slabs might be a good illustration. Plasticity analysis can be made to agree with the existing laboratory data only if the tensile strength is considered to be about $f_c'/200$, which is about 20-times less than the correct value of tensile strength. The proper conclusion from such an agreement should not be that plasticity of concrete works, but that it does *not* work, and that fracture mechanics, is, therefore, necessary. Obviously, the small value of nominal stress at failure must be due to the fact that the existing laboratory test data do not pertain to the initial horizontal portion of the diagram in Figure 1.19.

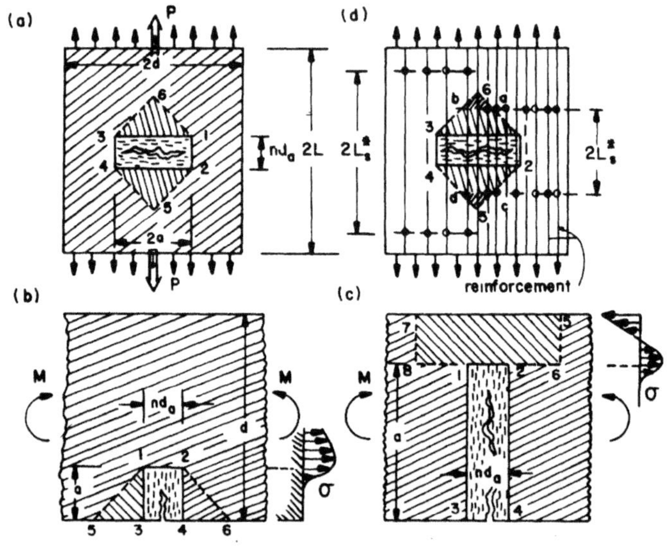

Fig. 1.20. Approximate analysis of energy release for various crack band situations.

Let us now try to derive a simple function to describe the typical transition curve in Figure 1.19. Consider first a center-cracked rectangular panel (Figure 1.20a) of thickness b, width $2d$, and a sufficiently large length $2L$. The panel is loaded by vertical normal stresses σ at top and bottom. The crack band is horizontal, symmetrically located, and has length $2a$ and width $w_c = nd_a$ ($n \simeq 3$, d_a = aggregate size). Before cracking, the strain energy density in the panel is uniform and equals $\sigma^2/2E$. The formation of the crack band may be imagined, as an approximation, to relieve stress and strain energy from the area 1254361 in Figure. 1.20(a), in which the "stress diffusion" lines $\overline{25}$, $\overline{45}$, $\overline{16}$, $\overline{36}$ have a certain fixed slope k_1 (close to 1). The energy release is

$$W = W_1 + W_2, \quad W_1 = 2k_1 a^2 b \frac{\sigma^2}{2E}, \quad W_2 = 2nd_a ab \frac{\sigma^2}{2E}. \tag{1.45}$$

Cracking is imagined to occur at fixed top and bottom boundaries such that the contribution of the work of load σ on the boundaries is zero. The potential energy release rate of the panel then is $\partial W/\partial a$, and the energy criterion in Eq. 1.33 reads $\partial U/\partial a = 2G_f b - \partial W/\partial a = 2bG_f - 2(2k_1 a + nd_a)b\sigma^2/2E = 0$. After substituting $G_f = nd_a(1 - E/E_t)f_t'^2/2E$ (Eq. 1.15), p may be solved from this equation; this yields $\sigma = Af_t^*$ with

$$f_t^* = \frac{f_t'}{\sqrt{1 + C\lambda}}, \quad \lambda = \frac{d}{d_a} (\geqslant n \simeq 3) \tag{1.46}$$

where

$$A=\sqrt{1+\frac{E}{-E_t}},\quad C=\frac{2k_1}{n}\frac{a}{d}. \tag{1.47}$$

A and C are constants when geometrically similar beams are considered. They are independent of the size. f_t^* may be called the *size-reduced strength*. It is a characteristic of the entire structure (and must be distinguished from f_{eq}, which is a characteristic of one finite element).

As a second example, consider a crack band of length a and width $w_c = nd_a$ ($n \simeq 3$) in a rectangular unreinforced beam of thickness b and depth d, subjected to bending moment M. First consider that $a \ll d$ (short cracks). The formation of the crack band may be imagined, as an approximation, to relieve the strain energy from the area 1264351 in Figure 1.20b, where the "stress diffusion" lines $\overline{15}$ and $\overline{26}$ have a certain empirical slope k_1 close to 1. Before cracking, the strain energy density at the tensile face of beam is $\sigma_1^2/2E$ where $\sigma_1 = 6M/bd^2$, and the same value approximately applies over the whole region 1264351 if $a \ll d$. Thus, the total energy release is $W = W_1 + W_2$, $W_1 = k_1 a^2\sigma_1^2/2E$, $W_2 = nd_a a\sigma_1^2/2E$. The potential energy release rate of the beam is $\partial W/\partial a$, and the energy criterion in Eq. 1.33 reads $\partial U/\partial a = bG_f - \partial W/\partial a = bG_f - b(2k_1 a + nd_a)(6M/bd^2)^2/2E = 0$. Substitute $G_f = nd_a(1 - E/E_t)f_t'^2/2E$ and evaluate the derivative $\partial W_1/\partial a$ of constant M. Then substitute $M = \sigma_N(d-a)^2/c_1$ (where c_1 = const. = 6 for elastic strength analysis, and $c_1 = 4$ for plastic strength analysis), and express σ_N from the resulting equation; this yields $\sigma_N = Af_t^*$ where f_t^* is given by Eq. 1.46 with

$$A=\frac{c_1}{6}\left(\frac{d}{d-a}\right)^2\sqrt{1+\frac{E}{-E_t}},\quad C=\frac{2k_1}{n}\frac{a}{d}. \tag{1.48}$$

Again C and A are constants when geometrically similar beams are considered.

Thirdly, consider the same beam but $a - d \ll d$ (short ligament); Figure 1.20c. Let $U = bG_f a - (M\theta/2) - W_0$. Here, W_0 = strain energy of beam if no crack existed, which is independent of a, and θ = additional rotation caused by crack band. Since the force resultants of the bending stresses over the ligament are zero, these stresses should affect only a region of size $d-a$, according to St. Venant's principle. It may be imagined that the localized bending moment M transmitted through the ligament $d-a$ affects the region 1265781 in Figure 1.20c, with segments $\overline{18}$ and $\overline{26}$ equal to $k_0(d-a)$ where k_0 = empirical constant, close to 1. Approximately, $\theta = [2k_0(d-a) + nd_a]M/EI_1$ where $I_1 = b(d-a)^3/12$ = inertia moment of the ligament section. Substitute $G_f = (1 -$

$E/E_t)nd_a f_t'^2/2E$, and evaluate the derivative $\partial(M\theta/2)/\partial a$ at constant M. Inserting the result, as well as the relation $M = \sigma_N(d-a)^2/c_1$, into the condition $\partial U/\partial a = G_f - \partial(M\theta/2)/\partial a = 0$, and solving σ_N from the resulting relation, it follows that $\sigma_N = Af_t^*$ where f_t^* is again given by Eq. 1.46, in which

$$A = \frac{c_1}{6}\sqrt{1 + \frac{E}{-E_t}}\ ; \quad C = \frac{4k_0}{3}\frac{d-a}{d}. \tag{1.49}$$

Eq. (1.46) can be derived for various other situations, e.g., edge-cracked panels, crack band in infinite medium, double-cantilever specimen, etc. The solutions are approximate in the evaluation of energy release; however, this causes uncertainty only in the constants k_1 and k_0, but not in the form of Eq. 1.46.

It appears that Eq. 1.46 might be of general applicability. This can be verified by a dimensional analysis. Let the geometry of a given two-dimensional structure of thickness b be characterized by some set of dimensions d, l_1, l_2, $l_3, \ldots, l_n$, and consider all geometrically similar structures such that the ratios $\xi_i = l_i/d$ $(i = 1, 2, \ldots, n)$ are the same, so that size of the structure may be characterized by one characteristic dimension d. From the preceding examples, note that fracture needs to be described by two independent parameters – length a of the crack band, and width nd_a of the cracking front. It may be noted further (e.g., from Eq. 1.45) that the strain energy relieved by cracking may be expressed as $W = W_1 + W_2$ where W_1 is the strain energy relieved from the outside of the crack band, which is proportional to a^2 (e.g., strain energy contained in triangular areas 136 and 245 in Figure 1.20a), and W_2 is the strain energy relieved from within the area and_a occupied by the crack band. To nondimensionalize these variables, one may set $a^2 = \alpha_1^2 d^2$ and $and_a = \alpha_2 d^2$ where α_1 and α_2 are the nondimensional parameters

$$\alpha_1 = \frac{a}{d}, \quad \alpha_2 = \frac{and_a}{d^2} \tag{1.50}$$

representing the nondimensional length and the nondimensional area of the crack band. The energy release by crack band formation may now be generally expressed as

$$W = f(\xi_i, \alpha_1, \alpha_2)\left(\frac{P}{bd}\right)^2 \frac{d^2 b}{2E} \tag{1.51}$$

where P is the given applied force or loading parameter, and function f depends on the shape of the structure and of the crack band, but is independent of size d. The condition of crack band propagation is

$\partial W/\partial a = G_f b$, and differentiating equation (1.51) at constant ξ_i (similar structures) leads to $(f_1/d + f_2 n d_a/d^2)P^2/2bE = G_f b$ in which the notations $f_1 = \partial f/\partial \alpha_1$ and $f_2 = \partial f/\partial \alpha_2$ have been adopted. Setting $G_f = nd_a(E^{-1} - E_t^{-1})f_t'^2/2$ (Eq. 1.15), $P = \sigma_N bd$, and $d = \lambda d_a$, yield the relation $\sigma_N = Af_t^*$, where f_t^* is again given by Eq. 1.46 and

$$\lambda = \frac{d}{d_a}, \quad A = \sqrt{\frac{1}{f_2}\left(1 + \frac{E}{-E_t}\right)}, \quad C = \frac{f_1}{nf_2} \tag{1.52}$$

where A and C are constant as the structure size is varied.

To sum up, the essential property which has led to Eq. 1.46 is the dependence of energy release on both the crack band area and the crack band length. If the energy release depended only on the crack band length ($f_2 = 0$), one would get $\sigma_N = (2G_f E/f_1)^{1/2}/\sqrt{d}$, which is the size dependence of linear fracture mechanics. If it depended only on the crack band area ($f_1 = 0$), one would get the size dependence of plasticity (σ_N = const.).

It may be concluded that Eq. 1.46 is of general applicability, as long as the two nondimensional parameters α_1 and α_2 (and no further parameters) are needed, and suffice, to characterize fracture.

For a small size relative to the size of aggregate, $\lambda \to 0$, $f_t^* \to f_t'$, and $\sigma_N = Af_t'$. For a very large size, $\lambda \to \infty$, the relation

$$f_t^* \simeq f_t'/\sqrt{C\lambda} \quad (\lambda \to \infty). \tag{1.53}$$

holds. Thus, Eq. 1.46 asymptotically approaches the size effect of linear fracture mechanics. Fracture-insensitive behavior is also a special case of Eq. 1.46; $C \simeq 0$.

Eq. 1.46 may be checked against the test data of Walsh [56] who tested geometrically similar three-point bent specimens of various beam depths d. His test results for six different concretes are plotted in Fig. 1.21 as Y vs. λ where $Y = (f_t'/\sigma_N)^2$. In such plots Eq. 1.46 is a straight line of slope C/A^2 and Y-intercept $1/A^2$. The regression lines corresponding to Eq. 1.46 are plotted in Fig. 1.21. It is seen that they agree reasonably well with the data. For strength theory, the regression lines would have to be horizontal, which is certainly not the case, and for linear fracture mechanics, the regression lines would have to pass through the origin, which is also evidently not the case.

Reinforcement located near the fracture front may have influence, too. To examine it, consider the same center-cracked rectangular panel as before (Figure 1.20d), reinforced by vertical steel bars which are spaced uniformly and so closely that a smeared modeling is possible. The panel is loaded on top and bottom by uniform normal stress σ. Before cracking,

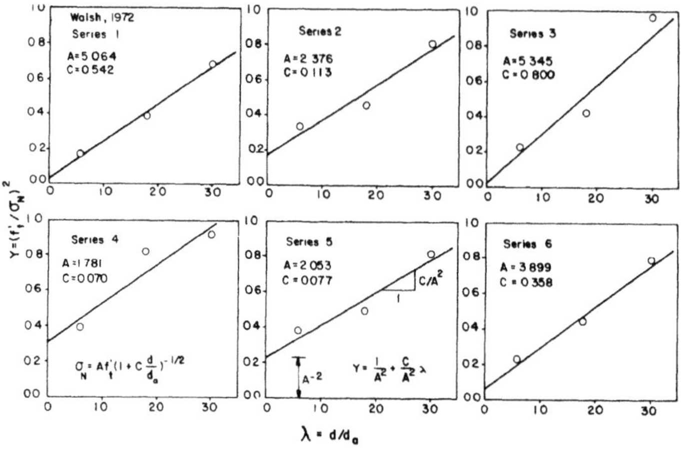

Fig. 1.21. Regression analysis of Walsh's (1972) fracture data.

the stress in concrete is $\sigma_c = \kappa_c\sigma$ where $\kappa_c = E_c/[E_c(1-p)+E_s p]$; $p=$ steel ratio; E_c, $E_s =$ Young's elastic moduli of concrete and steel. When cracks form, the steel bars slip near the cracks, as discussed before (Eq. 1.27). For the same reasons as before, the frictional slip may be replaced over the actual bond-slip length $2L_s$ by free (frictionless) slip over a modified, free bond-slip length $2L_s^*$ (Eq. 1.27), and assume perfect bond beyond this length (Figure 1.20d). Consider that $L_s^* = L_s/2$.

Formation on fracture relieves the stress in concrete from the region 1254361 in Figure 1.20(d), in which the "stress-diffusion" lines 16, 25, 36, 35 have a certain constant slope k_1. The stress relief is, however, complete only if this region is entirely within the free bond slip length $2L_s^*$; Figure 1.20(d). If crack band length a is so large that this region reaches beyond the free bond slip length, then the tensions in the steel bars introduce tensile stress into concrete within the triangular regions 5ab and 6cd in Figure 1.20(d). The value of stress in steel within the slip region, σ_{sL}, is less than (but probably close to) the stress that the steel carried before cracking, i.e., $\sigma_{sL} < \kappa_s\sigma$ where $\kappa_s = L_s/[E_c(1-p)+E_s p]$. Thus, the stress resultant per unit area, applied on these triangular regions is $< p\kappa_s\sigma$, which produces in concrete within the triangular regions the stress σ_c' such that $\sigma_c' < \kappa_c(p\kappa_s\sigma)$. This gives

$$\sigma_c' = c_t p\kappa_s\kappa_c\sigma \tag{1.54}$$

where c_t is a coefficient less than 1 but probably close to 1. The strain energy release from the panel of thickness b may now be expressed as

$$W = \left[\left(k_1a^2 + and_a\right)\frac{(\kappa_c\sigma)^2}{2E_c} - \frac{H_a}{k_1}\left(k_1a - 2L_s^*\right)^2\frac{\sigma_c'^2}{2E_c}\right]b \tag{1.55}$$

where $H_a = 1$ if $k_1 a > 2L_s^*$, and $H_a = 0$ if $k_1 a \leqslant 2L_s^*$.

In the energy balance, the energy consumed by bond slip should be included. The maximum slip of bars is at the crack axis and is roughly $(f_t'/E_c)L_s$. At the ends of length $2L_s$ the slip is zero, and so the mean slip is about $f_t'L_s/2E_c$. The bond stress is roughly U_b' per unit length of bar, as determined from pull-out tests. The number of steel bars per unit cross section of panel is p/A_b where A_b = cross-section area of one bar, and $L_s \simeq 2L_s^*$. So the work of bond stresses over length L_s per unit advance of the crack band is

$$W_b' = \frac{p}{A_b}\frac{f_t'L_s^*}{E_c}U_b'b \tag{1.56}$$

where b = panel thickness. In Eqs. 1.56 and 1.55, further substitution leads to $L_s^* = L_s/2 = c_L A_b/2U_b'$ where $c_L = \sigma_s - \sigma_s'$ as defined in Eq. 1.26.

The energy balance condition for crack band advance may now be written as $bG_f + W_b' = \partial W/\partial a$, where $G_f = nd_a(1 - E_c/E_t)f_t'^2/2E_c$. Differentiating Eq. 1.55 and substituting, we obtain $\sigma_c = A'f_t'$ where $\sigma_c = \kappa_c\sigma$ and

$$f_t^* = \frac{f_t'}{\sqrt{1 + C'\lambda}}, \quad \lambda = \frac{d}{d_a}(\geqslant n \simeq 3) \tag{1.57}$$

$$A' = \sqrt{A_1/B_1}, \quad C' = C_1/B_1, \quad A_1 = 1 - \frac{E_c}{-E_t} + \frac{pc_L}{nd_a f_t'},$$

$$B_1 = 1 + H_a\frac{2c_L}{nd_a}\frac{A_b}{U_b'}(c_t p\kappa_s)^2, \quad C_1 = \frac{2k_1}{n}\left[1 - 2H_a(c_t p\kappa_s)^2\right]\frac{a}{d}. \tag{1.58}$$

Consider now geometrically similar panels (same a/d), with same steel ratio p, and same bars, i.e., same A_b (also, $c_L \simeq$ constant). Then, Eq. 1.57 indicates the same type of dependence on structure size, λ, as Eq. 1.46, except that coefficient C' is larger than C. This shifts the asymptotic declining straight line in the plot of $\log f_t^*$ versus $\log\lambda$ to the right; see Figure 1.22a. If the steel bar size is increased with the structure size, the size effect becomes somewhat more pronounced since f_t^* decreases as A_b increases.

The size effect in reinforced structures is seen to be less pronounced for smaller sizes of the structure, but for large enough structure sizes the size effect becomes just as significant as for unreinforced structures since the asymptotic slope remains $-1/2$. This is, however, true only if the reinforcement remains elastic. For a long enough crack band, the opening in the center of its length becomes sufficiently large to cause the steel to

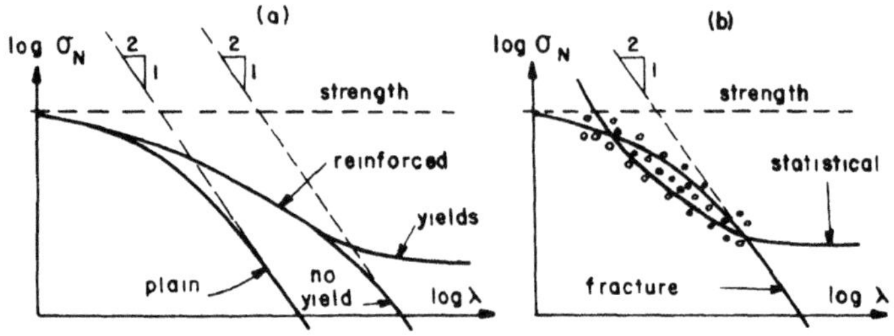

Fig. 1.22. (a) Structural size effects in reinforced concrete structures, and (b) statistical size effect of strength in structures.

yield, and that will completely alter the size effect. If the steel bars are yielding, the strains are so large that all resistance of concrete is lost, and the load is resisted by reinforcement alone. In that case, the value of σ becomes size independent. Therefore, in reinforced structures, the plot of $\log f_t^*$ versus $\log \lambda$ eventually stops decreasing and approaches a horizontal line. However, this limiting plastic value might be too low for practical purposes.

Finally, it is instructive to compare the results to the well-known statistical size effect. Concrete is heterogeneous, and the strength varies randomly throughout a concrete structure. This variation is independent of structure size. The stress gradient, on the other hand, normally varies inversely with the structure size, and the region of peak stress becomes larger in a larger structure. Therefore, the chance of encountering low strength in the peak stress region is higher in a larger structure, and so the apparent strength must decline with structure size. However, the decline stops when the peak stress region becomes much larger than the low strength regions. Therefore, all theories of the statistical size effect produce a plot of $\log \sigma_N$ versus $\log \lambda$ which tends to a horizontal asymptote. This is completely different from the fracture-mechanics size effect (Fig. 1.22b), except when yielding of reinforcement makes the response plastic.

It seems that many observed size effects in concrete structures should have been explained by fracture mechanics rather than statistical variation of strength. The dependence of the apparent bending strength on the depth of plain concrete beams is a blatant example. If the test data do not cover a very large range of λ, both theories seem to work. This may be misleading for extrapolations.

1.5 Applications and practical analysis

Diagonal shear failure of beams. Eq. 1.46 may be applied to introduce the size effect into various existing strength-based formulas for failure of concrete structures. For example, the ACI or CEB-FIP code formulas for the diagonal shear failure of beams with longitudinal reinforcement but without web reinforcement involve no size effect. In an on-going study at Northwestern University, J.K. Kim and Bažant analyzed failure data for over 300 beams which have been tested in various laboratories throughout the world and were reported in the literature. After determining and optimizing an approximate semiempirical formula (similar to that in ACI Code) for the nominal shear stress σ_N at failure as a function of the shear span and of the longitudinal reinforcement ratio, the dependence of σ_N on the size parameter $\lambda = d/d_a$ has been analyzed statistically, using Eq. 1.46 for regression analysis of existing data of diagonal shear failure of beams without web reinforcement, notably the data by Kani, Leonhardt, Bhal, Walraven, Taylor, Rüsch and Swamy [77–83].

Although most existing test data involve very small beam depths, there exist a few data which involve beams of various depths. One result of the ongoing study is the diagram in Figure 1.23 [84]. It is seen that the data points are not well approximated by a horizontal line, which would mean the absence of fracture mechanics type size effect; the data points agree well with the function in Eq. 1.46, plotted as the solid curve. This is better illustrated by statistical regression analysis in Figure 1.24, in which $Y = \sigma_N^{-2}$ is plotted versus λ. Function $f_t^*(\lambda)$ appears in this plot as a straight line of the equation $Y = \bar{a} + \bar{b}\lambda$ where $\bar{a} = (Af_t')^{-2}$, $\bar{b} = C\bar{a}$. Absence of the size effect would mean a horizontal regression line in Figure 1.24, and this is clearly contradicted by the data. Despite a large scatter, the data points exhibit an upward straight-line trend. Thus, the existing data clearly confirm a significant size effect and justify Eq. 1.46.

Based on this analysis, it seems that most of the code formulas for predicting the strength of structural members would be improved by replacing in them f_t' with f_t^*. A more detailed investigation is needed, however, and coefficient C needs to be determined for each case.

Reinhardt [85,86] has recently studied some of these data, and found that they reasonably agree with a linear fracture mechanics size effect ($C\lambda \gg 1$). This type of size effect would correspond in Figure 1.24 to a straight regression line passing through the origin, and in Figure 1.23, this would correspond to an inclined straight regression line rather than a curvilinear regression. It is seen from this figure that such trends are not confirmed when all available test data are considered. The size effect of linear fracture mechanics would be too strong. Clearly, the bulk of existing test results for diagonal shear failure indicates the need for a nonlinear fracture theory.

As an example of finite element fracture analysis, the results obtained by Cedolin and Bažant [60] may be presented for the shear failure of the panel sketched in Figure 1.25, which is reinforced only by horizontal steel bars concentrated near the bottom of the panel, and is loaded by a

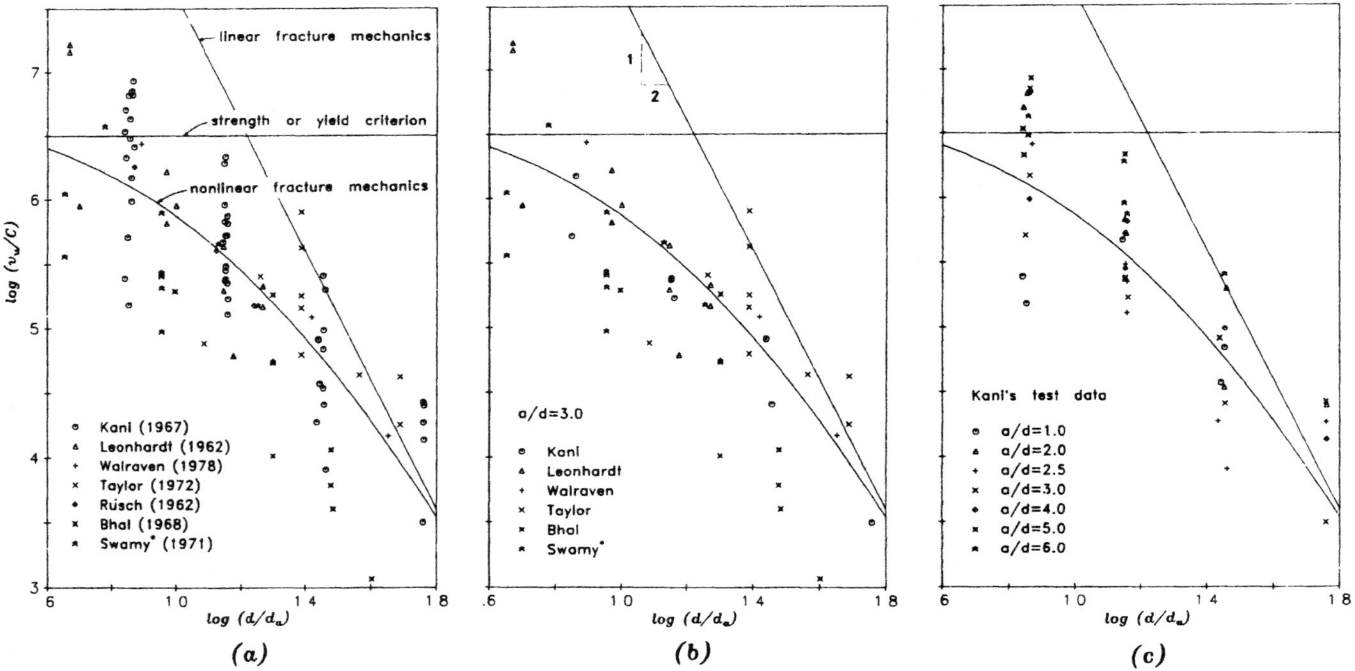

Fig. 1.23. Analysis of test data from various laboratories on diagonal shear failure of beams without web reinforcement (after Bažant and Kim, 1983).

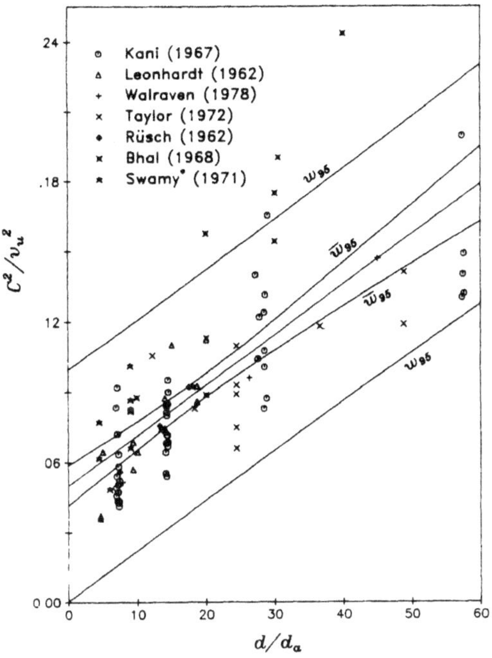

Fig. 1.24. Same data as in fig. 23a plotted in linear regression.

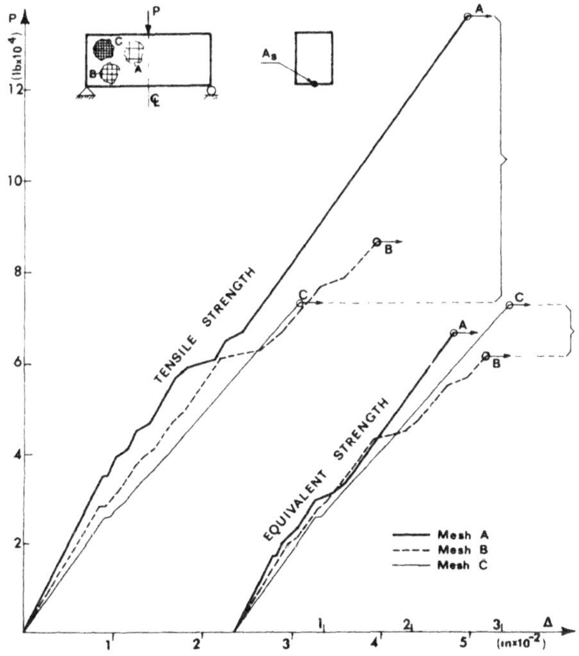

Fig. 1.25. Finite element analysis of shear failure of deep beams (after Bažant and Cedolin, 1982, 1983).

vertical force at midspan. The crack band path is not known is advance but is to be found. Among the elements sharing a side with the crack front element, the crack band was assumed to spread into that element in which the principal tensile stress is the largest. The analysis was carried out both for the actual tensile strength f_t' and for the equivalent strength f_{eq}'. The displacement at the loading point was introduced in small increments. At each load step, Newton-Raphson iterative procedure was used to redistribute the unbalanced nodal forces due to cracking until a stable crack band configuration was reached. Linear elastic behavior was assumed for concrete in compression.

The analysis was carried out for three different meshes (A, B and C in Figure 1.25), in which the finite element sizes are in the ratio 4 : 2 : 1. The load-deflection curves obtained for these three meshes are plotted in Figure 1.25. Even though this problem is less sensitive to the value of tensile strength than others, the deflection curves are more consistent for the equivalent strength criterion. The element size effect is largest for the value of the load at which the cracking zone reaches a certain fixed distance from the top. This distance was fixed as one-half of the size of the element in the crudest mesh (A), which then equals a distance of one element for mesh B, and of two elements for mesh C. The loads for which the cracking zone reaches this distance from the top are indicated in Figure 1.25 by horizontal arrows. For the equivalent strength criterion, they differ from each other much less than they do for the fixed tensile strength criterion. Furthermore, it was found (see Closure of [59]) that the crack patterns for meshes A, B and C are rather different and, in particular, the cracking zone for the finest mesh is not diffuse but localizes into narrow, separate crack bands of single element width at the front. This behavior is obtained, however, only when the loading steps are taken to be so small that no more than one element cracks during the first iteration of each loading step.

Deflections of cracking beams. As another application, consider deflections of unprestressed reinforced concrete beams. Their deflections are

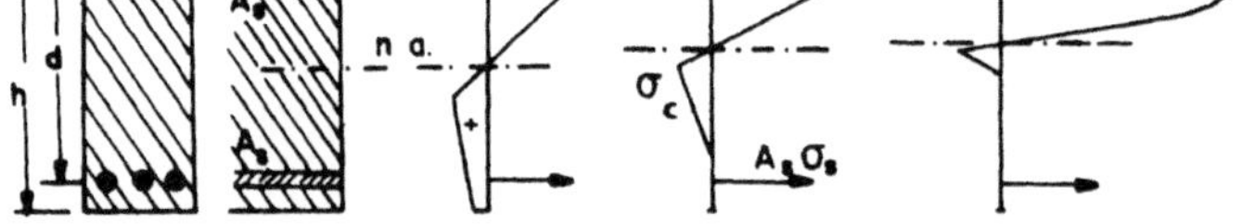

Fig. 1.26. (a) Stress-distribution in cross sections of beams with tensile-strain softening: (b) calculated beam deflections (Bažant and Oh, 1983) and their comparison with tests by Gerstle et al. (1964, 1965), Burns and Siess (1966), and Hollington (1970).

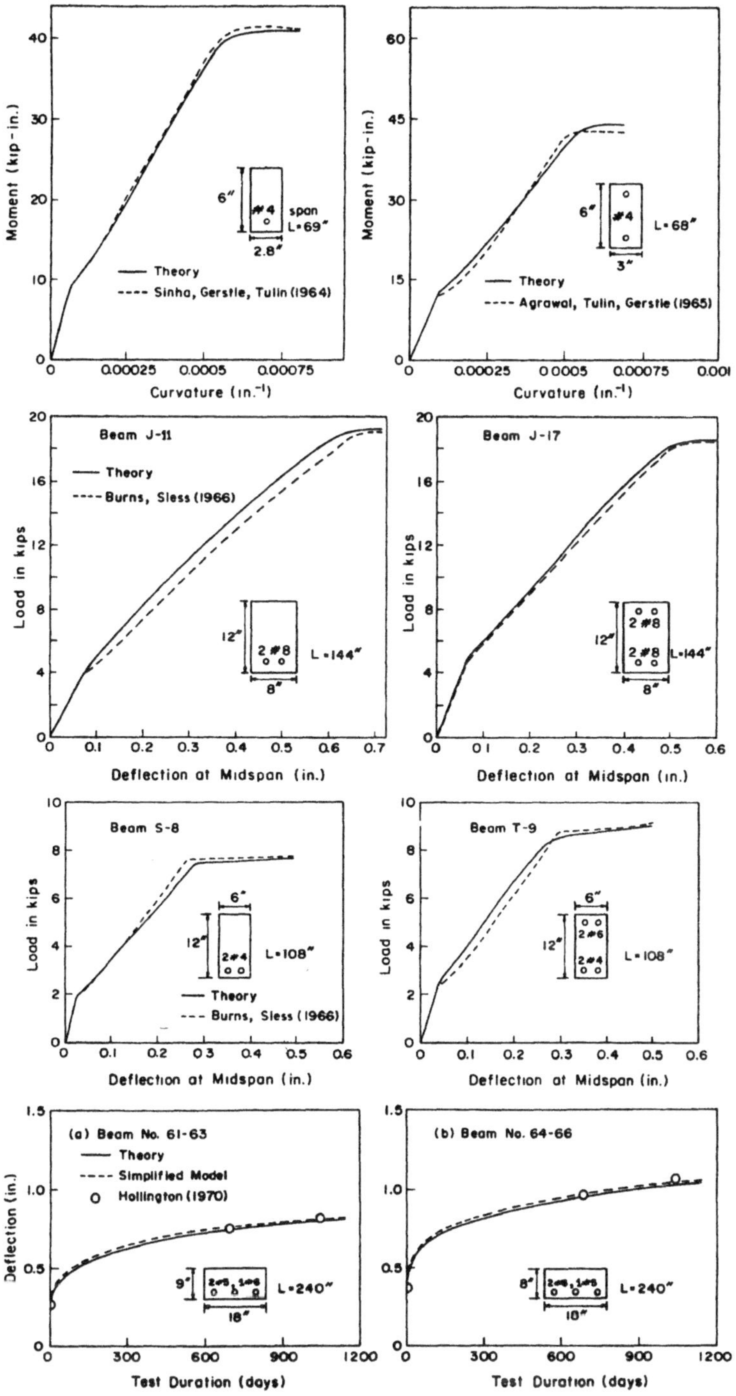
Moment (kip-in.)
Curvature (in.⁻¹)
6″
#4
span
L=69″
2.8″
Theory
Sinha, Gerstle, Tulin (1964)
6″
#4
L=68″
3″
Theory
Agrawal, Tulin, Gerstle (1965)
Load in kips
Deflection at Midspan (in.)
Beam J-11
Theory
Burns, Sless (1966)
12″
2 #8
L=144″
8″
Beam J-17
2 #8
2 #8
L=144″
Beam S-8
6″
12″
2#4
L=108″
Theory
Burns, Sless (1966)
Beam T-9
2#6
2#4
L=108″
Deflection (in.)
Test Duration (days)
(a) Beam No. 61-63
Theory
Simplified Model
Hollington (1970)
9″
L=240″
18″
(b) Beam No. 64-66
8″
L=240″
18″

Fig. 1.26 (b).

considerably less than those calculated under the assumption of no cracking; obviously the beams crack, even under service loads. On the other hand, if one assumes a no-tension material (which is the accepted approach to strength analysis, justified by the fact that strength is determined by the weakest cross section rather than the overall mean behavior), then the calculated deflections are much larger than the measured ones. This phenomenon, which is usually referred to as "tension-stiffening," is due to progressive microcracking of concrete in the tensile zone of beam, and so it should be possible to obtain correct deflections using our present bilinear stress-strain relation with strain softening (Figure 1.26).

This was done in [87]. The value of E_t was predicted from Eq. 1.21. On the compression side, Saenz' expression for strains due to uniaxial compression was assumed. The uniaxial stress-strain diagram of steel was considered as elastic–perfectly plastic, and the average strain of steel was assumed to be equal to the average strain of concrete at the same level. Based on these assumptions, the typical distributions of normal stress in concrete were as shown in Figure 1.26a.

As far as the cracking front is concerned, it is assumed here that the cracking does not localize and the strain everywhere follows the kinematic constraint expressed by the usual assumption that the cross sections remain plane and normal. The localization of cracking into certain cross sections is assumed to be prevented by tensile reinforcement and by the constraint provided to the cracking front by the compression zone (for deep beams, or for shear failures, the absence of localization cannot be assumed, of course).

Calculations based on the foregoing assumptions were compared [87] with the measurements of beam curvatures and deflections reported in the literature [89–92]. No fitting of data was attempted, i.e., no material parameter was adjusted to improve the fit. The comparisons are shown in Figure 1.26. The agreement is good, much better than the corresponding case when the tensile resistance of concrete is neglected. It was also shown [87] that the calculations give essentially the same results as the well-known Branson's empirical formula [76] within its range of validity (service loads). This success indirectly lends further support to our stress-strain relation for fracture.

The foregoing analysis of deflections, which applies to short time loading, was further extended to long-time loading. The effective modulus method was used to take concrete creep into account. The creep properties were predicted on the basis of double power law [93]. Since this creep law does not apply for the tensile strain-softening range, and since creep in this range may be expected to be larger, the tensile creep deformations were multiplied by an empirical coefficient c_ϕ the value of which was determined so as to get the best fit of measured long-time

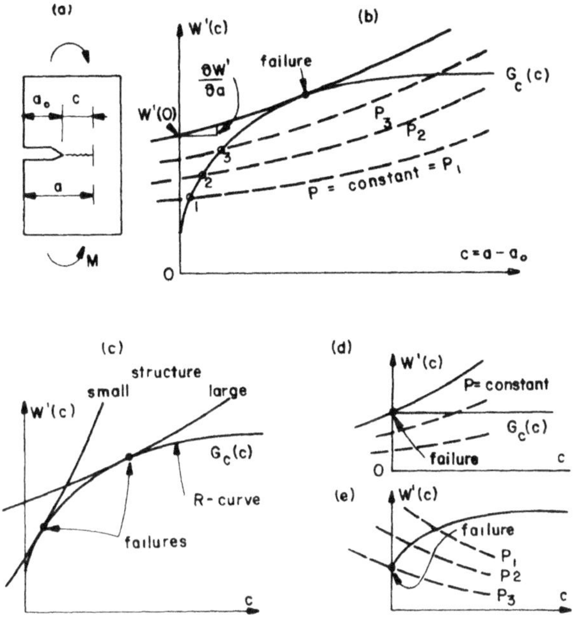

Fig. 1.27. Determination of crack extension from R-curves.

deflections ($c_\phi \simeq 3$). No other material parameter was adjusted. The results of calculations were compared [87] with the long-time deflections of beams tested by Hollington [94]; the agreement was again good (Figure 1.26). Furthermore, the long-time deflections have been calculated for typical singly and doubly reinforced beams, and were compared with an ACI empirical formula. The agreement was good again [87], and in particular, the strong effect of compression reinforcement on creep deflections of cracked beams, as known from tests, was predicted correctly.

Application of R-curves. For practical nonlinear fracture analysis, it is helpful to use the approach of resistance curves, or R-curves [95,41]. R-curve is the curve of effective or apparent fracture energy G_c as a function of the extension a of a crack from a notch (Figures 1.27a and 1.27b). For materials which deviate from linear fracture mechanics, it is found that the amount of energy per unit crack extension, which flows into the fracture process zone, increases as the crack extends until it stabilizes at some constant limiting value, G_f, unless some boundaries or loads are near the fracture front.

As shown in the foregoing (Figure 1.6), in the crack band approach, the R-curve can be predicted from the tensile strain-softening stress–strain relation. The predictions are different for different body geometries, or different loads and loading paths. It appears, however, that the shape of the R-curve (Figure 1.27) is nearly the same for most situations. Even though this can be exactly true only for infinitely short extensions of a crack from a notch, the R-curve can be assumed, for most purposes, to be approximately a fixed material property, independent of body geometry and the type of loading [41], as proposed in 1961 by Krafft et al. [95]. This assumption was proven to be rather successful for ductile fracture of metals [41]. Following the work in [96], a simple form of fracture analysis of concrete using the R-curve approach will be outlined.

Consider that the effective fracture energy, G_c, depends on c where $c = a - a_0$, a = length of crack with the notch, a_0 = length of the notch; $G_c = G_c(c)$. The energy that must be supplied to the structure is $U = \int G_c \mathrm{d}a - W(a)$ (if the thickness of the body is $b = 1$), where W = release of strain energy from the body, and G_c effective fracture energy at distance a from the notch. An equilibrium state of crack occurs when no energy needs to be supplied to change a by δa, i.e., when $\delta U = 0$ or $\delta U = (G_c - W')\delta a = 0$ or

$$W'(a) = G_c(c) \tag{1.59}$$

where $W'(a) = \partial W / \partial a$ = energy release rate, and $c = a - a_0$. The equilibrium crack state is stable if the second variation $\delta^2 U$ is positive, i.e., $\delta^2 U = \frac{1}{2}(\partial G_c/\partial a - \partial^2 W/\partial a^2)\delta a^2 > 0$ or $\partial G_c(c)/\partial c > \partial W'(a)/\partial a$. The limit of stability, i.e., failure, occurs if

$$\frac{\partial W'(a)}{\partial a} = \frac{\partial G_c(c)}{\partial c}. \tag{1.60}$$

For most structural situations, the strain energy release rate increases as the crack grows, i.e. $W'(a) > 0$. By elastic structural analysis, the curve $W_1'(a)$ corresponding to a unit load can be calculated. Then, for any load P, $W'(a) = P^2 W_1'(a)$. In Figure 1.27, we sketch the curves $W'(a)$ for a set of increasing P-values, P_1, P_2, P_3,... According to Eq. 1.59, the equilibrium states of crack extension for various loads are given by the intersections of these curves with the curve $G_c(c)$. These states are, according to Eq. 1.60, stable if the slope of the $G_c(c)$-curve is larger than the slope of the $W'(a)$-curve (see points 1, 2, 3 in Figure 1.27b). As the crack grows, the difference between the slopes $\partial G_c/\partial a$ and $\partial W'/\partial a$ gradually diminishes until the slopes become equal at the failure point (Figure 1.27); this point represents a critical state (or failure state) according to Eq. 1.60. Beyond this point, the crack extension is unstable. In the case that $W' < 0$ for all a, Eq. 1.60 is always satisfied. The crack is then stable for all a (Figure 1.27e).

In the case that G_c = constant = G_f, Eq. 60 reads $0 > \partial W'/\partial a$. This condition can never be satisfied if W' increases with a (Figure 1.27d). Thus, the fact that a stable crack growth from a notch exists implies that G_c cannot be constant but must increase.

For an approximate estimate of the values of $W'(a)$, one may use linear fracture mechanics provided that crack length a is interpreted as the equivalent length that gives the same remote stress field as the crack band. It might be difficult to actually determine this equivalent crack length for a given situation, but for practical purposes this is not necessary since the actual crack length at failure need not be known.

In a recent study [96], R-curves were used to analyze fracture test data from the literature [44,45,47–49,51,53,56] obtained with bent specimens or center-cracked specimens (the same data as used in Figures 1.5 and 1.6). The R-curve was assumed in the form

$$G_c(c) = G_f\left(1 - \beta_c e^{-c/c_1 d_a}\right) \qquad (c = a - a_0) \tag{1.61}$$

where G_f, β_c and c_1 are constants to be found by fitting test data, and d_a = maximum aggregate size. A more general expression, namely $G_c = G_f[1 - \beta_f \exp(-c/c_1 d_a)^q]^r$ was also studied but $q = r = 1$ was found to be about optimum. The energy release rates, $W_1'(a)$, for unit load $(P = 1)$, were calculated as $W_1'(a) = K_1^2/E$ from the existing analytical expressions for the stress intensity factor K_1, as listed for these specimens in Tada's manual [97]. The computational algorithm was as follows:

(1) Set the values of G_f, β_f, c_1, Δa $(= 0.01\ a_0)$, and set $a = a_0$.

(2) Increment a, replacing a with $a + \Delta a$. Set $c = a - a_0$.

(3) For each a, calculate $W_1'(a)$, and since $W' = W_1' P^2$ determine the load corresponding to a as $P = [G_c(c)/W_1'(a)]^{1/2}$. If this value of P is larger than the previous P-value, return to 2.

(4) Now $\partial G_c/\partial c < \partial W'/\partial a$, i.e. the specimen fails. Set $P_{max} = P$. (One could interpolate for the exact a at which $\partial G_c/\partial c = \partial W'/\partial a$ but this is not necessary if Δa is as small as $0.01\ a_0$.) Evaluate deviation of given test data from the theory as $\Delta P_{error} = P_m - P_{max}$ where P_m = measured value of maximum load.

(5) Repeat steps (1) to (4) for another case of the same test series (e.g., another notch depth a_0 or another beam depth H), and accumulate the sum $\Phi = \Sigma(\Delta P_{error}/P_0)^2$ where P_0 = prediction of failure load according to the bending strength theory (based on the ligament section).

The foregoing algorithm (computer subroutine) is then used with a library optimization algorithm, such as Marquardt-Levenberg's, to vary the values of G_f, β_f and c, until those values which give min Φ are determined. In this manner, it was found [96] that the values $\beta_f = 0.72$ and $c_1 = 1.85$ are nearly optimum for all concretes tested, while the values of G_f vary substantially from concrete [96]. Note that the optimum

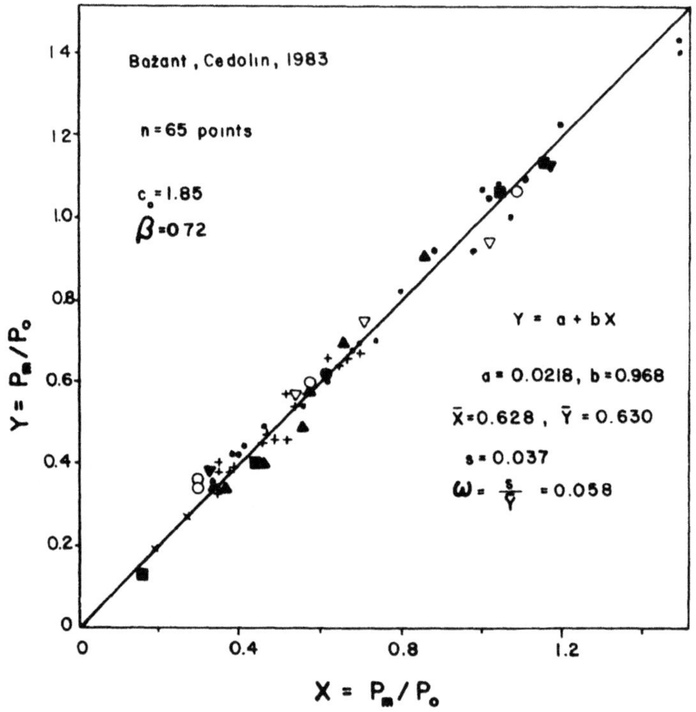

Fig. 1.28. Statistical comparison of measured failure loads with R-curve calculations for various test data from literature (after Bažant and Cedolin, 1983).

G_f values for the R-curve approach are not the same as those for the finite element crack band approach.

To get an idea of the error magnitude, Figure 1.28 reproduces from [96] a plot of Y versus X where $Y = P_t/P_0$, $X = P_m/P_0$, P_m = measured P_{max}, $P_t = P_{max}$ obtained from the R-curve analysis for the optimum material parameters determined as just described. This figure includes the data from [44,45,47–49,51,53,56]. By regression analysis, the standard deviation at the data centroid is found to be 0.037, and the coefficient of variation 5.8%. This is quite satisfactory and proves the applicability of the R-curve approach. Note that the error for the crack band finite element model (Figure 1.9) is about the same.

The R-curve analysis may be simplified if the energy release rate $W'(a)$ may be approximated by a straight line within the R-curve region, and if the R-curve is described by a parabola

$$G_c(c) = G_f\left[1 - \beta_f\left(\frac{c}{c_1 d_a} - 1\right)^2\right] \qquad (0 \leqslant c \leqslant c_1 d_a). \tag{1.62}$$

Eqs. 1.59 and 1.60 then become $P^2 W_1'' = \partial G_c/\partial c$ and $P^2(W_1' + W_1'' c) =$

$G_c(c)$. Substituting Eq. 1.62, we may reduce these two equations to a quadratic equation for c:

$$\left(\frac{c}{c_1 d_a} - 1\right)^2 - \frac{2}{c_1 d_a}\left(\frac{W_1'}{W_1''} + c\right)\left(\frac{c}{c_1 d_a} - 1\right) = \frac{1}{\beta_f}. \tag{1.63}$$

Solving c from this equation, $G_c(c)$ may be evaluated and the failure load may be calculated as $P_{max} = G_c(c)/(W_1' + W_1''c)$. For this solution, it suffices to determine from linear fracture analysis the values of W_1' and W_1'' at $a = a_0$ (and for unit load).

If $W'_1(a)$ is approximated by a parabola, an algebraic equation of fourth degree must be solved to determine c.

In another recent study [98], the R-curve approach has been used under the assumption that the nonlinear zone is negligibly small compared to specimen dimensions, crack length and ligament size. In this case, the stress field near the nonlinear zone may be solved considering that the crack tip is surrounded by an infinite elastic medium.

1.6 Crack development

Crack spacing at uniform strain. Crack spacing has a major influence on the crack width, which in turn, affects structural performance, including shear transfer, tensile stiffness, energy absorption capability, ductility and corrosion resistance. Recently, fracture mechanics energy analysis has been used to derive formulas for the spacing of parallel cracks produced by tension [88]. Only some simple solutions will be indicated here.

Consider one steel bar of diameter D embedded in the axis of concrete cylinder of diameter b (Figure 1.29). If the cross-section area of concrete per bar is replaced by a circular area, this situation may be used also as an approximate model for concrete reinforced by a regular array of parallel bars subjected to a uniform uniaxial tension. Consider equidistant cracks normal to the bar, denote their initial spacing as 2s (Figure 1.99), and try to determine the formation of further cracks that halve the spacing to s. In the light of the preceding theory, there are now two criteria to consider:

(1) The strength criterion decides whether the strain can exceed the strain value ε_p for peak stress, f_t'. If there is no bond slip, the strength criterion simply is

$$\varepsilon_s \geqslant f_t'/E_c \qquad \text{(strength, no slip)}. \tag{1.64}$$

If there is bond slip, and the ultimate bond force U_b' per unit length of

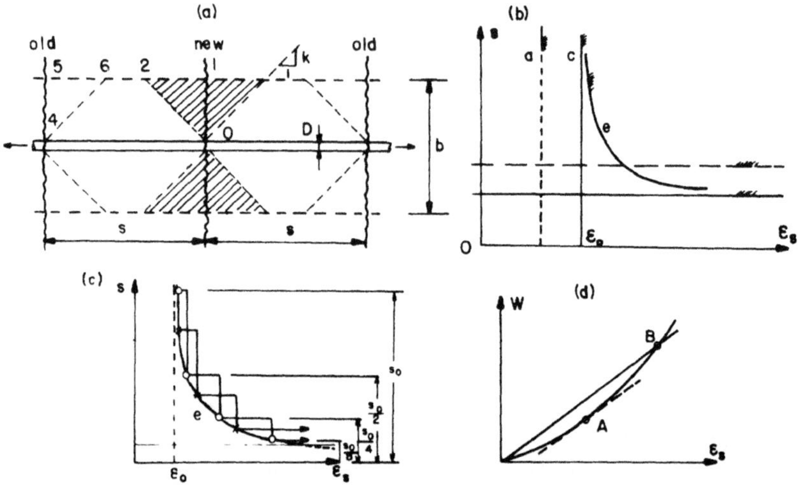

Fig. 1.29. (a) Parallel cracks normal to a reinforced bar, and (b) limits on crack spacing and strain at cracking. (a-strength, no slip; b-strength, slip; c-energy, no slip; d-energy, slip; e-curve for more accurate solution) (after Bažant and Oh, 1983).

bar is constant and is given, there results the equilibrium condition $U'_b L_b = A_c f'_t$ where $A_c = (b^2 - D^2)\pi/4$ = area of concrete, and L_b = length over which there is slip. Since $L_b \leqslant s$, it follows that

$$s \geqslant \frac{\pi f'_t (b^2 - D^2)}{4U'_b} \qquad \text{(strength, slip)}. \tag{1.65}$$

(2) For the energy criterion, the release of strain energy caused by cracking may be estimated by imagining that the formation of the crack relieves the strain energy from the shaded triangular region in Figure 1.29, limited by "stress-diffusion" lines of empirical slope k, close to 1. Assuming that $2s \geqslant k(D - b)$, these triangular areas do not overlap, and the volume per crack obtained by rotating these areas about the bar axis is $V_1 \simeq \pi b^3/12k$, in which D^3 is neglected in comparison with b^3. The strain energy density before cracking is $\sigma_1^2 V_1/2E_c$ where $\sigma_1 = E_c \varepsilon_s$ if there is no bond slip. The strain energy release due to formation of the new cracks between the original ones is $\Delta W \simeq V_1 \sigma_1^2/2E_c$. For the cracks to form, this must equal or exceed the energy consumed by the crack formation, which is $G_f \pi b^2/4$, D^2 being again neglected compared to b^2. This yields the condition

$$\varepsilon_s \geqslant \left(\frac{6kG_f}{E_c b}\right)^{1/2} \qquad \text{(energy, no slip)}. \tag{1.66}$$

A similar formula may be derived for the case $2s < k(D\text{-}b)$.

If there is bond slip, then $\sigma_1 = U_b' s/(\pi b^2/4)$. Using this to express ΔW_1, as before, the following condition is obtained:

$$s \geqslant \left(\frac{3\pi^2 k E_c G_f b^3}{8U_b'} \right)^{1/2} \qquad \text{(energy, slip)}. \qquad (1.67)$$

Eqs. 1.64 to 1.67 give boundaries of halving of crack spacing in the plot of spacing s versus steel strain ε_s; see Figure 1.29. In [88], more involved formulas based on more realistic assumptions are given; for them, the boundaries in Figure 1.29 become smoothly curved giving a one-to-one correspondence between ε_s and s (curve e in Figure 1.29b and 1.29c). Only with such a formulation it is possible to describe successive halvings of crack spacing at increasing strain [88]; Figure 1.29c.

In Eqs. 1.66 and 1.67, the energy criterion is used in an unorthodox manner. Instead of using the energy balance condition for small increments (or rates), the energy balance condition was used for the transition from no crack to a complete crack. This always indicates the cracking to occur somewhat later in the loading process than does the incremental energy condition (at point B instead of point A in Figure 1.27d). It seems that when analyzing cracks that are not much longer than the aggregate size, it makes no sense to consider infinitesimal crack length increments since a continuum model makes no sense on that scale.

As the strain is increased, some mutually very remote cracks form first, and all subsequent crack formation may be assumed to evolve by means of halving of the spacing, to which Eqs. 1.64 to 1.67 apply. The manner of crack formation differs depending on whether the strength criterion or the energy criterion is fulfilled first. The strength criterion (Eq. 1.64 or 1.65) indicates merely that ε_s exceeds the strain ε_p for peak stress. Therefore, if only the strength criterion is fulfilled, it means that microcracking begins but complete cracks do not necessarily form. For that to happen, the energy criterion (Eq. 1.66 or 1.67) must become also satisfied. In this case, the crack formation is obviously gradual, static.

If the energy criterion (Eq. 1.66 or 1.67) is satisfied first, cracks cannot begin to form, and so they cannot form at all. Then, if the strength criterion (Eq. 1.64 or 1.65) becomes satisfied later, there is an excess energy available for crack formation. The excess energy gets converted into kinetic energy, and so the cracks form suddenly, dynamically (emitting sound), in the manner of snap-through instability.

The theory just briefly outlined permitted achieving satisfactory comparisons with the measurements of crack spacing s and crack width w. As an example, Figure 1.30, taken from [88], shows a comparison with Chi and Kirstein's data [100], in which the crack width is estimated as

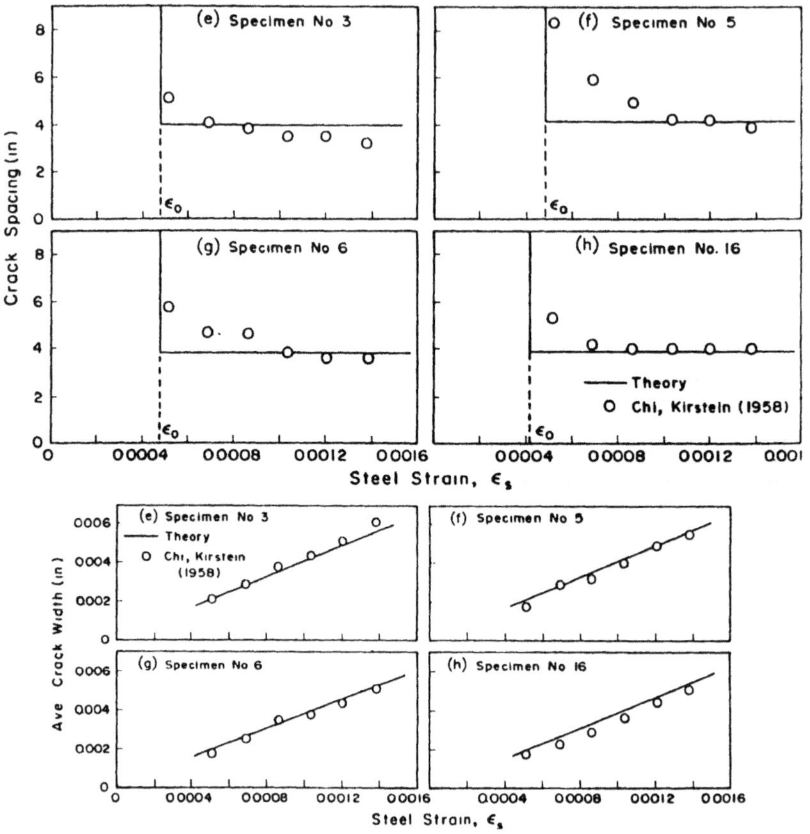

Fig. 1.30. Comparisons of theory with Chi and Kirstein's (1958) measurements of crack spacing and crack width (after Bažant and Oh, 1983).

$w = \varepsilon_s s$. Other data, e.g. those by Clark [101], Kaar and Mattock [102], Hognestad [103], and Mathey and Watstein [104] have been also successfully fitted [88].

Drying shrinkage cracks. Due to relatively small tensile strength and large shrinkage strains, drying typically produces cracks in concrete. If they are densely spaced and hair-thin they do little damage; however, long-time deformations are usually greatly affected. In particular, the drying creep as well as shrinkage cannot be realistically predicted without taking the effect of cracking and its evolution into account [105,106].

Consider just one typical problem: the initial spacing of drying cracks at the surface of a concrete halfspace. Using diffusion theory with given diffusivity of moisture in concrete, one can calculate profiles of specific moisture content of concrete at various times after the start of drying.

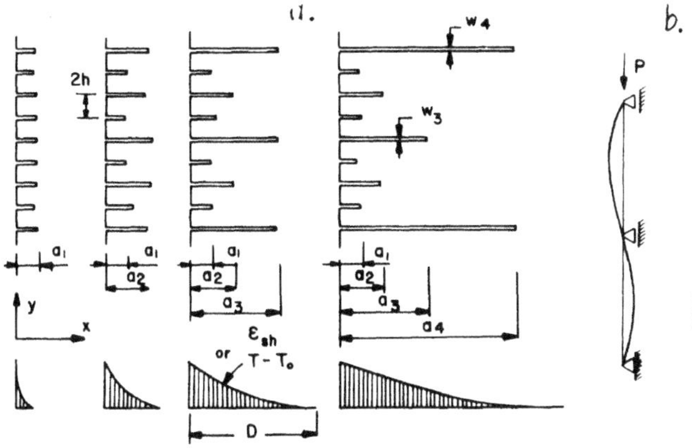

Fig. 1.31. System of parallel shrinkage or thermal cracks in a halfspace, and illustrations of constraint on instability mode by analogy with beam.

From these, one can get the profiles $\varepsilon_{sh}(x)$ of the free (unrestrained and incompatible) shrinkage strains. Assume that the environmental humidity is constant, and the moisture transmission at the surface is perfect. Then these profiles may be regarded as approximately parabolic.

Let x, y, z be Cartesian coordinates, x being normal to the halfspace surface. The stresses produced by the shrinkage strains may be solved from the conditions $(\sigma_y^0 - \nu\sigma_z^0)/E_{eff} - \varepsilon_{sh} = 0$ and $\sigma_z^0 = \sigma_y^0$, where $\nu =$ Poisson's ratio of concrete and $E_{eff} = E/(1+\phi) =$ effective modulus for elastic deformation plus creep, $\phi = \phi(t, t_0) =$ creep coefficient, which is a function of drying duration $t - t_0$ and age t_0 at drying start. Thus we obtain $\sigma_y^0 = \sigma_z^0 = \varepsilon_{sh}(x)E_{eff}/(1-\nu)$.

Assume now that a sudden formation of cracks normal to y-axis (Figure 1.31) reduces stress σ_y^0 to 0, i.e., the stress change is $\Delta\sigma_y = -\sigma_y^0$. Let $\Delta\sigma_z$ be the change of σ_z, and let $\Delta\varepsilon_y^m$ be the change of strain in concrete between the cracks. From Hooke's law $\Delta\varepsilon_y^m = (-\sigma_y^0 - \nu\Delta\sigma_z)/E$ and $\Delta\varepsilon_z = (\Delta\sigma_z + \nu\sigma_y^0)/E = 0$, the solution of which is $\Delta\sigma_z = -\nu\sigma_y^0$, $\Delta\varepsilon_y^m = -(1-\nu^2)\sigma_y^0/E$. The loss of strain energy due to cracking per unit area of halfspace surface is $\Delta W_1 = \int_0^D(\sigma_y^0 + \frac{1}{2}\Delta\sigma_y)\Delta\varepsilon_y^m dx$ or $\Delta W_1 = \int_0^D(\sigma_y^0/2)(1-\nu^2)(\sigma_y^0/E)dx = [E_{eff}/(1-\nu)]^2\int_0^D \varepsilon_{sh}^2(x)dx(1-\nu^2)/2E$. Assuming the profile of $\varepsilon_{sh}(x)$ to be parabolic, we may substitute $\varepsilon_{sh}(x) = (1 - x/D)^2\varepsilon_{sh}^0$ for $x \leqslant D$, and $\varepsilon_{sh}(x) = 0$ for $x \geqslant D$, where $\varepsilon_{sh}^0 =$ constant $=$ shrinkage strain at halfspace surface, and $D =$ penetration depth of drying (which is a function of drying duration $t - t_0$). Integration then yields $\Delta W_1 = 0.1\ DE[\varepsilon_{sh}^0/(1+\phi)]^2(1+\nu)/(1-\nu)$.

Stresses σ_y^0 actually are not reduced completely to zero. Therefore, the actual potential energy release will be $r\Delta W_1$ where r is a certain fraction

($0 < r < 1$) which can be determined only by exact solution of the stress field. Probably r is between 0.6 and 0.8. The energy balance during crack formation requires that $r\Delta W_1 s \geqslant G_f a$, where a = crack depth. This yields the condition

$$s \geqslant \frac{10(1-\nu)G_f}{r(1+\nu)E}\left(\frac{1+\phi}{\varepsilon_{sh}^0}\right)^2 \frac{a}{D} \qquad \text{(energy)} \tag{1.68}$$

derived, without creep, in [99]. Note that factor $1/(1+\phi)$ is applied to ε_{sh}^0, not E. The cracks, however, form early in the drying process, and then $\phi \simeq 0$.

The ratio a/D, as well as the ratio a/s, can be determined by linear fracture analysis; this was done by finite elements [74], and yielded a/D as a certain function of D/s. For short cracks, one may use $a/D \simeq 1.5$ and $a \simeq 2s$ as very crude estimates.

For typical properties of concrete, Eq. 1.68 yields $s \simeq 5$ cm, and for typical properties of hardened portland cement paste, $s \simeq 3$ mm. The corresponding crack widths, calculated as $w \simeq s\varepsilon_{sh}^0$, are 0.03 mm for concrete, and 0.004 mm for cement paste [105]. Cracks as fine as this are obviously not visible. Moreover, they are so fine that they cannot be continuous. So, the drying cracks must begin as microcrack zones, which means that concrete may still transmit substantial normal tensile stress across the cracks. This is not necessarily so, however, at a later stage of drying, as it will be shown in the next section.

The strength criterion simply requires that $\sigma_y \geqslant f_t'$ or

$$\varepsilon_{sh}^0 \geqslant \frac{f_t'}{E}(1-\nu)(1+\phi) \qquad \text{(strength).} \tag{1.69}$$

This criterion decides whether progressive microcracking can start but not whether complete cracks form.

If the surface shrinkage ε_{sh}^0 is sufficiently large, Eq. 1.69 is satisfied before Eq. 1.68 and then the initial drying cracks form gradually, statically. If ε_{sh}^0 is not large enough, the cracks can never initiate, regardless of the penetration depth D. However, some other disturbance can make the strength criterion satisfied, and then the cracks form dynamically.

In testing shrinkage and creep at drying, deleterious cracking may be avoided on very thin specimens if the environmental humidity is varied gradually and sufficiently slowly. Formulas indicating the maximum admissible drying rate have been developed [105]. Also, a formula for cracking of a tubular drying specimen has been derived, using the global energy balance for complete crack formation [105]. Another interesting question is a possible coupling between drying and cracking due to the effect of cracking on diffusivity.

Crack system instability. Whereas the cracks in a tensioned reinforced bar may become denser as the loading proceeds, in some other situations, they may become sparser. This is so for cracks growing toward a compression region, a typical example of which is the system of parallel equidistant drying cracks or cooling cracks growing perpendicularly to the surface of a halfspace (Fig. 1.31).

In general, the work, U, that needs to be done to produce cracks of lengths a_i in an elastic body may be expressed as [107,99]

$$U = W(a_1, a_2, \ldots, a_n;\ D) + \sum_{i=1}^{N} \int_0^{a_i} G_{f_i}\, da_i' \tag{1.70}$$

where W = strain energy, G_{f_i} = fracture energy = material property (which could depend on a_i), D = loading parameter which represents here the penetration depth of drying. The equilibrium state of cracks is characterized by a zero value of the first variation of U, i.e.,

$$\delta U = \sum_{i=1}^{m} \left(W_{,i} + G_{f_i}\right)\delta a_i + \sum_{j=m+1}^{k} W_{,j}\delta a_j = 0 \tag{1.71}$$

where $W_{,i} = -\partial W/\partial a_i$ = energy release rates; $i = 1,\ldots,m$ are the cracks which grow ($\delta a_i > 0$); $j = m+1,\ldots,k$ are those which close ($\delta a_i < 0$); and $i = k+1,\ldots,N$ are those which neither grow nor close ($\delta a_i = 0$). Since Eq. 1.71 must be satisfied for any admissible δa_i, it follows, for equilibrium (nondynamic) crack extensions,

$$\text{for } \delta a_i > 0:\quad -W_{,i} = G_{f_i}; \qquad \text{for } \delta a_i < 0:\quad W_{,i} = 0. \tag{1.72}$$

The first condition is the well-known Griffith failure criterion [40], identical to Eq. 1.33. The cases $-W_{,i} > G_{f_i}$ and $-W_{,i} < 0$ cannot happen for equilibrium states because Eq. 1.71 could give $\delta U < 0$. So it follows that only the following crack length variations are admissible:

$$\begin{aligned} &\text{for } -W_{,i} = G_{f_i}: && \delta a_i \geqslant 0 \\ &\text{for } 0 < -W_{,i} < G_{f_i}: && \delta a_i = 0 \\ &\text{for } W_{,i} = 0: && \delta a_i \leqslant 0 \end{aligned} \tag{1.73}$$

Equivalent statements can be made in terms of the stress intensity factors $K_i = (-W_{,i}/E)^{1/2}$.

The question of stability of the states which satisfy the equilibrium conditions (Eq. 1.71) is decided by the second variation of W [107,99]:

$$\delta^2 U = \frac{1}{2} \sum_{i=1}^{n} \sum_{j=1}^{n} U_{ij} \delta a_i \delta a_j \begin{cases} > 0 \text{ for all admissible } \delta a_i \ldots \text{stable} \\ = 0 \text{ for some admissible } \delta a_i \ldots \text{critical} \\ < 0 \text{ for some admissible } \delta a_i \ldots \text{unstable} \end{cases} \tag{1.74}$$

in which $U_{ij} = \partial^2 U / \partial a_i \partial a_j$. When G_{f_i} are independent of a_i (the usual case, except if one uses the R-curve approach), $U_{ij} = W_{ij} = \partial^2 W / \partial a_i \partial a_j$. So, stability is decided not by the energy release rates but by their derivatives. Eq. 1.74 further implies

$$\det_n(U_{ij}) \begin{cases} > 0 \text{ for all } n,\ \delta a_i \text{ all admissible} \ldots \text{stable} \\ = 0 \text{ for some } n, \text{ admissible } \delta a_i \ldots \text{critical} \\ < 0 \text{ for some } n, \text{ admissible } \delta a_i \ldots \text{unstable} \end{cases} \tag{1.75}$$

where $\det_n$ = principal minors of matrix U_{ij} of sizes $n \leqslant N$. The eigenvector δa_i corresponding to the critical state is determined from the equation system [107,99]:

$$\sum_{j=1}^{n} U_{ij} \delta a_j = 0. \tag{1.76}$$

Consider now the parallel cracks that are initially of equal length (Figure 1.31), and examine when the increments of a_i can become different, alternating between δa_1 and δa_2. In the 2×2 matrix U_{ij} $(= W_{ij})$, there is the need to check the sign of $U_{22} = \det_1(U_{ij}) = \partial^2 U / \partial a_2^2$, and of $\det(U_{ij})$. If $\det(U_{ij}) = 0$, then $U_{12}U_{21} = U_{11}U_{22}$ or $U_{12}^2 = U_{11}^2$ (since $a_1 = a_2$), and the eigenvector is then given by $U_{11}\delta a_1 + U_{12}\delta a_2 = 0$ and $U_{21}\delta a_1 + U_{22}\delta a_2 = 0$. Thus, if $\partial U / \partial a_1 > 0$, it follows that $\delta a_2 / \delta a_1 = -W_{11}/W_{21} = -W_{12}/W_{22}$. Now, since the energy release rates $(-\partial W / \partial a_i)$ should decrease with increasing a_i if the crack system is stable before the point of instability, it is expected that $W_{22} > 0$ and $W_{12} > 0$, and numerical calculations confirm that [107,99]. Thus, either δa_2 or δa_1 must be negative. A negative δa_i is, however, impossible if $-\partial W / \partial a_i = G_f$ (growing crack). So, this type of instability cannot happen.

This interesting situation is analogous to that of buckling of a continuous beam with two spans (Figure 1.31) which are constrained so that they can buckle only to the left. Without those one-sided constraints, the spans would buckle in opposite directions (the first critical state). The

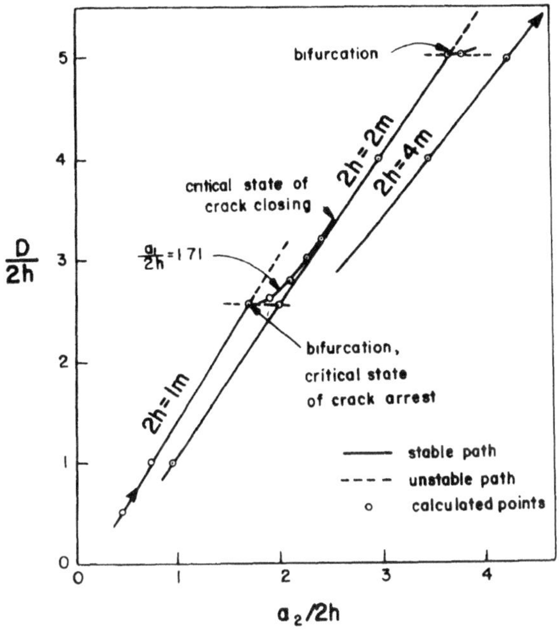

Fig. 1.32. Evolution of crack lengths in a system of parallel cracks in Fig. 31 (after Bažant, Ohtsubo, Aoh, 1979).

presence of one-sided constraints makes the first critical state inapplicable and causes that the beam can buckle only in the second critical mode, in which both spans deflect to the left. The same must happen for the crack system, i.e. instability can arise only due to the second critical condition, $W_{22} = 0$. In that case, since the eigenvector is given by $W_{22}\delta a_2 + W_{21}\delta a_1 = 0$, it follows that $W_{21}\delta a_1 = 0$, and since $W_{21} \neq 0$ according to numerical calculations, $\delta a_1 = 0$, while δa_1 cannot be negative since $-\partial W/\partial a_t = G_f$. So, there exists an instability such that $\delta a_1 > 0$ and $\delta a_2 = 0$ at constant D.

These conclusions and further, more detailed analysis [99] indicate the following evolution of the parallel crack system. Equally spaced drying cracks extend at first at equal length ($a_1 = a_2$) as the drying front advances into the halfspace. The (2×2) determinant of $\partial^2 U/\partial a_1 \partial a_2$ vanishes first, but this does not represent any critical state. Subsequently, $\partial^2 U/\partial a_2^2$ vanishes, and this does represent a critical state at which each other crack gets arrested. Further growth of cracks at equal length is impossible since it is unstable (as $\partial^2 U/\partial a_2^2 < 0$), even though the energy balance condition (Eq. 1.71) is satisfied. Rather, cracks a_2 suddenly jump ahead at constant a_1 and constant D (Figure 1.32). Then cracks a_2 extend again gradually as the drying proceeds (D increases). At the same time, cracks a stop growing and $\partial U/\partial a_1$ gradually diminishes as a_2 and D increase, until $\partial U/\partial a_1$ becomes zero; this represents a second critical state, which can be shown to correspond to a vanishing (2×2) determinant of U_{ij}. At this point, cracks a_2 suddenly close over a certain part of their length at constant a_2 and D. Thereafter, cracks a_2 grow at increasing D. Because cracks a_1 are closed, this is equivalent to the initial situation with equal crack lengths, but the spacing is doubled. The process then gets repeated at doubled spacing, quadrupled spacing, etc. (For a precise numerical calculation of this process, see [99].)

The foregoing behavior is true for a two-dimensional problem. In three dimensions, the behavior is similar but the cracks form hexagonal prisms rather than parallel planes. Some hexagon sides successively close and the hexagon sizes multiply, as the remaining cracks open ever more widely during drying. Drying mud-flats (in dried lakes) demonstrate this behavior [108,109].

The closing of cracks during the progress of drying has the effect that the spacing and the width of the open cracks doubles at each critical state. Numerical calculations [99,74] show that the lengths of the leading cracks and the spacing of the open cracks fluctuate within the limits

$$0.67\,D \leqslant a \leqslant 0.77\,D, \quad 0.39\,D \leqslant s \leqslant 0.61\,D \tag{1.77}$$

where the first limit corresponds to the start of closing of every other crack, and the second limit to the completion of their closing. According to these inequalities, the average crack length and average crack spacing are

$$a \simeq 0.72\,D, \quad s \simeq 0.69\,a \simeq 0.5\,D. \tag{1.78}$$

As the open cracks are getting more widely spaced, their opening w increases. Roughly, $w = s\varepsilon_{sh}$ (s = spacing). The drying cracks become visible when, roughly, $s\varepsilon_{sh} \geqslant 0.2$ mm, and considering that $\varepsilon_{sh} = 0.0006$, this happens for $s \geqslant 30$ cm, $D \geqslant 60$ cm and $a \geqslant 43$ cm. The required value D is so large that cracks caused by drying in massive concrete walls cannot become visible except after many years.

From the foregoing conclusion, it follows that the drying cracks normally are too narrow to form distinct, sharp, and continuous cracks. Rather, what has been referred to as cracks must be cracking bands (or microcrack bands). The foregoing conclusion does not apply in other situations, e.g., when cracking releases flexural strain energy. For example, thin-wall tubular specimens exposed to drying may develop longitudinal cracks in a radial plane. It was shown [105] that these cracks form when the shrinkage strain reaches the value

$$\varepsilon_{sh} = \left(\frac{1-\nu}{1+\nu} \frac{12 G_f}{\pi r E_{ef}} \right)^{1/2} \tag{1.79}$$

where r = radius up to midthickness of tube wall. For a typical cement paste specimen of $r = 7.5$ mm, the critical shrinkage strain is achieved by a drop from 100% to 90% relative humidity [105].

In various types of tests of concrete or cement paste specimens, e.g., in measurements of shrinkage and drying creep as a material property, sorption isotherms, internal surface areas, etc., it is important to make sure that no tensile cracks or microcracks be produced by drying (or cooling). This problem was analyzed in detail in [105] from the viewpoints of both linear and nonlinear diffusion theories. If one considers a linear time variation of surface humidity, a perfect moisture transmission at the surface, and a planar wall of thickness b, then the maximum rate of change of environmental relative humidity h_e to assure a crack-free state is found to be [105]

$$\mathrm{Max}\left(\frac{\mathrm{d}h_e}{\mathrm{d}t}\right) = 0.14\frac{C}{b^2} \tag{1.80}$$

where C = moisture diffusivity. For typical properties of structural concrete, this gives the rate 2.2% humidity change per year for a 6-inch thick slab, which means that prevention of cracking cannot be assured in normal specimens. The bulk of existing test data grossly violates this condition. For an 0.75 mm thick wall, the maximum rate is 10% humidity change per hour, and so tests of material properties at drying must be carried out on extremely thin specimens, made of cement paste or fine mortar, if cracking should be avoided. (Note, however, that cracking can be also avoided by applying sufficient biaxial compression parallel to the surface, and then large specimens are usable.)

The gradual closing of some cracks at the expense of a wider opening of others exists in various other situations. For example, it has been demonstrated [75] that equally spaced bending cracks growing toward the neutral axis in an unreinforced or reinforced beam subjected to bending or eccentric compression can exhibit this type of behavior. However, it is also found that here this question is only academic since this type of behavior is possible only for reinforcement percentages below about 0.18%, which do not represent realistic situations. For normal reinforced beams and plates, the bending cracks do not exhibit the instability just analyzed and maintain the same spacing, which is analyzed in a preceding section from another viewpoint.

Crack path and shear fracture. The question of determining the crack path is rather difficult and the present knowledge is fragmentary. For many situations, it seems, one may assume that the crack propagates in such a direction that a Mode I situation would prevail at fracture front, (i.e., the stress and displacement fields would be symmetric). The Mode I

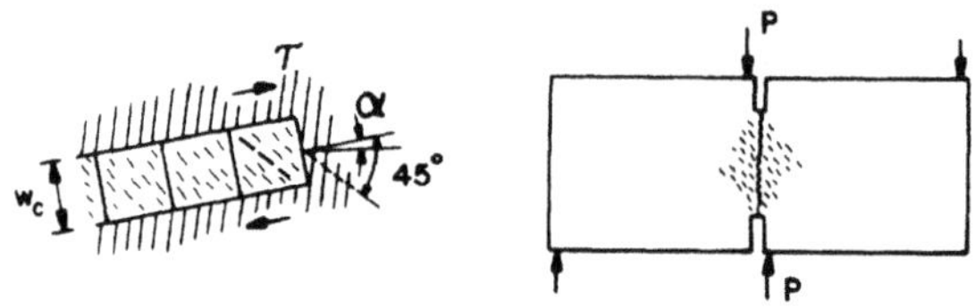

Fig. 1.33. Shear fracture as a band of inclined microcracks.

situation is justified by the experimental fact that, due to crack surface roughness, the ratio of crack slip increment to crack opening increment is nearly zero at very small openings [28]. However, if Mode I always applied, then shear fracture would be impossible, yet it exists in certain situations, e.g., that pictured in Figure 1.33 where the shear zone is narrow and concentrated. If the fracture would propagate in Mode I, it would extend in 45° inclined directions which would quickly bring the crack front into a zone of small principal tensile stress and arrest the crack.

Shear fractures probably develop as a band of 45° inclined microcracks, which themselves are of Mode I (symmetric) type (Figure 1.33). This concept of shear fracture can be modeled by finite elements using the blunt crack band approach. In this case, the fracture energy for the band of inclined cracks would probably be about the same as for tensile (Mode I) crack band, i.e., equal to G_f. However, the multiaxial stress state at the crack front needs to be taken into account. Thus, if the frontal element is subjected to pure shear, Eq. 1.22 ($\sigma_3 = -\sigma_1$) yields the equivalent strength in diagonal tension:

$$f'_{eq_{II}} = \frac{1}{\sqrt{1+2\nu'}} f'_{eq_I} \tag{1.81}$$

where f'_{eq_I} is the equivalent strength in pure tension (Mode I), as given by Eq. 1.22 with $r_f = 1$.

1.7 General model for progressive fracturing

Microplane Model. For some types of loading, especially the dynamic ones, it may happen that a principal tensile stress of direction z causes only partial cracking and fracture is completed later by superimposing another loading of different principal strain direction. For such situations, one needs a softening stress-strain relation that can be applied for general loading paths, in particular, loading paths with rotating principal stress directions. For this purpose, the stress-strain relation used before (Eq. 1.12) needs to be generalized.

An attractive method to do this is the microplane model developed in [110,111]. This model is defined by specifying the constitutive properties by a relation between the stresses and strains acting within the microstructure on a plane of any orientation. This involves no tensorial invariance restrictions. These restrictions can then be satisfied by a suitable combination of planes of various orientations, e.g., in the case of isotropy, each orientation must be equally frequent.

The idea of defining the inelastic behavior independently on planes of different orientation within the material, and then in some way superimposing the inelastic effects from all planes, has a long history. It appeared in Taylor's work [112] on plasticity of polycrystalline metals. Batdorf and Budianski [113] formulated the slip theory of plasticity, in which the stresses acting on various planes of slip are obtained by resolving the macroscopic applied stress, and the plastic strains (slips) from all planes are then superimposed. The same superposition of inelastic strains was used in the so-called multilaminate models of Zienkiewics et al. [114] and Pande et al. [115,116], and in many works on plasticity of polycrystals. Recently, a model of this type was developed to describe tensile strain-softening due chiefly to microcracking [110]. While in previous works dealing with plasticity of polycrystals, the stresses on various microplanes were assumed to be equal to the resolved macroscopic stress, this new model uses a similar assumption for part of the total strains.

The resultants of the stresses acting on the weak planes over unit areas of the heterogeneous material will be called the microstresses, s_{ij}, and the strains of the heterogeneous material accumulated from the deformations on the microplanes will be called the microstrains, e_{ij}. With regard to the interaction between the micro- and macro-levels, one may introduce the following basic hypotheses [110,111].

Hypothesis I. The tensor of macroscopic strain, ε_{ij}, is a sum of a purely elastic macrostrain ε^{a}_{ij} that is unaffected by cracking, and an inelastic macrostrain e_{ij} which reflects the stress relaxation due to cracking, i.e.,

$$\varepsilon_{ij} = \varepsilon^{a}_{ij} + e_{ij} \tag{1.82}$$

Here, Latin lower case subscripts refer to Cartesian coordinates x_i (i = 1, 2, 3).

Hypothesis II. The normal microstrain e_n which governs the progressive development of cracking on a microplane of any orientation is equal to the resolved macroscopic strain tensor e_{ij} for the same plane, i.e.,

$$e_n = n_i n_j e_{ij} \tag{1.83}$$

in which n_i = direction cosines of the unit normal $\boldsymbol{n}$ of the microplane, and the repeated Latin lower case subscripts indicate a summation over 1, 2, 3.

Hypothesis III. The stress relaxation due to all microcracks normal to $\boldsymbol{n}$ is characterized by assuming that the microstress s_{n} on the microplanes of any orientation is a function of the normal microstrain e_{n} on the same plane, i.e.,

$$s_{\mathrm{n}} = \frac{2\pi}{3} F(e_{\mathrm{n}}). \tag{1.84}$$

The factor $(2\pi/3)$ is introduced just for convenience, as it will later cancel out.

The last hypothesis is similar to that made for shear microstresses and microstrains in the slip theory of plasticity. Hypothesis II is however opposite. There are three reasons for hypothesis II:

(1) Using resolved stresses rather than strains on the microplanes would hardly allow describing strain-softening, since, in this case, there are two strains corresponding to a given stress but only one stress corresponding to a given strain.

(2) The microstrains must be stable when the macrostrains are fixed. It has been experienced numerically that, in the case of strain-softening, the model becomes unstable if resolved stresses rather than strains are used.

(3) The use of resolved strains, rather than stresses, seems to reflect the microstructure of a brittle aggregate material better. The use of resolved stresses is reasonable for polycrystalline metals in which local slips scatter widely while the stress is roughly uniformly distributed throughout the microstructure. By contrast, in a brittle aggregate material consisting of hard inclusions embedded in a weak matrix, the stresses are far from uniform, having sharp extremes at the locations where the surfaces of aggregate pieces are nearest. The deformation of the thin layer of matrix between two aggregate pieces, which yields the major contribution to inelastic strain, is determined chiefly by the relative displacements of the centroids of the two aggregate pieces, which roughly correspond to the macroscopic strain. The microplanes may be imagined to represent the thin layers of matrix and the bond planes between two adjacent aggregate pieces (Figure 1.34), since microcracking is chiefly concentrated there.

In Hypothesis III, the relaxation of shear microstresses s_{nt} caused by the shear and normal microstrains e_{nt} and e_{n} is neglected. This assumption is probably quite good for very small crack openings, since it has been deduced from test data on shearing of cracks in concrete that no relative shear displacements on the rough interlocked cracks is possible before a certain finite crack opening is produced, and that the shear

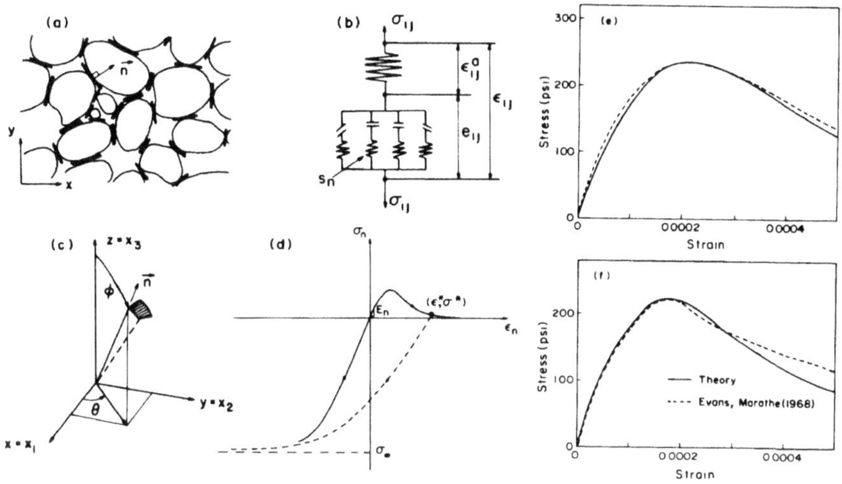

Fig. 1.34. Explanations for microplane model (a–d), and comparison with tests of Evans and Marathe (1968) (e–f).

stiffness of the cracks decreases rather slowly as the crack gradually opens. One must admit, however, that Eq. 1.84 (Hypothesis III) is also useful for its simplicity. It would be much more complicated to assume a general relation between the normal and shear microstresses and microstrains on each plane.

The virtual work of stresses per unit volume may be written, according to Eq. 1.82, as $\delta W = \sigma_{ij}\delta\varepsilon_{ij} = \sigma_{ij}\delta\varepsilon^{a}_{ij} + \sigma_{ij}\delta e_{ij}$. Summing the virtual work due to $\delta\varepsilon^{a}_{ij}$ and δe_{ij}, it gives $\delta W = \sigma^{a}_{ij}\delta\varepsilon^{a}_{ij} + s_{ij}\delta e_{ij}$, in which s_{ij} is the macrostress tensor resulting from s_n on all planes, and σ^{a}_{ij} is the stress tensor corresponding to ε^{a}_{ij}. Since both expressions for δW must hold for any $\delta\varepsilon^{a}_{ij}$ and any δe_{ij}, $s_{ij} = \sigma^{a}_{ij} = \sigma_{ij}$. Equilibrium conditions may be expressed by means of the principle of virtual work:

$$\delta W^{c} = \tfrac{4}{3}\pi\sigma_{ij}\delta e_{ij} = 2\int_{S} s_{n}\delta e_{n} f(\boldsymbol{n})\,\mathrm{d}S \tag{1.85}$$

in which S represents the surface of a unit hemisphere, the factor $(4\pi/3)$ is due to integrating over the surface of a sphere of radius 1, and $\mathrm{d}S = \sin\delta\,\mathrm{d}\theta\,\mathrm{d}\phi$ (Figure 1.34c). Note that there is no need to integrate over the entire surface of the sphere, since the values of σ_n or e_n are equal at any two diametrically opposite points on the sphere. Function $f(\boldsymbol{n})$ introduces the relative frequency of planes of various orientations $\boldsymbol{n}$, contributing to inelastic stress relaxation.

Substituting Eqs. 1.83 and 1.84 into Eq. 1.85 yields $\sigma_{ij}\delta e_{ij} =$

$\int_S F(e_n)n_i n_j \delta e_{ij} f(\boldsymbol{n})\mathrm{d}S$, and because this must hold for any δe_{ij} it gives

$$\sigma_{ij} = \int_0^{2\pi}\int_0^{\pi/2} F(e_n)n_i n_j f(\boldsymbol{n}) \sin\phi \mathrm{d}\phi \mathrm{d}\theta. \tag{1.86}$$

Furthermore, according to Eq. 1.83, $\mathrm{d}F(e_n) = F'(e_n)\mathrm{d}(e_n) = F'(e_n)n_k n_m \mathrm{d}e_{km}$, and thus the differentiation of Eq. 1.86 finally yields

$$\mathrm{d}\sigma_{ij} = D^{\mathrm{c}}_{ijkm}\mathrm{d}e_{km} \tag{1.87}$$

in which [110,111]

$$D^{\mathrm{c}}_{ijkm} = \int_0^{2\pi}\int_0^{\pi/2} a_{ijkm}F'(e_n)f(\boldsymbol{n}) \sin\phi \mathrm{d}\phi \mathrm{d}\theta, \quad \text{with} \quad a_{ijkm} = n_i n_j n_k n_m. \tag{1.88}$$

D^{c}_{ijkm} may be called the tangent stiffnesses of the microplane system. Noting that the sequence of subscripts of D^{c}_{ijkm} is immaterial, it is seen that there are only six independent values of incremental stiffnesses. Eq. 1.88 applies to initially anisotropic solids. For isotropic solids, substitute $f(\boldsymbol{n}) = 1$.

The mathematical structure of the present model may be geometrically visualized with the rheologic model in Figure 1.34b.

The compliance corresponding to the additional elastic strain σ^{a}_{ij} must satisfy isotropy conditions, and so

$$C^{\mathrm{a}}_{ijkm} = \frac{1}{9K^{\mathrm{a}}}\delta_{ij}\delta_{km} + \frac{1}{2G^{\mathrm{a}}}\left(\delta_{ik}\delta_{jm} - \tfrac{1}{3}\delta_{ij}\delta_{km}\right) \tag{1.89}$$

in which K^{a} and G^{a} are certain bulk and shear moduli which cannot be less than the actual initial bulk and shear moduli K and G. For fitting of test data, it was assumed, with success, that $1/G^{\mathrm{a}} = 0$ [110].

Recalling Eq. 1.82 (and Figure 1.34b), the incremental stress-strain relation may be written as

$$\mathrm{d}\sigma_{ij} = D_{ijkm}\mathrm{d}\varepsilon_{km}, \quad \text{with} \quad \left[D_{ijkm}\right] = \left[(\mathbf{D}^{\mathrm{c}^{-1}})_{ijkm} + C^{\mathrm{a}}_{ijkm}\right]^{-1}. \tag{1.90}$$

Applying Eq. 1.88 to elastic deformations (with $f(\boldsymbol{n}) = 1$), one finds that the matrix in Eq. 1.88 always yields Poisson's ratio $\nu = 1/4$. This is because the microplane shear stiffnesses are neglected. Since $\nu = 1/4$ is not quite true for concrete, the additional elastic strain must be used to make a correction. Now, determine the value of K^{a} needed to achieve the desired Poisson's ratio ν. Let superscripts c and a distinguish between the values corresponding to D^{c}_{ijkm} and C^{a}_{ijkm}. For uniaxial stress, it is found that $\varepsilon_{11} = \sigma_{11}/9K^a + \sigma_{11}/E^c$ and $\varepsilon_{22} = \sigma_{11}/9K^a - \nu^c\sigma_{11}/E^c$ in which $\nu^c = 1/4$ and $E^c = 2\pi E_n/5$, $E_n = F'(0)$ = initial normal stiffness for the microplane. Since $\varepsilon_{22} = -\nu\varepsilon_{11}$, it follows that [110,111]:

$$K^a = \frac{1+\nu}{9(\nu^c - \nu)} E^c \qquad (\text{for } \nu \leqslant \nu^c). \tag{1.91}$$

This is, of course, under the assumption that $1/G^a = 0$.

The stress–strain relation for the microplanes, relating σ_n to ε_n, must describe cracking all the way to complete fracture, at which σ_n reduces to zero. In view of the kinematics visualized in Figure 34d, it is clear that σ_n as a function of ε_n must first rise, then reach a maximum, and then gradually decline to zero. The final zero value is chosen to be attained asymptotically, since no precise information exists on the final strain at which $\sigma_n = 0$, and since a smooth curve is convenient computationally. The following expressions were used in computations [110] (Figure 1.34d):

$$\text{for } e_n > 0\text{: } \sigma_n = E_n e_n e^{-(k e_n^p)}, \quad \text{for} \quad e_n \leqslant 0\text{:} \quad \sigma_n = E_n e_n \tag{1.92}$$

in which E_n, k and p are positive constants; $k = 1.8 \times 10^7$, $p = 2$.

The integral in Eq. 1.88 has to be evaluated numerically, approximating it by a finite sum [110,111]

$$D^c_{ijkm} = \sum_{\alpha=1}^{N} w_\alpha \left[a_{ijkm} F'(e_n) \right]_\alpha. \tag{1.93}$$

in which α refers to the values at certain numerical integration points on a unit hemisphere (i.e., certain directions), and w_α are the weights associated with the integration points. Since in finite element programs for incremental loading, the numerical integration needs to be carried out a great number of times, a very efficient numerical integration formula is needed. For the slip theory of plasticity, the integration was performed using a rectangular grid in the $\theta - \phi$ plane. This formula is, however, computationally inefficient since the integration points are crowded near the poles, and since in the $\theta - \phi$ plane, the singularity arising from the poles takes away the benefit from a use of high-order integration formula.

Optimally, the integration points should be distributed over the spherical surface as uniformly as possible. A perfectly uniform subdivision is obtained when the microplanes normal to the α-directions are the faces of a regular polyhedron. A regular polyhedron with the most faces is the icosahedron, for which $N = 10$ (half the number of faces); such a formula was proposed by Albrecht and Collatz [118].

TABLE 1.2
Direction cosines and weights for 2×21 points (integrates exactly 9-th degree polynomials).

α	x_1^α	x_2^α	x_3^α	w^α
1	0.1875924741	0	0.9822469464	0.01984126984
2	0.7946544723	−0.5257311121	0.3035309991	0.01984126984
3	0.7946544723	0.5257311121	0.3035309991	0.01984126984
4	0.1875924741	−0.8506508084	−0.4911234732	0.01984126984
5	0.7946544723	0	−0.6070619982	0.01984126984
6	0.1875924741	0.8506508084	−0.4911234732	0.01984126984
7	0.5773502692	−0.3090169944	0.7557613141	0.02539682540
8	0.5773502692	0.3090169944	0.7557613141	0.02539682540
9	0.9341723590	0	0.3568220897	0.02539682540
10	0.5773502692	−0.8090169944	−0.1102640897	0.02539682540
11	0.9341723590	−0.3090169944	−0.1784110449	0.02539682540
12	0.9341723590	0.3090169944	−0.1784110449	0.02539682540
13	0.5773502692	0.8090169944	−0.1102640897	0.02539682540
14	0.5773502692	−0.5	−0.6454972244	0.02539682540
15	0.5773502692	0.5	−0.6454972244	0.02539682540
15	0.3568220898	−0.8090169944	0.4670861795	0.02539682540
17	0.3568220898	0	−0.9341723590	0.02539682540
18	0.3568220898	0.8090169944	0.4670861795	0.02539682540
19	0	−0.5	0.8660254038	0.02539682540
20	0	−0.5	−0.8660254038	0.02539682540
21	0	1	0	0.02539682540

Numerical experience revealed, however, that 10 points are not enough when strain-softening takes place; it was found that the strain-softening curves calculated for uniaxial tensile stresses oriented at various angles with regard to the α-directions significantly differ from each other, even though within the strain-hardening range the differences are not very large. Therefore, more than 10 points are needed, and then a perfectly uniform spacing of α-directions is impossible.

Bažant and Oh [119] developed numerical integration formulas with more than 10 points, which give consistent results even in the strain-softening range. The most efficient formulas, with a nearly uniform spacing of α-directions, are obtained by certain subdivisions of the faces of an icosahedron and/or a dodecahedron [119]. Such formulas do not exhibit orthogonal symmetries. Other formulas which do were also developed [119]. Taylor series expansions on a sphere were applied and the weights w_α were solved from the condition that the greatest possible number of terms of the Taylor series expansion of the error would cancel out. The angular directions of certain integration points were further determined so as to minimize the error term of the expansion. Formulas involving 16, 21, 33, 37 and 61 points were derived, with errors of 10th, 12th and 14th order. Tables 1.2 and 1.3 define two of these numerical integration formulas, with 21 and 25 points, one without, and one with

TABLE 3
Direction cosines and weights for 2×25 points (integrates exactly 11-th degree polynomials) *.

α	x_1^α	x_2^α	x_3^α	w^α
1	1	0	0	0.01269841058
2	0	1	0	0.01269841058
3	0	0	1	0.01269841058
4	0.7071067812	0.7071067812	0	0.02257495612
5	0.7071067812	−0.7071067812	0	0.02257495612
6	0.7071067812	0	0.7071067812	0.02257495612
7	0.7071067812	0	−0.7071067812	0.02257495612
8	0	0.7071067812	0.7071067812	0.02257495612
9	0	0.7071067812	−0.7071067812	0.02257495612
10	0.3015113354	0.3015113354	0.9045340398	0.02017333557
11	0.3015113354	0.3015113354	−0.9045340398	0.02017333557
12	0.3015113353	−0.3015113354	0.9045340398	0.02017333557
13	0.30151113354	−0.3015113354	−0.9045340398	0.02017333557
14	0.3015113354	0.9045340398	0.301513354	0.02017333557
15	0.3015113354	0.9045340398	−0.3015113354	0.02017333557
16	0.3015113354	−0.9045340398	0.3015113354	0.02017333557
17	0.3015113354	−0.9045340398	−0.3015113354	0.02017333557
18	0.9045340398	0.3015113354	0.3015113354	0.02017333557
19	0.9045340398	0.3015113354	−0.3015113354	0.02017333557
20	0.9045340398	−0.3015113354	0.3015113354	0.02017333557
21	0.9045340398	−0.3015113354	−0.3015113354	0.02017333557
22	0.5773502692	0.5773502692	0.5773502692	0.02109375117
23	0.5773502692	0.5773502692	−0.5773502692	0.02109375117
24	0.5773502692	−0.5773502692	0.5773502692	0.02109375117
25	0.5773502692	−.05773502692	−0.5773502692	0.02109375117

$\beta = 25.239401°$
* Note added in proof: This formula was previously obtained by McLaren and is given by Stroud [178] with better accuracy than here.

orthogonal symmetry. These formulas give sufficient accuracy for most practical purposes. For crude calculations, a formula with 16 points [119] may sometimes also suffice. The directions of integration points are illustrated in Fig. 1.35. Also shown are the stress-strain diagrams calculated with the formula for uniaxial tension applied in various directions with regard to the integration points (directions a, b, c, d,...); the spread of the curves characterizes the range of error.

The following numerical algorithm may be used for the microplane model:

(1) Determine $e_n^{(u)}$ from Eqs. 1.82 and 1.83 for all direction $\alpha = 1,\ldots,N$. In the first iteration of the loading step, use ε_{ij} for the end of the previous step, and in subsequent iterations, use the value of ε_{ij} determined for the mid-step in the previous iteration. In structural analysis, repeat this for

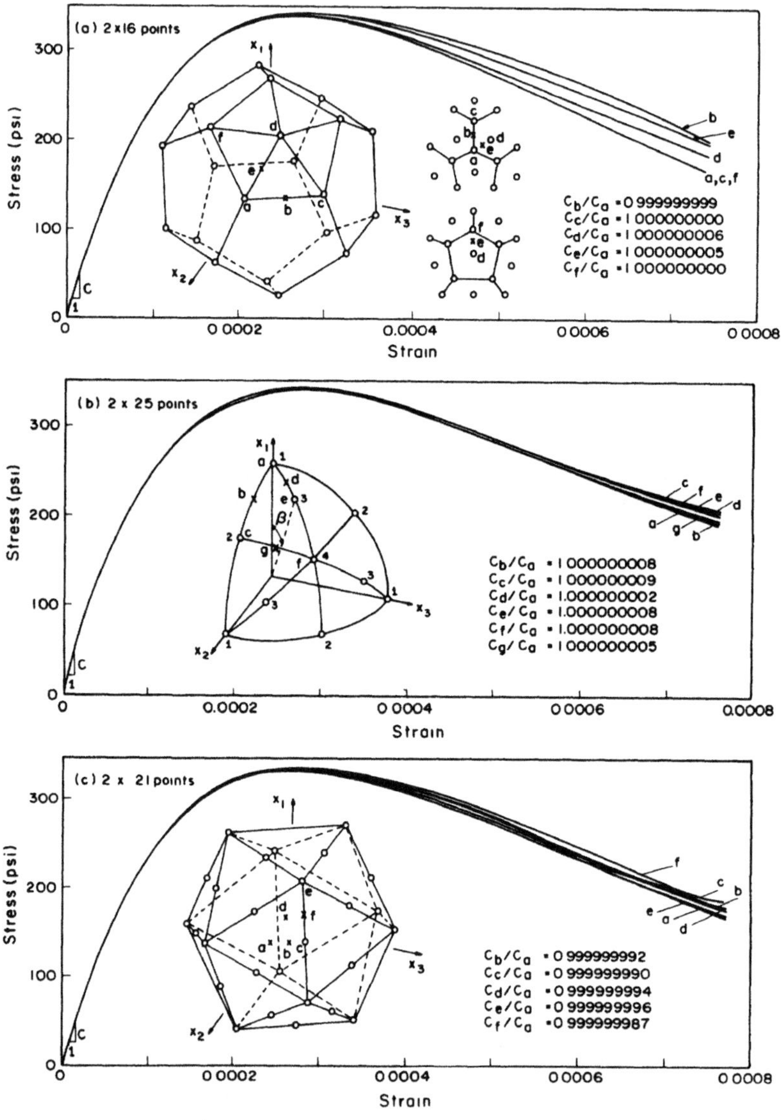

Fig. 1.35. Directions of integration points for some integration formulas for the surface of a sphere, and response curves for universal tension in direction a, b, c, d, ... (after Bažant and Oh, 1983).

all finite elements and for all integration points within each finite element.

(2) For all directions $n^{(\alpha)}$, evaluate $F'(e_n)$ for use in Eq. 1.93. Also check for each direction whether unloading occurs, as indicated by violation of the condition $s_n \Delta e_n \geqslant 0$. If violated, replace $F'(e_n)$ with the

unloading stiffness (which may be approximately taken as E_n; however, a better expression exists [120]).

(3) Evaluate D^c_{ijkm} from Eq. 1.88 and D_{ijkm} from Eq. 1.90. In structural analysis, repeat this for all elements and all integration points in each element. When solving stress-strain curves, calculate then the increments of unknown stresses and unknown strains from Eq. 1.90. In structural analysis, solve (by the finite element method) the increments of nodal displacements from the given load increments, and subsequently calculate the increments of ε_{ij} and σ_{ij} for all elements and all integration points in each element. Then advance to the next iteration of the same loading step, or advance to the next loading step.

In simulating uniaxial tensile loading of fixed direction, the unloading criterion is not important since the only unloading occurs at moderate compressive stresses, for which a perfectly elastic unloading may be assumed.

The microplane model can be calibrated by comparison with direct tensile tests which cover the strain-softening response. Such tests, which can be carried out only in a very stiff testing machine and on sufficiently small test specimens, have been performed by Evans and Marathe [21] as well as others [22–25]. Optimal values of the three parameters of the model, E_n, k, and p, have been found [110] so as to achieve the best fits of the data of Evans and Marathe. Some of these fits are shown as the solid lines in Figure 1.34, and the data are shown as the dashed lines. A better test of the model would, of course, be a tensile test under rotating principal stress directions, but such tests have not yet been performed.

Note that, in this theory, one has only two material parameters, E_n and k, to determine by fitting test data. Trial-and-error approach is sufficient for that.

Shear in cracked concrete. The microplane model just described appears capable of modeling also the resistance of cracks in concrete for shear, characterized by crack friction (aggregate interlock effect) and dilatancy. For this purpose, the model needs to be enhanced by more realistic $\sigma_n - \varepsilon_n$ curves for unloading, and if cyclic shearing is considered, then also for reloading. This was done in [120].

Test data are available only for shear loading of blocks (Figure 1.36a) that have been previously fully cracked in tension [28,121–126]. Even though a finite separation is evident from the relative displacement of the blocks, the resistance to shear is not zero, not even at the beginning of shear. Obviously, there must be some contacts between the opposite surfaces even after the tensile stress normal to the crack has already been reduced to zero.

Using the microplane model, we treat the distinct crack in a rectangular test specimen as a band of certain finite width, w_c. This is probably

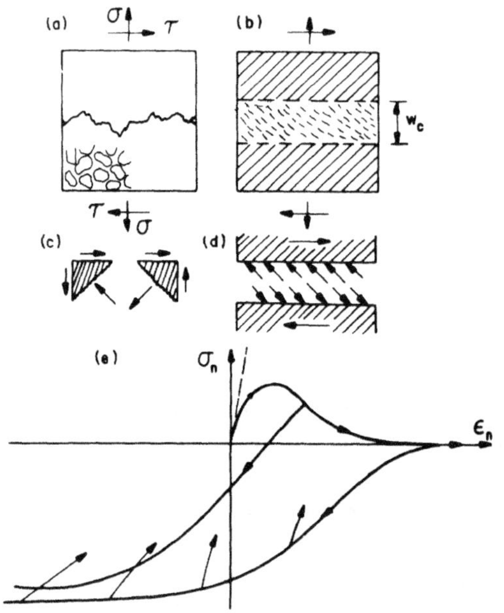

Fig. 1.36. Shear resistance of crack modeled by crack band.

not too unrealistic in view of the roughness of the crack surface, as well as the fact that concrete near the crack must have been microcracked during its previous tensile loading that produced the crack. In numerical simulation, one starts with an intact concrete and implements first uniaxial tensile loading in direction z until the stress σ_z is reduced to zero (in practice, to 0.001 f_t'). Subsequently, either a shear stress τ_{xz} or a shear strain γ_{xz}, depending on the conditions of simulated test, is gradually applied in small increments; see Figure 1.36b. Doing this, the normal strain ε_n on the microplanes inclined $+45°$ (Figure 1.36c) is increased, and so σ_n remains zero on these microplanes. However, ε_n on the microplanes inclined $-45°$ (Figure 1.36c) is decreased, and so contraction (unloading) occurs on those microplanes. For contraction, the normal stiffness is non-zero. Therefore, shear produces in the crack band a set of inclined compression forces illustrated in Figure 1.36d. These forces have a component along the crack, representing crack friction, and a component normal to the crack, representing the pressure opposing dilatancy. If such a pressure is not generated by the support conditions, then a simultaneous expansion (dilatancy) occurs so as to reduce the normal force component to zero.

Figure 1.36e shows the unloading σ_n–ε_n curves that have been used in [120], in which analytical expressions for these unloading curves may be found. Typical response curves which have been simulated with the microplane model are shown in Figure 1.37, where they are compared

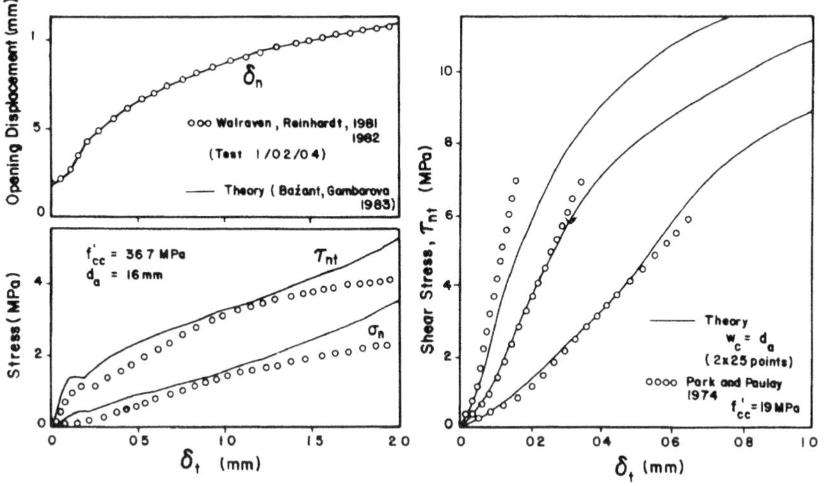

Fig. 1.37. Comparison of crack band model with measurements by Walraven and Reinhardt (1981, 1982) and Park and Paulay (1974) (after Bažant and Gambarova, 1983).

with the data points obtained in Paulay and Loeber tests [121]. In [120], it was shown that this model can fit essentially all the existing data on aggregate interlock or crack shear.

Thus we have a model which correctly describes both the tensile strain-softening up to full fracture and the crack shear in fully fractured concrete. It may be expected that the model would also represent the shear resistance of partially cracked concrete, and perhaps also shear fracture. Another possible use is biaxial (spatial) nonproportional shear loading, i.e., loading of the crack plane by shear stresses τ_{xz} and τ_{yz} (or shear strains γ_{xz} and γ_{yz}) which do not increase in proportion.

If a simultaneous representation of both tensile fracture and crack shear is not needed, one may of course use simpler models consisting in a relation between the stress and relative displacement on the crack. For a realistic description, one must use (in 2 dimensions) both the normal and tangential relative displacements δ_n and δ_t across a crack, and describe the normal and shear stresses on the crack, σ_n and σ_{nt}, as a function of δ_n and δ_t. In an incremental form, such a relation may be written as

$$\begin{Bmatrix} d\sigma_{nn} \\ d\sigma_{nt} \end{Bmatrix} = \begin{bmatrix} B_{nn} & B_{nt} \\ B_{tn} & B_{tt} \end{bmatrix} \begin{Bmatrix} d\delta_n \\ d\delta_t \end{Bmatrix} \tag{1.94}$$

in which $B_{nn}, \ldots, B_{tt}$, called crack stiffness coefficients, depend on δ_n and δ_t. A model of this type was developed and callibrated by test data in [28].

Experimental evidence [28,121–127] reveals that the crack stiffness coefficients are extremely variable depending on the normal and tangential displacements across the crack. For very small openings, the cracks present a very large resistance to shear, while for large crack openings, this resistance may become quite small. At small crack openings, even a small tangential displacement produces large compressive stresses if dilation of the crack is prevented, while for a wide crack even large tangential displacements do not produce large compressive stresses.

Considering a system of dense, parallel and equidistant cracks, and superimposing the deformations on the crack and those due to solid concrete between the cracks, one can obtain the flexibility matrix of cracked concrete. For small crack openings, this matrix can be written in the following explicit form [28]:

$$\begin{Bmatrix} d\varepsilon_n \\ d\varepsilon_t \\ d\gamma_{nt} \end{Bmatrix} = \begin{bmatrix} C_{11} + A|\sigma_{nt}|^p\sigma_{nn}^{-2}, & C_{12}, & \pm Ap|\sigma_{nt}|\sigma_{nm}^{-1} \\ C_{21}, & C_{22}, & 0 \\ \pm B|\sigma_{nt}|^{p+1}\sigma_{nn}^{-2}, & 0 & C_{33} + 3(p+1)|\sigma_{nt}|\sigma_{nn}^{-1} \end{bmatrix} \begin{Bmatrix} d\sigma_n \\ d\sigma_t \\ d\sigma_{nt} \end{Bmatrix} \tag{1.95}$$

in which A, B, p = material constants which depend on crack spacing; $C_{11}, \ldots, C_{33}$ = flexibility coefficients for the concrete between the cracks, which should properly be larger than those for intact concrete and should reflect the damage done to concrete during previous tensile loading which produced the cracks. The $\pm$ signs refer to shear to the left and to the right. Note that this matrix contains large off-diagonal terms which determine normal stress produced by the shear strain in confined concrete, as well as the shear deformation caused by a change in normal strain at constant shear stress. These effects are neglected in those models in which crack response to shear is described only by assuming a finite, smaller than elastic, shear stiffness.

As a cruder approximation, the crack stiffness matrix can be obtained by assuming frictional dilatant slip with a constant friction coefficient k and a constant dilatancy ratio α_d, defining the magnitude of the ratio of normal to tangential relative displacements across the crack. The result is [29]

$$\begin{Bmatrix} d\sigma_n \\ d\sigma_t \\ d\sigma_{nt} \end{Bmatrix} = \begin{bmatrix} 1 & \nu & \pm 2\alpha_d \\ \nu & \dfrac{E_c}{E^*} + \nu^2 & \pm 2\alpha_d\nu \\ \pm k & \pm k\nu & (\pm 2\alpha_d)(\pm k) \end{bmatrix} \begin{Bmatrix} d\varepsilon_n \\ d\varepsilon_t \\ d\varepsilon_{nt} \end{Bmatrix}. \tag{1.96}$$

It is interesting that this matrix is singular, which is a consequence of the friction relation. However, this singularity does not cause problems in reinforced concrete analysis because the deformation is stabilized by reinforcement or the boundary restraint.

Finally, it should be pointed out that if the slip of cracks should be prevented by reinforcement, then the reinforcement required to balance given loads is larger than that when crack friction is neglected. This is because the tensile reinforcement must balance not only the applied loads but also the normal component of friction on the cracks [117,73]. Consideration of friction leads to a reinforcement design for which the maximum deformations of orthogonally reinforced concrete, which occur in the diagonal directions, are significantly less than for the design in which friction is neglected.

1.8 Conclusion

In conclusion of the present work, it may be emphasized that fracture mechanics offers a realistic and consistent approach to the analysis of cracking in concrete structures. The form of the fracture mechanics to be used must be nonlinear, taking into account the existence at the fracture front of a finite nonlinear zone in which the material undergoes progressive microcracking. This type of fracture modeling is objective in that it is independent of the chosen mesh, and it gives a correct structural size effect. The effect of structure size on the apparent (nominal) strength of concrete is the most important salient feature of fracture mechanics. The present crack band theory gives the size effect as a gradual transition from failures governed by the strength limit to failures governed by fracture energy, with linear fracture mechanics as the limit for very large structures. It appears that introduction of such a size effect into various existing provisions of design codes, such as the ACI Code [65] or the CEB-FIP Model Code [47] would improve the predictions of structural response, particularly when the structure to be designed is much larger than the laboratory structures that were used to verify and calibrate the code provisions. In the writer's opinion, improvement in regard to the size effect should be taken as one important goal for further code development.

References

1. Bažant, Z.P., and Oh. B.H., Crack Band Theory for Fracture of Concrete, Materials and Structures (RILEM, Paris), Vol. 16, pp. 155–177 (1983) (based on [130]).
2. Bažant, Z.P., and Kim, S.S., Plastic-Fracturing Theory for Concrete, Journal of the Engineering Mechanics Division, ASCE, Vol. 105, No. EM3, Proc. Paper 14653, pp. 407–428 (1979).
3. Bažant, Z.P., Crack Band Model for Fracture of Geomaterials, Proc., 4th Intern. Conf. on Numerical Methods in Geomechanics, held in Edmonton, Alberta, Canada, ed. by Z. Eisenstein, Vol. 3 (1982), pp. 1137–1152.

4. Bažant, Z.P., Instability, Ductility and Size Effect in Strain-Softening Concrete, J. of the Engineering Mechanics Division ASCE, Vol. 102, No. EM2, pp. 331–344 – Paper 12042 (1976).
5. Bažant, Z.P., and Panula, L., Statistical Stability Effects in Concrete Failure, J. of the Engineering Mechanics Division, ASCE, Vol. 104, No. EM5, pp. 1195–1212, Paper 14074 (1978).
6. Bažant, Z.P., and Cedolin, L., Blunt Crack Band Propagation in Finite Element Analysis, Journal of the Engineering Mechanics Division, ASCE, Vol. 105, No. EM2, Proc. Paper 14529, pp. 297–315 (1979).
7. Cedolin, L., and Bažant, Z.P., Effect of Finite Element Choice in Blunt Crack Band Analysis, Computer Methods in Applied Mechanics and Engineering, Vol. 24, No. 3, pp. 305–316 (1980).
8. Rashid, Y.R., Analysis of Prestressed Concrete Pressure Vessels, Nuclear Engng. and Design, Vol. 7, No. 4, pp. 334–344 (1968).
9. Mindess, S., and Diamond, S., A Preliminary SEM Study of Crack Propagation in Mortar, Cement and Concrete Research, Vol. 10, pp. 509–519 (1980).
10. Cedolin, L., Dei Poli, S., and Iori, L. Experimental Determination of the Fracture Process Zone in Concrete, Cement and Concrete Research, Vol. 13, pp. 557–567 (1983).
11. Cedolin, L., dei Poli, S., and Iori, L., Experimental Determination of the Stress–Strain Curve and Fracture Zone for Concrete in Tension, Proc., Int. Conf. on Constitutive Laws for Engineering Materials, ed. by C. Desai, University of Arizona, Tucson (1983).
12. Barenblatt, G.I., The Formation of Equilibrium Cracks During Brittle Fracture, General Ideas and Hypothesis. Axially–Symmetric Cracks, Prikladnaya Matematika i Mekhanika, Vol. 23, No. 3, pp. 434–444 (1959).
13. Dugdale, D.S., Yielding of Steel Sheets Containing Slits, J. Mech. Phys. Solids, Vol. 8, pp. 100–108 (1960).
14. Kfouri, A.P., and Miller, K.J., Stress Displacement, Line Integral and Closure Energy Determinations of Crack Tip Stress Intensity Factors, Int. Journal of Pres. Ves. and Piping, Vol. 2, No. 3, pp. 179–191 (1974).
15. Kfouri, A.P., and Rice, J.R., Elastic/Plastic Separation Energy Rate for Crack Advance in Finite Growth Steps, in Fracture 1977 (Proc. of the 4th Intern. Conf. on Fracture, held in Waterloo, Ontario, June 1977), ed. by D.M.R. Taplin, University of Waterloo Press, Vol. 1, pp. 43–59 (1977).
16. Knauss, W.C., On the Steady Propagation of a Crack in a Viscoelastic Sheet; Experiments and Analysis, in The Deformation in Fracture of High Polymers, ed. by H.H. Kausch, Pub. Plenum Press, pp. 501–541 (1974).
17. Wnuk, M.P., Quasi-Static Extension of a Tensile Crack Contained in Viscoelastic Plastic Solid, Journal of Applied Mechanics, ASME, Vol. 41, No. 1, pp. 234–248 (1974).
18. Hillerborg, A., Modéer, M., and Petersson, P.E., Analysis of Crack Formation and Crack Growth in Concrete by Means of Fracture Mechanics and Finite Elements, Cement and Concrete Research, Vol. 6, pp. 773–782 (1976).
19. Petersson, P.E., Fracture Energy of Concrete; Method of Determination, Cement and Concrete Research, Vol. 10, 1980, pp. 78–89, and Fracture Energy of Concrete: Practical Performance and Experimental Results, Cement and Concrete Research, Vol. 10, pp. 91–101 (1980).
20. Suidan, M., and Schnobrich, W.C., Finite Element Analysis of Reinforced Concrete, Journal of the Structural Division, ASCE, Vol. 99, No. ST10, Proc. Paper 10081, pp. 2109–2122 (1973).
21. Evans, R.H., and Marathe, M.S., Microcracking and Stress-Strain Curves for Concrete

in Tension, Materials and Structures (RILEM, Paris), No. 1, Jan.–Feb., pp. 61–64 (1968).
22. Heilmann, H.G., Hilsdorf, H.H., and Finsterwalder, K., Festigkeit und Verformung von Beton unter Zugspanungen, Deutscher Ausschuss für Stahlbeton, Heft 203, W Ernst & Sohn, West Berlin (1969).
23. Rüsch, H., and Hilsdorf, H., Deformation Characteristics of Concrete under Axial Tension, Voruntersuchungen, Bericht Nr. 44, Munich (1963).
24. Hughes, B.P., and Chapman, G.P., The Complete Stress-Strain Curve for Concrete in Direct Tension, Bulletin RILEM, No. 30, pp. 95–97 (1966).
25. Petersson, P.E., Crack Growth and Development of Fracture Zones in Plain Concrete and Similar Materials, Doctoral Dissertation, Lund Institute of Technology, Lund, Sweden (1981).
26. Reinhardt, H.W., and Walraven, J.C., Cracks in Concrete Subject to Shear, J. of the Structural Division ASCE, Vol. 108, No. ST1, Paper 16802, pp. 207–224 (1982).
27. ASCE State-of-the-Art Report on Finite Element Analysis of Reinforced Concrete, prepared by a Task Committee chaired by A. Nilson, Am. Soc. of Civil Engrs., New York (1982).
28. Bažant, Z.P., and Gambarova, P., Rough Cracks in Reinforced Concrete, Journal of the Structural Div., Proc. ASCE, Vol. 106, No. ST4, April 1980, pp. 819–842, Paper 15330; Discussion pp. 2579–2581, pp. 1377–1388 (1981).
29. Bažant, Z.P., and Tsubaki, T., Slip-Dilatancy Model for Cracked Reinforced Concrete, Journal of the Structural Division, ASCE, Vol. 106, No. ST9, Paper No. 15704, pp. 1947–1966 (1980).
30. Kupfer, H.B., and Gerstle, K.H., Behavior of Concrete under Biaxial Stress, Journal of the Engineering Mechanics Division, ASCE, Vol. 99, No. EM4, Proc. Paper 9917, pp. 853–866 (1973).
31. Liu, T.C.Y., Nilson, A.H., and Slate, F.O., Biaxial Stress–Strain Relations for Concrete, Journal of the Structural Division, ASCE, Vol. 98, No. ST5, Proc. Paper 8905, pp. 1025–1034 (1972).
32. Rosenthal, I., and Glucklich, J., Strength of Plain Concrete under Biaxial Stress, ACI Journal, pp. 903–914 (1970).
33. Kachanov, L.M., Time of Rupture Process Under Creep Conditions, Izv. Akad, Nauk, SSR, Otd. Tekh, Nauk, No. 8, pp. 26–31 (1958).
34. Janson, J., and Hult, J., Fracture Mechanics and Damage Mechanics, – A Combined Approach, Journal de Mécanique Appliquée, Vol. 1, No. 1, pp. 69–84 (1977).
35. Løland, K.E., Continuous Damage Model for Load-Response Estimation of Concrete, Cement and Concrete Research, Vol. 10, pp. 395–402 (1980).
36. Lorrain, M., On the Application of the Damage Theory to Fracture Mechanics of Concrete, A State-of-the-Art Report, Civil Engineering Department, I.N.S.A., 31077 Toulouse, Cedex, France.
37. Mazars, Mechanical Damage and Fracture of Concrete Structures, 5th International Conference on Fracture, Edited by D. François, Cannes, France, Vol. 4, pp. 1499–1506 (1981).
38. Arrea, M., Ingraffea, A.R., Mixed-Mode Crack Propagation in Mortar and Concrete, Report No. 81–83, Dept. of Structural Engineering, Cornell University, Ithaca, N.Y., Feb. (1982).
39. Knott, J.F., Fundamentals of Fracture Mechanics, Butterworths, London, England (1973).
40. Parker, A.P., The Mechanics of Fracture and Fatigue, E. & F.N. Spon, Ltd. – Methuen, London (1981).
41. Broek, D., Elementary Engineering Fracture Mechanics, Noordhoff International Publishing, Leyden, Netherlands (1974).

42. Mindess, S., The Application of Fracture Mechanics to Cement and Concrete: A Historical Review, Chapter I in 'Fracture Mechanics of Concrete', ed. by F.H. Wittmann, Elsevier, The Netherlands (1983).
43. Brown, J.H., Measuring the Fracture Toughness of Cement Paste and Mortar, Magazine of Concrete Research, Vol. 24, No. 81, pp. 185–196 (1972).
44. Carpinteri, A., Experimental Determination of Fracture Toughness Parameters K_{IC} and J_{IC} for Aggregative Materials, Advances in Fracture Research, (Proc., 5th International Conference on Fracture, Cannes, ed. by D. François, Vol. 4, pp. 1491–1498 (1981).
45. Carpinteri, A., Static and Energetic Fracture Parameters for Rocks and Concretes, Report, Instituto di Scienza delle Costruzioni-Ingegneria, University of Bologna, Italy (1980).
46. Entov, V.M., and Yagust V.I., Experimental Investigation of Laws Governing Quasi-Static Development of Macrocracks in Concrete,: Mechanics of Solids (translation from Russian), Vol. 10, No. 4, pp. 87–95 (1975).
47. Gjørv, O.E., Sørensen, S.I., and Arnesen, A., Notch Sensitivity and Fracture Toughness of Concrete, Cement and Concrete Research, Vol. 7, pp. 333–344 (1977).
48. Huang, C.M.J., Finite Element and Experimental Studies of Stress Intensity Factor for Concrete Beams, Thesis Submitted in Partial Fulfillment of the Requirements for the Degree of Doctor of Philosophy, Kansas State University, Kansas (1981).
49. Kaplan, M.F., Crack Propagation and the Fracture of Concrete, American Concrete Institute Journal, Vol. 58, No. 11 (1961).
50. Kesler, C.E., Naus, D.J., and Lott, J.L., Fracture Mechanics – Its Applicability to Concrete, International Conference on the Mechanical Behavior of Materials, Kyoto, August (1971).
51. Mindess, S., Lawrence, F.V., and Kesler, C.E., The J-Integral As a Fracture Criterion for Fiber Reinforced Concrete, Cement and Concrete Research, Vol. 7, pp. 731–742 (1977).
52. Naus, D.J., Applicability of Linear-Elastic Fracture Mechanics to Portland Cement Concretes, Thesis Submitted in Partial Fulfillment of the Requirements for the Degree of Doctor of Philosophy, University of Illinois at Urbana-Champaign (1971).
53. Shah, S.P., and McGarry, F.J., Griffith Fracture Criterion and Concrete, Journal of the Engineering Mechanics Division, ASCE, Vol. 97, No. EM6, Proc. Paper 8597, pp. 1663–1676 (1971).
54. Sok, C., Baron, J., and François, D., Mécanique de la Rupture Appliquée au Béton Hydraulique, Cement and Concrete Research, Vol. 9, pp. 641–648 (1979).
55. Swartz, S.E., Hu, K.K., Fartash, M., and Huang, C.M.J., Stress Intensity Factors for Plain Concrete in Bending – Prenotched Versus Precracked Beams, Report, Department of Civil Engineering, Kansas State University, Kansas (1981).
56. Walsh, P.F., Fracture of Plain Concrete,: The Indian Concrete Journal, Vol. 46, No. 11, pp. 469, 470, and 476 (1979).
57. Wecharatana, M., and Shah, S.P., Resistance to Crack Growth in Portland Cement Composites, Report, Department of Material Engineering, University of Illinois at Chicago Circle, Chicago, Illinois (1980).
58. Bažant, Z.P., and Oh, B.H., Rock Fracture via Stress-Strain Relations, Concrete and Geomaterials, Report No. 82-11/665r, Northwestern University, Evanston, Illinois (1982); also to appear in ASCE Journal of Engineering Mechanics, Vol. 110 (1984).
59. Bažant, Z.P., and Cedolin, L., Fracture Mechanics of Reinforced Concrete, Journal of the Engineering Mechanics Division, ASCE, Vol. 106, No. EM6, Proc. Paper 15917, December 1980, pp. 1287–1306; with Discussion and Closure in Vol. 108, EM., pp. 464–471 (1982).
60. Bažant, Z.P., and Cedolin, L., Finite Element Modeling of Crack Band Propagation, Journal of Structural Engineering, ASCE, Vol. 109, No. ST2, pp. 69–92 (1983).

61. Flanagan, D.P., and Belytschko, T., A Uniform Strain Hexahedron and Quaderlateral with Orthogonal Hourglass Control, Int. J. for Numerical Methods in Engineering, Vol. 17, pp. 679–706 (1981).
62. Marchertas, A., Bažant, Z.P., Belytschko, T., and Fistedis, S.H., Extension of HCDA Safety Analysis of Large PCRV Containment Structures, Preprints 4th Intern. Conf. on Struct. Mech. in Reactor Technology, San Francisco, Paper E4/1 (1977).
63. Marchertas, A.H., Beltyschko, T.B., Bažant, Z.P., Transient Analysis of LMFBR Reinforced/Prestressed Concrete Containment, Trans, 5th Intern, Conf. on Struct. Mech. in Reactor Tech., Vol. H, Paper H4/1, West Berlin, ed. by T.A. Jaeger and B.A. Boley, publ. by North Holland (1979).
64. Marchertas, A.H., Fistedis, S.H., Bažant, Z.P., and Belytschko, T., Analysis and Application of Prestressed Concrete Reactor Vessels for LMFBR Containment, Nuclear Engng. and Design, Vol. 49 pp. 155–173 (1978).
65. Building Code Requirements for Reinforced Concrete (ACI Standard 318–377), Am. Concrete Institute, Detroit (1977).
66. Marchertas, A.H., Kulak, R.F., and Pan. Y.C., Performance of Blunt Crack Approach Within a General Purpose Code, in Nonlinear Numerical Analysis of Reinforced Concrete, ed. by L.E. Schwer, Am. Soc. of Mech. Engrs., New York 1982, (presented at ASME Winter Annual Meeting, Phoenix, pp. 107–123 (1982).
67. Bažant, Z.P., Pfeiffer, P., Finite Element Crack Band Analysis, in preparation.
68. Bažant, Z.P., Pfeiffer, P., Marchertas, A.H., Blunt Crack Band Propagation in Finite Element Analysis for Concrete Structures, Preprints 7th Int. Conf. on Structural Mechanics in Reactor Technology, Chicago (1983).
69. Rice, J.R., Mathematical Analysis in the Mechanics of Fracture, in Fracture, an Advance Treatise, H. Liebowitz, ed. Vol. 2, Academic Press, New York, N.Y., pp. 191–250 (1968).
70. Haugeneder, E., A Note of Finite Element Analysis of Blunt Crack Band Propagation, Proc., Intern. Conf. on Constitutive Equations for Engineering Materials, ed. by C. Desai, University of Arizona, Tucson, pp. 561–564 (1983).
71. Pan, Y.C., Marchertas, A.H., Kennedy, J.M., Finite Element of Blunt Crack Band Propagation, A Modified J-Integral Approach, Preprints, 7th Intern. Conf. on Structural Mechanics in Reactor Technology, Paper H, Chicago (1983).
72. Rice, J.R., The Localization of Plastic Deformation, Preprints of the 14th IUTAM Congress (Int. Union of Theor. and Appl. Mech.), held in Delft, Netherlands, Edited by W.T. Koiter, North Holland Publishing Co, Amsterdam, pp. 207–220 (1976).
73. Bažant, Z.P., Tsubaki, T., and Belytschko, T.B., Concrete Reinforcing Net: Safe Design, Journal of the Structural Division, Proc. ASCE Vol. 106, No. ST9, pp. 1899–1906, Paper 15705 (1980).
74. Bažant, Z.P., and Wahab, A.B., Instability and Spacing of Cooling or Shrinkage Cracks, Journal of the Engineering Mechanics Division, ASCE, Vol. 105, No. EM5, Proc. Paper 14933, pp. 873–889 (1979),
75. Bažant, Z.P., and Wahab, A.B., Stability of Parallel Cracks in Solids Reinforced by Bars, International Journal of Solids and Structures, Vol. 16, pp. 97–105 (1980).
76. Branson, D.E., Design Procedures for Computing Deflection, ACI Journal, Vol. 65, No. 9, pp. 730–742 (1968).
77. Kani, G.N.J., Basic Facts Concerning Shear Failure, Part I and Part II, J. of ACI, Vol. 63, No. 6, pp. 675–692 (1966).
78. Leonhardt, F., and Walther, R., Beiträge zur Behandlung der Schubprobleme im Stahlbetonbau, Beton-u Stahlbetonbau, Vol. 56, No. 12 (1961), Vol. 57, No. 2, 3, 6, 7, 8, (1962), Vol. 58, No. 8, 9 (1963).
79. Bhal, N.S. Über den Einfluss der Balkenhöhe auf Schubtragfähihkeit von einfeldrigen Stahlbetonbalken mit und ohne Schubbewehrung, Dissertation, Unversität Stuttgärt (1968).

80. Walraven, J.C., The Influence of Depth on the Shear Strength of Lightweight Concrete Beams without Shear Reinforcement, Stevin Laboratory Report No. 5-78-4, Delft University of Technology (1978).
81. Taylor, H.P.J., The Shear Strength of Large Beams, J. of the Structural Division ASCE, Vol. 98, pp. 2473–2490 (1972).
82. Rüsch, M., Haugli, F.R., and Mayer, M., Schubversuche an Stahlbeton Rechteckbalken mit Gleichmässig Verteilter Belastung, Deutscher Ausschuss fur Stahlbeton, Heft 145, W. Ernst u. Sohn, West Berlin (1962).
83. Swamy, R.N., and Qureshi, S.A., Strength, Cracking and Deformation Similitude in Reinforced T-Beams under Bending and Shear, Part I and II, J. of Am. Concrete Inst., Vol. 68, No. 3, pp. 187–195 (1971).
84. Bažant, Z.P., Kim. J.K., Size Effect in Shear Failure of Longitudinally Reinforced Beams, Am. Concrete Institute Journal Vol. 81 (1984), in press.
85. Reinhardt, H.W., Masstabeinfluss bei Schubversuchen im Light der Bruchmechanik (Size Effect in Shear Tests in the Light of Fracture Mechanics), Beton-und Stahlbetonbau, Vol. 7, No. 1, pp. 19–21 (1981).
86. Reinhardt, H.W., Similitude of Brittle Fracture of Structural Concrete, Proc. IABSE Colloquium Advanced Mechanics of Reinforced Concrete, Delft, pp. 201–210 (1981).
87. Bažant, Z.P., Oh, B.H., Deflections of Cracked Reinforced Concrete Beams, Am. Concrete Institute Journal, Vol. 81 (1984), in press.
88. Bažant, Z.P., Oh, B.H., Spacing of Cracks in Reinforced Concrete, J. of Engng. Mech. ASCE, in press.
89. Agrawal, G.L., Tulin, L.G., and Gerstle, K.H., Response of Doubly Reinforced Concrete Beams to Cyclic Loading, ACI Journal, Proc. Vol. 63, No. 7, pp. 823–835 (1965).
90. Bruns, N.H., and Siess, C.P., Plastic Hinging in Reinforced Concrete, J. of the Structural Division ASCE, Vol. 92, No. ST5, pp. 45–64 (1966).
91. Burns, N.H., and Siess, C.P., Repeated and Reversed Loading in Reinforced Concrete, J. of the Structural Division ASCE, Vol. 92, No. ST5, pp. 65–78 (1966).
92. Sinha, B.P., Gerstle, K.H., and Tulin, L.G. Response of Singly Reinforced Beams of Cyclic Loading, ACI Journal, Proc. Vol. 61, No. 8, pp. 1021–1038 (1964).
93. Bažant, Z.P., and Panula, L., Practical Prediction of Time-Dependent Deformations of Concrete, Materials and Structures (RILEM, Paris), Vol. 11, pp. 307–328, 415–434 (1978), Vol. 12, pp. 169–183 (1979).
94. Hollington, M.R., A Series of Long-Term Tests to Investigate the Deflection of Representative Precast Concrete Floor Components, Technical Report TRA 442, Cement and Concrete Association (London) (1970).
95. Krafft, J.M., Sullivan, A.M. Boyle. R.W., Effect of Dimensions on Fast Fracture Instability of Notched Sheets, Cranfield Symposium, Vol. I, pp. 8–28 (1961).
96. Bažant, Z.P., Cedolin, L., Approximate Linear Analysis of Concrete Fracture by R-Curves, Journal of Structural Engineering ASCE, Vol. 110, No. 5T6, June (1984).
97. Tada, H., Paris, P.C., and Irwin, G.R., The Stress Analysis of Cracks Handbook, Del Research Corp., Hellertown, Pa. (1973).
98. Wnuk, M., Bažant, Z.P., and Law, E., Stable growth of Fracture in Brittle Aggregate Materials, Theoretical and Applied fracture Mechanics, Vol. 2 (1984) in press.
99. Bažant, Z.P., Ohtsubo, H., and Aoh, K., Stability and Post-Critical Growth of a System of Cooling or Shrinkage Cracks, International Journal of Fracture, Vol. 15, No. 5., pp. 443–456 (1979).
100. Chi, M., and Kirstein, A.F., Flexural Cracks in Reinforced Concrete Beams, Journal, American Concrete Institute, Proc., Vol. 54, No. 10, pp. 865–878 (1958).
101. Clark, A.P., Cracking in Reinforced Concrete Flexural Member, Journal, American Concrete Institute, Proc., Vol. 52, No. 8, pp. 851–862 (1956).

102. Kaar, P.H., and Mattock, A.H., High Strength Bars as Concrete Reinforcement, Part 4. Control of Cracking, Journal, Portland Cement Association Research and Development Laboratories, Vol. 5, No. 1, pp. 15–38 (1963).
103. Hognestad, E., High Strength Bars as Concrete Reinforcement, Part 2. Control of Flexural Cracking, Journal, Portland Cement Association Research and Development Laboratories, Vol. 4, No. 1, pp. 46–63 (1962).
104. Mathey, R.G., And Watstein, D., Effect of Tensile Properties of Reinforcement on the Flexural Characteristics of Beams, Journal, American Concrete Institute, Proc. Vol. 56, No. 12, pp. 1253–1273 (1960).
105. Bažant, Z.P., and Raftshol, W.J., Effect of Cracking in Drying and Shrinkage Specimens, Cement and Concrete Research, Vol. 12, pp. 209–226 (1982).
106. Bažant, Z.P., Mathematical Models for Creep and Shrinkage of Concrete, in Creep and Shrinkage in Concrete Structures, ed. by Z.P. Bažant and F.H. Wittmann, J. Wiley & Sons, London, pp. 163–285 (1982).
107. Bažant, Z.P., and Ohtsubo, H., Stability Conditions for Propagation of a System of Cracks in a Brittle Solid, Mechanics Research Communications, Vol. 4, No. 5, pp. 353–366 (1977).
108. Lachenbruch, A.H., Journal of Geophysical Research, Vol. 66, p. 4273 (1961).
109. Lister, C.R.B., Geophysical Journal of the Royal Astronomical Society, Vol. 39, pp. 465–509 (1974).
110. Bažant, Z.P., and Oh, B.H., Model of Weak Planes for Progressive Fracture of Concrete and Rock, Report No. 83-2/448m, Center for Concrete and Geomaterials, Northwestern University, Evanston, Il. (1983).
111. Bažant, Z.P., and Oh, B.H., Microplane Model for Fracture Analysis of Concrete Structures, Proc., Symp. on the Interaction of Nonnuclear Munitions with Structures, U.S. Air Force Academy, Colorado Springs, May 1983, ed. by C.A. Ross, publ. by McGregor & Werner, Inc., Wash. D.C.
112. Taylor, G.I., Plastic Strain in Metals, J. Inst. Metals, Vol. 63, pp. 307–324 (1983).
113. Batdorf, S.B., and Budiansky, B., A Mathematical Theory of Plasticity Based on the Concept of Slip, NACA TN1871 (1949).
114. Zienkiewics, O.C., and Pande, G.N., Time-Dependent Multi-laminate Model of Rocks – A Numerical Study of Deformation and Failure of Rock Masses, Int. Journal of Numerical and Analytical Methods in Geomechanics, Vol. 1, pp. 219–247 (1977).
115. Pande, G.N., and Sharma, K.G., Multi-Laminate Model of Clays – A Numerical Evaluation of the Influence of Rotation of the Principal Stress Axes, Report, Department of Civil Engineering, University College of Swansea, U.K., 1982; see also Proceedings, Symposium on Implementation of Computer Procedures and Stress-Strain Laws in Geotechnical Engineering, ed. by C.S. Desai and S.K. Saxena, held in Chicago, Aug. 1081, Acorn Press, Durham, N.C., pp. 575–590 (1981).
116. Pande, G.N., and Xiong, W., An Improved Multi-laminate Model of Jointed Rock Masses, Proceedings, International Symposium on Numerical Models in Geomechanics, ed. by R. Dungar, G.N. Pande, and G.A. Studer, held in Zurich, Sept. 1982, Balkema, Rotterdam, 1982, 218–226.
117. Bažant, Z.P., and Tsubaki, T., Concrete Reinforcing Net: Optimum Slip-Free Limit Design, Journal of the Structural Division, ASCE Vol. 105, No. ST2, Proc. Paper 14344, Feb. 1979, pp. 327–346; Discussion and Closure, pp. 1375–1383 (1981).
118 Albrecht, J., and Collatz, L., Zur numerischen Auswertung mehrdimensionaler Integrale, Zeitschrift fur Angewandte Mathematik und Mechanik, Band 38, Heft 1/2, Jan./Feb., pp. 1–15.
119. Bažant, Z.P., and Oh, B.H., Efficient Numerical Integration on the Surface of a Sphere, Report, Center for Concrete and Geomaterials, Northwestern University, Evanston, Ill. (1982).

120. Bažant, Z.P., and Gambarova, P., Crack Shear in Concrete: Crack Band Microplane Model, Journal of Engineering Mechanics ASCE, Vol. 110 (1984), in press.
121. Paulay, T., and Loeber, P.J., Shear Transfer by Aggregate Interlock, Am. Concr. Inst. Special Publ. SP42, Detroit, pp. 1–15 (1974).
122. Reinhardt, H.W., and Walraven, J.E., Crack in Concrete Subject to Shear, Journal of the Structural Division ASCE, Vol. 108, pp. 207–224 (1982).
123. Walraven, J.C., and Reinhardt, H.W., Theory and Experiments on the Mechanical Behavior of Cracks in Plain and Reinforced Concrete Subjected to Shear Loading, HERON Journal Vol. 26, No. 1A, Dept of Civil Eng. Delft University of Technology, Delft (1981).
124. Laible, J.P., White, R.N., and Gergely, P., Experimental Investigation of Shear Transfer across Cracks in Concrete Nuclear Containment Vessels, Am. Concrete Inst., Special Publ. SP53, pp. 203–226 (1977).
125. Mattock, A.H., The Shear Transfer Behavior of Cracked Monolithic Concrete Subject to Cyclical Reversing Shear, REport SM7404, Dept. of Civil Engng., Univ. of Washington, Seattle (1974).
126. Paulay, T., Park, R., and Phillips, M.H., Horizontal Construction Joints in Cast-in-Place Reinforced Concrete, in Shear in Reinf. Concrete, Vol. 2, Am. Concrete Inst. Special Publ. SP42, Detroit (1974).
127. Laible, J.P., White, R.N., and Gergely, P., Experimental Information on Shear Transfer Across Cracks in Concrete Nuclear Containment Vessels, Special Publ. SP53, Am. Concr. Inst., pp. 203–226, Detroit 1977.

Additional bibliography

128. Baumann, T., Zur Frage der Netzbewehrung von Flächentragwerken, Der Bauingenieur, Vol. 47, No. 10, pp. 367–377 (1971).
129. Bažant, Z.P., Mathematical Models for Creep and Shrinkage of Concrete, Chapter 7 in Creep and Shrinkage in Concrete Structures, ed. by Z.P. Bažant and F.H. Wittmann, J. Wiley & Sons, London, pp. 163–256.
130. Bažant, Z.P., and Oh, B.H., Strain Rate Effect in Rapid Triaxial Loading of Concrete, Vol. 108, pp. 764–782 (1982).
131. Bažant, Z.P., and Oh, B.H., Concrete Fracture via Stress-Strain Relations, Report No. 81-10/665c, Center for Concrete and Geomaterials, Technological Institute, Northwestern University, Evanston, Ill. (1981).
132. Bažant, Z.P. and Oh, B.H., Deformation of Cracked Net-Reinforced Concrete Walls, Journal of the Structural Engineering ASCE, Vol. 109, No. ST2, pp. 93–108 (1983).
133. Bhal, N.S., Über den Einfluss der Balkenhöhe auf Schubtragfähigkeit von einfeldrigen Stahlbetonbalken mit und ohne Schubbewehrung, Dissertation, Universität Stuttgart (1968).
134. Building Code Requirements for Reinforced Concrete ACI-318-77, ACI Committee 318, American Concrete Institute, Detroit, Mich. (1977).
135. CEB-FIP Model Code for Concrete Structures, Commité Eurointernational du Béton, CEB Bulletin No. 124/125-E, Paris (1978).
136. Cedolin, L., and Bažant, Z.P., Fracture Mechanics of Crack Bands in Concrete, Fracture Mechanic Methods for Ceramics, Rocks and Concrete, ed., S.W. Freiman and E.P. Fuller, Am. Soc. for Testing Materials STP745, pp. 221–236 (1981).
137. Cedolin, L., and Dei Poli, S., Finite Element Studies of Shear Critical R/C Beams, J. of the Engng. Mech. Div., ASCE, Vol. 103, No. EM3, pp. 395–410 (1979).
138. Cervenka, V., and Gerstle, K.H., Inelastic Analysis of Reinforced Concrete Panels, Publications, Intern. Assoc. for Bridge and Structural Engng., Zurich, Switzerland, Vol. 31, 1971, pp. 31–45 and Vol. 32, pp. 25–79 (1972).

139. Christensen, R.M., A Rate-Dependent Criterion for Crack Growth, Intern. J. of Fracture, Vol. 15, No. 1, Feb. 1979, pp. 3–21,; disc. Vol. 16, pp. R229–R232, R233–R237 (1980).
140. Chudnovsky, A., On the Law of Fatigue Crack Layer Propagation in Polymers, Polymer Engineering and Science, Vol. 22, No. 15, pp. 922–927 (1982).
141. Chudnovsky, A., Proceedings of NSF Workshop on Damage, held in General Butler State Park in May 1980, University of Cincinnati.
142. Crisfield, M.A., Local Instabilities in the Non-linear Analysis of Reinforced Concrete Beams and Slabs, Proc., Institution of Civil Engrs. Part 2, Vol. 73, pp. 135–145 (1982).
143. Crisfield, M.A., Accelerated Solution Techniques and Concrete Cracking, Comp. Methods in Appl. Mech. and Engng. Vol. 30 (1982).
144. Carpinteri, A., Notch-Sensitivity and Fracture Testing of Aggregate Materials, Engng, Fracture Mechanics, Vol. 16, No. 14, pp. 467–481 (1982).
145. Ingraffea, A.R., Numerical Modeling of Fracture Propagation, in Rock Fracture Mechanics, ed. by H.P. Rossmanith, publ. by The International Center for Mechanical Sciences, Udine, Italy (1983).
146. Isida, J., Elastic Analysis of Cracks and Stress Intensity Factors, Baifukan Publishing Col, Japan (1976).
147. Kachanov, M., Continuum Model of Medium with Cracks, J. of the Engng. Mech. Div. ASCE, Vol. 106, pp. 1039–1051 (1980).
148. Kachanov, M.L., A Microcrack Model of Rock Inelasticity, Mechanics of Materials, North Holland, Vol. 1, pp. 19–41 (1982).
149. Kani, G.N.J., Basic Facts Concerning Shear Failure, Part I and Part II, J. of ACI, Vol. 63, No. 6, pp, 675–692 (1966).
150. Leonhardt, F., and Walther, R., Beitrage nur Behandlung der Schubprobleme im Stahlbeton bau, Beton-u Stahlbetonban, Vol. 56, No. 12 (1961), Vol. 57, No. 2, 3, 6, 7, 8, (1962), Vol. 58, No. 8, 9 (1963).
151. Marchertas, A.H., Belytschko, T.B., Comparison of Transient PCRV Model Test Results with Analysis, Trans. 5th Int. Conf. on SMiRT, Paper H8/2, Berlin (1979).
152. Marchertas, A.H., and Belytschko, T.B., Transient Analysis of a PCRV for LMFBR Primary Containments, Special Issue on Mechanics of Applications to Test Breeder Reactor Safety, Nuclear Technology, Vol. 51, No. 3., pp. 433–442 (1980).
153. Margolin, L.G., Numerical Simulation of Fracture, Proceedings, INtern. Conf. on Constitutive Relations for Engineering Mechanics, ed. by C. Desai, University of Arizona, Tucson, pp. 567–572 (1983).
154. Mihashi, H., and Zaitzev, J.W., Statistical Nature of Crack Propagation, Chapter 4 in Report of RILEM Techn. Comm. 50-FMC, to appear
155. Morley, C.T., Yield Criteria for Elements of Reinforced Concrete Slabs, Introductory Report, Colloquium, Plasticity in Reinforced Concrete, Report of the Working Commission, Inter. Assoc. for Bridge and Struct. Engng., Vol. 8, pp. 35–47 (1979).
156. Nilson, A., Nonlinear Analysis of Reinforced Concrete by Finite Element Method, Am. Concrete Institute Journal, Vol. 65 (1968).
157. Paris, P.C., Fracture Mechanics in the Elastic Plastic Regime, Flaw Growth and Fatigue, ASTM Special Techn. Publ. 631, Am. Soc. for Testing Materials, Philadelphia, pp. 3–27 (1977).
158. Park, R., and Paulay, T., Reinforced Concrete Structures, J. Wiley & Sons, New York, (1975).
159. Paulay, T., and Loeber, P.J., Shear Transfer by Aggregate Interlock Shear in Reinforced Concrete, Special Publications SP-42, American Concrete Institute, Detroit, Mich., pp. 1–15 (1974).
160. Pietruszczak, S., and Mroz, Z., Finite Element Analysis of Deformation of Strain-Softening Materials, Intern. J. for Numerical Methods in Engineering, Vol. 17, pp. 327–334 (1981).

161. Rice, J.R., An Examination of the Fracture Mechanics Energy Balance from the Point of View of Continuum Mechanics, Proc. First International Conference on Fracture (held in Sandia) T. Yokobori, et al., eds. Japanese Soc. for Strength and Fracture of Materials, Tokyo, Japan, Vol. 1, pp. 309–340 (1965).
162. Rice, J.R., Mathematical Analysis in the Mechanics of Fracture, in Fracture an Advance Treatise, H. Liebowitz, ed., Vol. 2, Academic Press, New York, pp. 191–250 (1968).
163. Rüsch, E.H., Haugh, F.R., and Mayer, M., Schubversuche an Stahlbeton Rechteckbalken mit Gleichmassig verteilter Belastung, Deutscher Ausschuss für Stahlbeton, Heft 145, W. Ernst & Sohn, West Berlin (1962).
164. Saouma, V.E., Ingraffea, A.R., and Catalano, D.M., Fracture Thoughness of Concrete-K_{IC} Revisited, J. of the Energy Mech. Div. ASCE, Vol. 108, pp. 1152–1166 (1982).
165. Seaman, L., Curran, D.R. Shockey, D.A., Computational Models for Ductile and Brittle Fracture, J. of Applied Physica, Vol. 47, No. 11, 1976, pp. 4814–4826 (also L. Seaman, Proceedings, NSF Workshop on Damage and Fracture, Stone Mountain, Nov. 1982, ed. by A. Altun, Georgia Institute of Technology, Atlanta.
166. Stout, R.B., Deformation and Thermodynamic Response for a Dislocation Model of Brittle Fracture, Engineering Fracture Mechanics, to appear (also Report UCRL-87472, Lawrence Livermore Laboratory, 1982).
167. Stout, R.B., Thigpen, L., and Peterson, L., Modeling the Deformation of Materials Involving Microcracks Kinetics, Report UCRL-85477, Lawrence Livermore National Laboratory, Livermore, Cal., 1981 (to appear in Int. J. of Num. Methods in Geomechanics, 1983).
168. Stroud, A.H., Approximate Calculation of Multiple Integrals, Prentice Hall, Englewood Cliffs, New Jersey, pp. 296–302 (1971).
169. Swamy, R,N., and Qureshi, S.A., Strength, Cracking and Deformation Similitude in Reinforced T-Beams under Bending and Shear, Part I and Part II, J. of Am. Concrete Inst., Vol. 68, No. 3, pp. 187–195 (1971).
170. Taylor, H.P.J., The Shear Strength of Large Beams, J. of the Structural Division ASCE, Vol. 98, pp. 2473–2490 (1972).
171. Walraven, J.C., The Influence of Depth on the Shear Strength of Lightweight Concrete Beams without Shear Reinforcement,: Stevin Laboratory Report No. 5-78-4, Delft University of Technology (1978).
172. Watstein, D., and Mathey, R.G., Width of Cracks in Concrete at the Surface of Reinforcing Steel Evaluated by Means of Tensile Bond Specimens, Journal. American Concrete Institute, Proc. Vol. 56, No. 1, pp. 47–56.
173. Wecharatana, M., and Shah, S.P., Double Torsion Tests for Studying Slow Crack Growth of Portland Cement Mortar, Cement and Concrete Research, Vol, 10, pp. 833–844 (1980).
174. Wecharatana, M., and Shah, S.P. Slow Crack Growth in Cement Composites, J. of the Structural Division ASCE, Vol. 108, pp. 1400–1413 (1982).
175. Wittmann, F.H., Mechanics and Mechanisms of Fracture of Concrete, Advances in Fracture Research (Preprints, 5th Intern. Conf. on Fracture, in Cannes, March 1981), Vol. 4, Pergamon Press, Paris, pp. 1467–1487 (1981).
176. Zaitsev, Y.W., and Wittmann, F.H., Crack Propagation in a Two-Phase Material Such as Concrete, Fracture, 1977 (Proc. 4th Intern. Conf. on Fracture), University of Waterloo, Vol. 3, pp. 1197–1203 (1977).
177. Zech, B., and Wittmann, F.H., Variability and Mean Value of Strength of Concrete as a Function of Load, Am. Concrete Institute Journal, Vol. 77, No. 5, pp. 358–362 (1980).
178. Zech, B. and Wittmann, F.H., Influence of Rate of Loading on Strength of Concrete, Manuscript (1981).

A. CARPINTERI

2

Scale effects in fracture of plain and reinforced concrete structures

2.1 Introduction

With the advent of modern computational techniques, it is now possible to perform step-by-step stress and strain calculations of structures with complex geometries and materials that may be heterogeneous, anisotropic and non-linear. In the case of a linear elastic material and a proportional loading, the stress and strain relation at a point appears to be proportional up to failure. For an elastic-perfectly plastic material or elastic-softening material, stress and strain are not proportional and the critical condition depends on the failure mode. The collapse mechanism corresponds to plastic flow (i.e. constant stress and strain level tending to infinity) or to fracture (i.e. vanishing stress and strain level tending to infinity). It has been pointed out experimentally that materials with aggregates (e.g. concretes and rocks) follow elastic-softening constitutive laws such that stresses are relaxed beyond a certain strain level. Several theoretical laws have been proposed, including three-dimensional ones, to simulate softening, which, from a mechanical point of view, corresponds to unstable deformation [1]. On the other hand, it is not completely clear so far how stresses really relax, so that they totally vanish after the crack formation. In fact, fracture in solids has been neglected by the researchers in the field of Continuum Mechanics for a long time. The presence of a crack involves discontinuity and reduces the effectiveness of a material to transmit stress locally. The discipline of Fracture Mechanics considers cracks as inherent part of the system that can affect the strength of materials and structures [2].

Several authors [3–5] proposed softening constitutive laws connecting strain (or strain rate) with relaxed stress (or negative stress rate). Although these stress-strain laws can be conveniently used as computer input, they only describe progressive damage in concrete structures in an average, and do not address thc crack formation phenomenon, which is localized and anisotropic in nature. In other words, they tend to simulate only damage due to microcracking in the gross sense and the fracturing process is smeared over a region rather than being discrete.

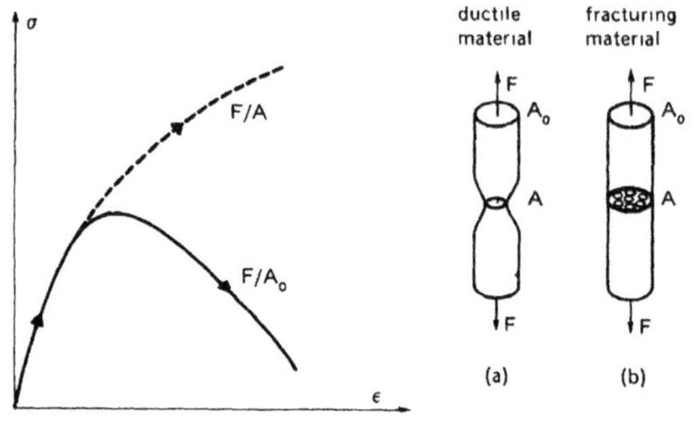

Fig. 2.1. Softening due to void formation.

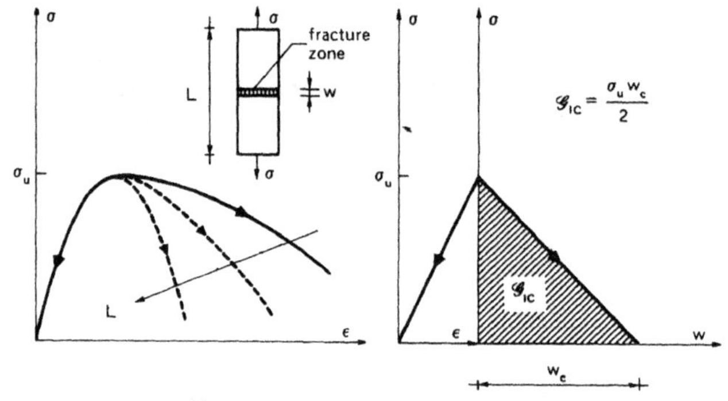

Fig. 2.2. Softening due to damage localization.

Descending stress-strain (or stress rate-strain rate) relationship always prevails in plasticity corresponding to the stage of necking in metals. However, if the stress were estimated in relation to the reduced area, the descending portion of the curve would disappear (Figure 2.1a). For the fracturing of materials a pseudo-necking probably occurs, i.e. the net area is reduced; in this case this is due to void and microcrack formation at the weakest cross-section [6] (Figure 2.1b).

Moreover, by observing tensile tests on concrete specimens [7], it is possible to assert that the damage zone is more and more localized as the loading capacity decreases. In fact, while the material within the fracture zone softens, Figure 2.2a, stress and strain in the undamaged portion of the material still behave in a proportional manner. As a result, strains accumulate in the fracture zone, while the remaining part of the structure unloads itself. Hillerborg, Modeer and Petersson [8] assumed that the original length of the fracture zone in the tensile direction is equal to zero, whereas it is different from zero while it is actually developing. Stress at the softening stage, therefore, will be a function of such a length

w (Figure 2.2b). This simple hypothesis explains the size effects for the descending branch of the $\sigma - \varepsilon$ curve which appears to vary with specimen size [6] (Figure 2.2a). Another remarkable consequence of this hypothesis, widely found in practice, is that it is possible to have similarity in the physical fracture behaviour when the value w_c of a total stress relaxation is proportional to the structural size L. In other words, the fracture energy $\mathcal{G}_{IC} \simeq \sigma_u w_c/2$ is then proportional to L (Figure 2.2b). In such a case, the global collapse dilatation $\varepsilon_c \simeq w_c/L$ becomes constant as the size L is varied.

Changing the abscissa as the unstable stage begins (Figure 2.2b) leads to important theoretical consequences, which will be emphasized subsequently. Dilatation ε, in fact, is a dimensionless quantity, while the width w of the fracture zone has the dimension of length $[L]$ *. Size (or scale) effects in Fracture Mechanics originate from this transition. In fact, the area under a $\sigma - \varepsilon$ curve represents energy dissipated per unit volume, thus having the physical dimension of *stress* $[FL^{-2}]$. It is well-known that the classical strength criteria such as those advanced by Beltrami [9] using the strain energy density, and Von Mises [10] using the distortion energy density are equivalent to imposing limits to a stress quantity. On the other hand, the area under a $\sigma - w$ curve (Figure 2.2b) represents energy dissipated per unit area, thus having the dimensions of *surface energy* $[FL^{-1}]$. More precisely, the area under the descending branch of the curve $\sigma - w$ represents the fracture energy $\mathcal{G}_{IC}$, i.e. the energy necessary to create a unit free surface. Such a quantity was introduced by Griffith [11] in 1921 and was called the *surface energy*, γ. He proposed it as a thermodynamic quantity capable of explaining the fracture phenomenon in elastic solids. In 1957 Irwin [2] introduced it again with the name of *critical energy release rate*. His contribution was to connect $\mathcal{G}_{IC}$ with the critical value of the stress-intensity factor, K_{IC}, a stress amplifying factor at the crack tip which, in many cases, can be easily expressed as a function of geometry and loads. In 1968, Rice [12] eventually extended this concept to elastic-plastic solids by formulating the *J-integral* Theory. All these energetic parameters γ, $\mathcal{G}_{IC}$ and J_{IC} have dimensions $[FL^{-1}]$. Such physical dimensions are unusual in the field of traditional Solid Mechanics. It is indeed the transition from a continuous to a discontinuous system which requires the introduction of such quantities. This is not just a mere chance. Another well-known critical parameter of Fracture Mechanics, the *strain energy density factor* S_{IC}, proposed by Sih [13] in

* J. Clerk Maxwell, the Scottish physicist and philosopher, was the first who employed symbols such as $[F]$, $[M]$, $[L]$, $[T]$, $[\Theta]$ to denote force, mass, length, time and temperature respectively. In his paper "On the Mathematical Classification of Physical Quantities", Proc. London Math. Soc., Vol. III, no. 34, p. 224, Mar. 1871, he formed products of powers of these symbols, which he called "dimensions".

1973, has the same physical dimension.

Bažant and Cedolin [14,15] implemented a numerical finite element model to describe the propagation of microcracked zones in concrete, where the collapsed elements become orthotropic with zero stiffness orthogonally to the crack surface. If a tensile strength crack propagation criterion, or rather a volumetric energy criterion, is applied, such a model becomes unrealistic in terms of varying the finite element size. It must then resort to devising superficial energy criteria. This means that, when separation phenomena are involved, it is necessary to utilize mechanical parameters which explicitly refer to developing free surfaces. Moreover, Bažant and Cedolin proved that the above-mentioned argument is valid only when the structural sizes are considerable, as in the case of dams or concrete nuclear vessels. However, when the sizes of the structural elements are small, e.g. beams or panels, the stress criterion tends to coincide with the energy criterion. In this case, traditional criteria may be applied. A transition from a tensile strength collapse to a crack propagation collapse has thus been numerically pointed out, when the scale only is changing.

Therefore, just as structural engineers should pay attention to the fact that, above certain sizes, stress (or volumetric) criteria are meaningless, so material scientists should pay attention to the fact that, below certain sizes, energy (or superficial) criteria in Fracture Mechanics are meaningless. Thus it would be absurd to carry out Fracture Mechanics tests by means of extremely small specimens. This is a well-known problem which has led many researchers to unsatisfactory results. In fact, if a parameter is measured relating to a certain phenomenon, while another phenomenon occurs instead of the supposed one, unexpected variations of the parameter will obviously result. This has often occurred in the fracture testing of metallic materials where geometry and size of specimen and crack depth [16] are varied. On the other hand, when collapse really occurs because of crack propagation, this condition is usually called *fracture* (or *notch*) *sensitivity*.

Petersson [7] utilized the above-mentioned crack model (Figure 2.2b) and performed numerical simulations of crack propagation. The result was that, when the structural sizes are increased, the material became more and more "brittle", i.e. the final collapse can be better described by means of Fracture Mechanics parameters. When the structural sizes are decreased, the achievement of an ultimate tensile strength induces the final collapse. Consequently, an embrittlement occurs by increasing the structural sizes, a fact which corresponds to the fundamental laws of physical similitude [17].

Above certain structural sizes and for particular values of steel percentage, even reinforced concrete may result to be sufficiently "brittle" and may be studied by using Fracture Mechanics concepts. On the other hand, in the Limit Analysis of the reinforced concrete beam cross-section the stretched part of concrete is conventionally assumed to be not traction bearing, while a perfectly plastic behaviour of the compressed part is hypothesized. Such analysis does not take cracks into consideration, which are usually developed in concrete and generally cause the collapse of the concrete-steel system. In fact, the Limit Analysis approach does not consider the stiffness variation and the concentration of stresses near the crack tip.

By tradition, the problems relating to cracked masonry or concrete constructions are studied on the basis of empirical parameters, such as the crack opening. Such parameters cannot be considered as absolute indications of crack stability, but only as a warning to incipient collapse. In fact, the crack opening will not be constant, but will generally vary on the crack surface. It will, however, depend on the sizes of the cracked structure.

A few authors have already analyzed the behaviour of cracked reinforced concrete by means of Fracture Mechanics [15,18–22]. Bažant [15] took into consideration the slippage of steel bars due to a crack; Petersson [20] studied the shear fracture in reinforced concrete beams; Ingraffea [21] applied a mixed mode fracture criterion and Francois et al. [22] analyzed the influence of reinforcement on fracture energy, utilizing very large and prestressed DCB-specimens.

In the following sections, a general formulation of the problem of sensitivity and stability of fracture in plain and reinforced cement composites is presented. It will be shown that these phenomena can be explained by *laws of scale*. The notch *sensitivity* effects on fracture testing of material with aggregates and varying specimen and crack sizes have been studied on the basis of Dimensional Analysis concepts. Such effects are due to the co-existence of two different failure modes induced by generalized forces with different physical dimensions, $[\sigma]=[FL^{-2}]$ and $[K_I]=[FL^{-3/2}]$, and because the specimen sizes are finite. The application of Buckingham's Theorem concerning physical similitude and scale modelling [23] allows the definition of a non-dimensional number s (the brittleness number), which governs the notch sensitivity phenomenon. Some recurring experimental inconsistencies are thus explained. The *stability* of the fracturing process is generally due to compressive loadings or to a sufficiently high degree of redundancy in the system. The adjunctive elements producing such redundancy are usually fibre or bar reinforcements. Two particular cases are considered: (1) a masonry wall, subjected to eccentric compressive axial force, and (2) a reinforced concrete beam element, subjected to bending moment. It is shown that the stability of the process of concrete fracture and steel plastic flow depends on the mechanical and geometrical (scale included) properties of the beam section. The investigation on reinforced concrete beams subjected to monotonic loadings, is then extended to the case of repeated loadings, giving special emphasis to shake-down and *hysteretic* effects due to fracture of concrete and to slippage or yielding of steel.

2.2 Dimensional analysis applied to plain and reinforced concrete structures

Many features of material with aggregates have discouraged the application of Fracture Mechanics to analyze plain and reinforced concrete structures. The biggest difficulties in explaining the experimental results and in extrapolating them to the structural design of large structures are due to the *heterogeneity* and to the *non-linearity* of concrete. In fact, Linear Elastic Fracture Mechanics is a valid theory only for macroscopically homogeneous and linear materials. Moreover, there are two disturbing phenomena, which occur during the fracture tests of cement materials and make the research-workers pessimistic about the application of Linear Elastic Fracture Mechanics to materials similar to rocks and concretes:

(1) the *slow crack growth*, which reveals itself prior to the unstable crack propagation and makes the real crack length (at the moment of final collapse) unknown to the experimenter and

(2) the *microcracking* at the crack tip, which is an energy absorption mechanism, as in the case of plastic flow for metallic materials.

Notch sensitivity is the condition in which a notched specimen of brittle material collapses, owing to the stress concentration effect of the notch. In other words, a material is notch sensitive when the net section rupture stress is lower than the ultimate tensile strength σ_u. Several authors have observed that such effect tends to disappear by decreasing the specimen size. Below a certain size threshold the net section rupture stress is coincident with the ultimate strength σ_u and any stress concentration and amplification effect appears to vanish [24]. Since the stress-intensity factor K_I is derived from a linear elastic analysis, the experimental determination of its critical value K_{IC} is therefore valid only when the crack tip process zone is small in relation with the specimen and crack sizes (*condition of small scale yielding*). For elastic-plastic materials, like ductile metallic alloys, the small scale yielding condition implies the notch sensitivity condition. Since the size of the crack tip plastic zone, under fracture conditions, is approximately independent of the specimen and crack geometry, it is possible to assert that the small scale yielding condition corresponds to a lower bound to the specimen and crack size. For linear elastic materials, the small scale yielding condition, which is always satisfied as the plastic zone size tends to zero, would in any case imply the notch sensitivity condition. This would be true if the tensile strength σ_u were infinite, as is often implicitly assumed in Linear Elastic Fracture Mechanics, i.e. if the unique potential collapse were that of fracture. Assuming the co-existence of two potential collapses, those caused by fracture and ultimate load, it is possible to show that, for brittle materials such as rocks, mortars and concretes, notch sensitivity is not an intrinsic material property, but it depends on the sizes of the structure where the crack is localized.

Buckingham's Theorem. Just as in Hydraulics a fluid flow and its characteristic dynamic phenomena can be reproduced by a scale model only if the model sizes are not lower than certain critical value, in Solid Mechanics, the mechanical behaviour of a structure and its collapse mechanisms can be reproduced by a scale model (specimen) only if the model sizes are not too small. This is what happens in the case of a cracked body with two different potential failure modes:

(1) the *collapse at ultimate strength*, induced by the highest normal stress σ in the body, considering the crack only as a weakening for the cross section and not as a stress concentrator and
(2) the *crack propagation collapse*, induced by the stress intensity factor K_I, supposing structure and loadings as being symmetrical with respect to the crack line.

The presence of two generalized forces with different physical dimensions, $[\sigma]=[FL^{-2}]$ and $[K_I]=[FL^{-3/2}]$, requires the application of Dimensional Analysis so that non-dimensional parameters are needed to decide the order of the two failure modes. Dimensional Analysis is a method by which information concerning a phenomenon is reduced by a dimensionally correct equation involving certain variables. With little effort, a formal solution to nearly any problem can be obtained. On the other hand, a complete solution is not obtained, nor is the inner mechanism of a phenomenon revealed by dimensional reasoning alone [25].

Buckingham's Theorem [23] concerning physical similitude and scale modelling will be applied to a simple case: a symmetrically reinforced body, symmetrically loaded with respect to the crack line, of homogeneous, isotropic and *elastic-perfectly plastic* or *elastic-brittle* material (Figure 2.3). In the first case, the yield strength σ_y will be used as reference, while in the second case the tensile strength σ_u will be taken into consideration. As Figure 2.4a makes evident the two cases considered are the limit cases of an *elastic-linear softening* material. All the following considerations could therefore be easily extended to a material of this general type.

Let q_0 be the load of final collapse (i.e. the maximum bearable load) in the case of the cracked structure of Figure 2.3. This load is a function of several variables:

$$q_0 = f(q_1, q_2, \ldots, q_n; r_1, r_2, \ldots, r_m), \tag{2.1}$$

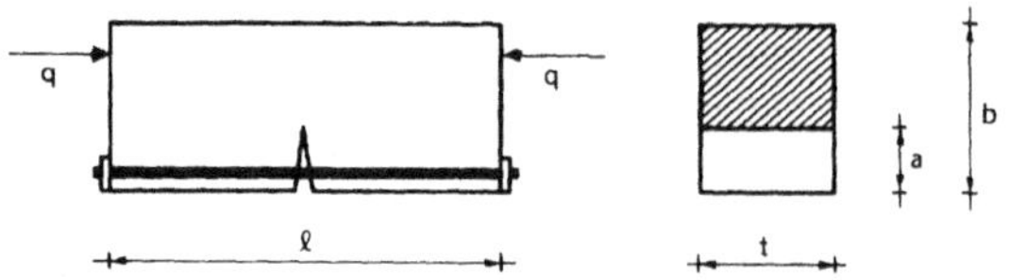

Fig. 2.3. Cracked reinforced body.

where q_i are physical quantities with different dimensions and r_i are non-dimensional numbers. Each quantity with certain dimensions appears just once in function f. For example, in the case of an elastic-work-hardening material, q_1 could be the ultimate strength σ_u and r_1 the ratio between the yield strength σ_y and the ultimate strength; q_2 could be a linear size b of the structure (Figure 2.3); $r_2, r_3, \ldots$ the ratios between the other sizes, which are characteristic for the considered structural geometry, and the size b; then q_3 could be the fracture toughness K_{IC} and so on.

Consider now two dimensionally independent quantities q_1 and q_2 (statical case). As clarified in Section 2.5 (Appendix), they can be considered as being fundamental, i.e. it is possible that the product $q_1^{\alpha_{10}} q_2^{\alpha_{20}}$, for suitable values of α_{10} and α_{20}, has the same dimensions as q_0. In the same way, the product $q_1^{\alpha_{13}} q_2^{\alpha_{23}}$ can have the same dimensions as q_3, for suitable values of α_{13} and α_{23}, and so on.... The function in equation (2.1) can be therefore transformed as follows:

$$\frac{q_0}{q_1^{\alpha_{10}} q_2^{\alpha_{20}}} = \varphi\left(q_1, q_2, \frac{q_3}{q_1^{\alpha_{13}} q_2^{\alpha_{23}}}, \ldots, \frac{q_n}{q_1^{\alpha_{1n}} q_2^{\alpha_{2n}}}; r_1, r_2, \ldots, r_m \right). \tag{2.2}$$

The function f becomes φ because of the performed non-dimensionalizations. If the unit of measure of q_1 changes, φ does not vary, being a non-dimensional number. Thus, φ is not really a function of q_1; analogously it is not a function of q_2. It is only a function of $(n-2+m)$ non-dimensional numbers:

$$\frac{q_0}{q_1^{\alpha_{10}} q_2^{\alpha_{20}}} = \varphi(N_3, N_4, \ldots, N_n; r_1, r_2, \ldots, r_m). \tag{2.3}$$

In the simple case of elastic-perfectly plastic material, it results (Figure 2.3):

$$q_0 = f\left(\sigma_y, b, K_{IC}, F_P; \frac{a}{b}, \frac{t}{b}, \frac{l}{b} \right), \tag{2.4}$$

where also the crack length a appears among the geometric parameters and F_P is the force of the plastic flow of the reinforcement (due to yielding or slippage). Considering σ_y and b as fundamental quantities, equation (2.4) becomes

$$\frac{q_0}{\sigma_y^\alpha b^\beta} = \varphi\left(\frac{K_{IC}}{\sigma_y b^{1/2}}, \frac{F_P}{\sigma_y b^2}; \frac{a}{b}, \frac{t}{b}, \frac{l}{b}\right), \tag{2.5}$$

where the function φ depends on the geometry of structure and on external loadings. As far as the most usual fracture test geometries are concerned, thickness t and length l are not always really present as arguments in function φ. When they are, however, it is possible to separate function φ into two functions φ_1 and φ_2:

$$\frac{q_0}{\sigma_y^\alpha b^\beta} = \varphi_1\left(s; \frac{a}{b}\right)\varphi_2\left(\frac{t}{b}, \frac{l}{b}\right). \tag{2.6}$$

The function φ_1, which is more interesting and governs the notch sensitivity phenomenon, is a function of the non-dimensional number (*brittleness number*):

$$s = \frac{K_{IC}}{\sigma_y b^{1/2}}, \tag{2.7}$$

and of the relative crack length a/b. Note that both the mechanical properties of the material and the size of the body appear in s. The true function φ_1 is very difficult to be obtained in the general case considered (Figure 2.3). However, if a simple series of structural geometries, without reinforcing elements, is considered, it is possible to theoretically justify the scale effects. In all the elementary cases considered by the author [16,17], in fact, the priority of fracture collapse with large structural sizes and the priority of plastic flow collapse (or ultimate strength collapse) with small structural sizes have been verified. As a general rule, fracture or rather the "separation" of a solid can be reproduced only above certain sizes, for cracks that are neither too small nor too large (Figures 2.5 to 2.7).

For the *tension test* (TT, Figure 2.5) the stress-intensity factor is given by

$$K_I = \sigma\sqrt{\frac{\pi}{2}a}\left(\sec\frac{\pi}{2}\frac{a}{b}\right)^{1/2}. \tag{2.8}$$

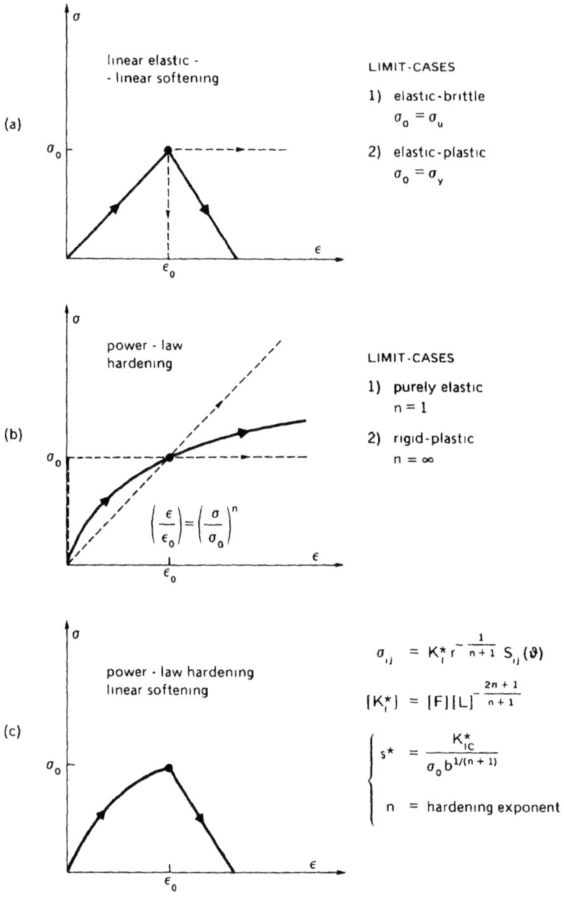

Fig. 2.4. Constitutive laws.

From equation (2.8), it is possible to obtain the crack propagation load

$$\frac{\sigma_{CR}}{\sigma_y} = s\left(\frac{\cos \pi/2\, a/b}{\pi/2\, a/b}\right)^{1/2}. \tag{2.9}$$

On the other hand, the load of plastic flow collapse, relating to the cracked section is

$$\frac{\sigma_{CR}}{\sigma_y} = 1 - \frac{a}{b}. \tag{2.10}$$

Such a load is really only an *upper bound*, because of the co-existence of

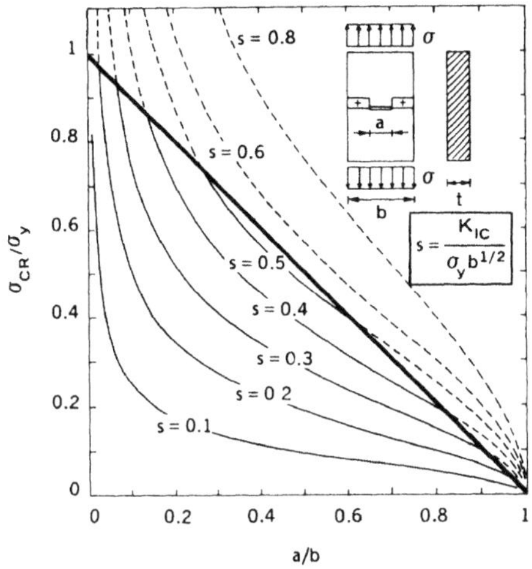

Fig. 2.5. Interaction between strength collapse and separation collapse (tension test).

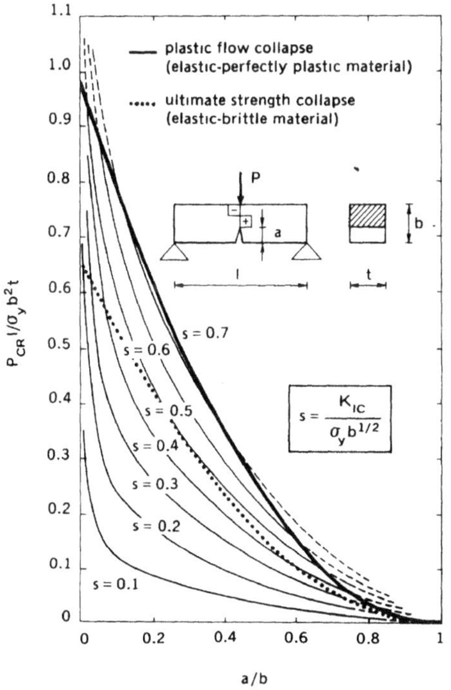

Fig. 2.6. Interaction between strength collapse and separation collapse (three points bending test).

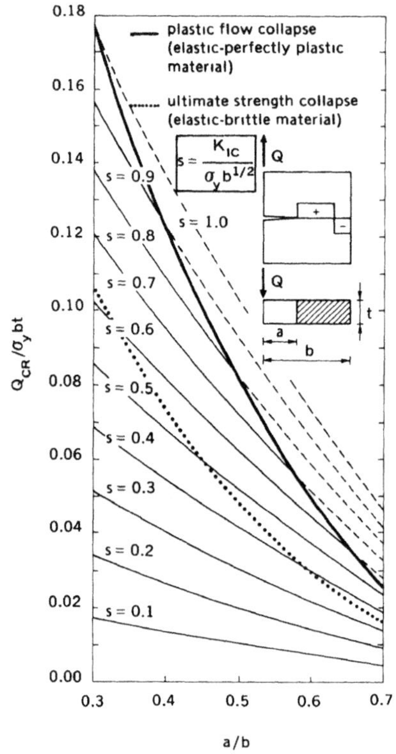

Fig. 2.7. Interaction between strength collapse and separation collapse (compact test).

potential fracture collapse. In Figure 2.5 the non-dimensional load of fracture collapse is reported as a function of the relative crack length a/b by varying the brittleness number s. The load of plastic flow collapse is also reported, and it is evident that fracture tests lose their meaning completely for s higher than the critical value $s_0 \cong 0.54$ as plastic flow collapse occurs in this case. For values of $s \lesssim 0.54$, fracture tests are significant only with cracks of intermediate length. For the *three points bending test* (TPB test, Figure 2.6) the stress intensity factor is

$$K_{\mathrm{I}} = \frac{Pl}{tb^{3/2}} f\left(\frac{a}{b}\right), \tag{2.11}$$

where

$$f\left(\frac{a}{b}\right) = 2.9\left(\frac{a}{b}\right)^{1/2} - 4.6\left(\frac{a}{b}\right)^{3/2} + 21.8\left(\frac{a}{b}\right)^{5/2} -$$

$$- 37.6\left(\frac{a}{b}\right)^{7/2} + 38.7\left(\frac{a}{b}\right)^{9/2}.$$

From equation (2.11) the crack propagation force can be obtained:

$$\frac{P_{CR}l}{\sigma_y b^2 t} = \frac{s}{f(a/b)}. \tag{2.12}$$

The force of plastic flow collapse obtained by a limit analysis is

$$\frac{P_{CR}l}{\sigma_y b^2 t} = \left(1 - \frac{a}{b}\right)^2, \tag{2.13}$$

while the force of ultimate strength collapse for an elastic-brittle material is

$$\frac{P_{CR}l}{\sigma_u b^2 t} = \tfrac{2}{3}\left(1 - \frac{a}{b}\right)^2. \tag{2.14}$$

In Figure 2.6 diagrams are reported which are analogous to the previous ones. In the case of elastic-perfectly plastic material, fracture tests are not significant for $s \gtrsim 0.73$. In fact, the fracture curve $s = 0.73$ is tangent to the plastic flow curve.
In the case of elastic-brittle material, fracture tests are not significant for $s \gtrsim 0.50$ ($s = K_{IC}/\sigma_u b^{1/2}$). Being $0.73 > 0.54$, it is possible to assert that, for simple geometric reasons, TPB tests are more favourable than TT, as far as the notch sensitivity for an elastic-perfectly plastic material is concerned. For an elastic-brittle material exactly the contrary can be asserted. For the *compact test* (CT, Figure 2.7) the stress-intensity factor is

$$K_I = \frac{Q}{tb^{1/2}} h\left(\frac{a}{b}\right), \tag{2.15}$$

where

$$h\left(\frac{a}{b}\right) = 29.6\left(\frac{a}{b}\right)^{1/2} - 185.5\left(\frac{a}{b}\right)^{3/2} + 655.7\left(\frac{a}{b}\right)^{5/2} - 1017\left(\frac{a}{b}\right)^{7/2} +$$

$$+ 638.9\left(\frac{a}{b}\right)^{9/2} \quad \text{for} \quad 0.3 \leqslant (a/b) \leqslant 0.7.$$

From equation (2.15) the crack extension force is

$$\frac{Q_{CR}}{\sigma_y bt} = \frac{s}{h(a/b)}, \tag{2.16}$$

while the plastic flow collapse at the weakened section due to the eccentric load occurs when

$$\frac{Q_{CR}}{\sigma_y bt} = \left[\left(1 + \frac{a}{b}\right)^2 + \left(1 - \frac{a}{b}\right)^2\right]^{1/2} - \left(1 + \frac{a}{b}\right). \tag{2.17}$$

Up to $s \simeq 1.02$, fracture tests are significant (Figure 2.7). The compact test is therefore reasonably favourable with respect to notch sensitivity. In other words, the fracture collapse can occur also with small specimens provided that the cracks are not too long. For $s = 0.9$, for example, the relative crack length must be $a/b \lesssim 0.4$ (Figure 2.7). It is curious to notice that the condition of small scale yielding, according to ASTM-E 399, requires exactly the contrary, i.e. sufficiently long cracks: $a/b > 0.45$.

Something analogous happens in Fluid Mechanics, where the turbulent *separation* phenomenon occurs only when the fluid flow exceeds certain size according to its viscosity and velocity. This size depends on the surface energy of the liquid. It is interesting to observe that viscosity (fluid) and fracture toughness (solid) play the same role in the physical similitude description [26]. It may happen then that forces having practically no effect on the prototype behaviour may affect the behaviour of the model. For example, surface tension does not influence ocean waves, but, if the waves in a model harbor are less than one inch long, their behavior would be dominated by surface tension. Disturbing influences of this type are called *scale effects*. Another analogous case is that of the capillary phenomenon. The difficulty in analyzing *separation collapse* with small structural sizes may be due to the influence of surface tension, which becomes more and more important as size decreases. In other words, if b is a characteristic structural size, for $b \to 0$ the volume tends to zero as b^3, while the external surface tends to zero as b^2. The external surface then results to be an infinitesimal quantity of lower rank with respect to the volume, and the forces associated to it become more and more relevant. Roughly speaking, a small specimen could therefore result to be bound by surface forces to a greater extent than a large structure of the same material.

For a *power-law hardening* material (Figure 2.4b), the stress, strain and displacement fields at the crack tip present the following forms respectively:

$$\sigma_{ij} = K_I^* r^{-1/(n+1)} S_{ij}(\vartheta),$$

$$\varepsilon_{ij} = \varepsilon_0 \left(\frac{K_I^*}{\sigma_0}\right)^n r^{-n/(n+1)} E_{ij}(\vartheta), \tag{2.18}$$

$$u_i = \varepsilon_0 \left(\frac{K_I^*}{\sigma_0}\right)^n r^{1/(n+1)} U_i(\vartheta).$$

A remark can be made from equation (2.18): *the plastic stress-intensity factor K_I^*, when the material is power-law hardening, has physical dimensions depending on the hardening exponent n*. More precisely, K_I^* has the dimensions $[FL^{-(2n+1)/(n+1)}]$. As a particular case, when the material is purely elastic (i.e. $n = 1$), the stress-intensity factor has the well-known dimensions $[FL^{-3/2}]$. On the other hand, according to the J-integral theory, it is interesting to observe that, for $n \to \infty$, K_{IC}^* tends to assume the physical dimensions $[FL^{-2}]$, i.e. the dimensions of a stress. Therefore, for a rigid-perfectly plastic material the fracture parameter K_{IC}^* coincides with the yield strength $\sigma_0 = \sigma_y$; it seems then that the considered model cannot predict another failure mode different from the plastic collapse. Increasing n from 1 to ∞, i.e. increasing the non-linearity of the material, a transition from a brittle fracture to a plastic flow collapse takes place. Such a transition is analogous to that previously described for an elastic-softening material (Figure 2.4a), which occurs when the structural sizes are decreased with geometrical similitude. On the basis of these considerations, a certain equivalence then appears to exist between an high non-linearity in the constitutive law of the material and a sufficiently small structural size [27].

It is possible then to define a *power-law hardening and linear softening* material (Figure 2.4c), a combination of the two elementary models previously considered (Figures 2.4a and b) which is able to describe most mechanical behaviours observed in practice. In this more general case, the dimensionless load of collapse will be a function of the hardening exponent n as well as of the statical brittleness number $s_s^* = K_{IC}^*/\sigma_0 b^{1/(n+1)}$ (or of the energy brittleness number $s_E^* = J_{IC}/\sigma_0 b = w_c/2b$). A strength collapse should then occur only for points $P(n, s^*)$ which are sufficiently far from the reference point $P_0(1, 0)$, being $n \geqslant 1$ and $s^* > 0$. In other words, an interaction between non-linearity and size should happen.

Finally, consider the cracked body of Figure 2.3 again. If the function in equation (2.5) is differentiated with respect to crack length a, it gives [28]:

$$\frac{\left(\dfrac{\partial q_0}{\partial a}\right)}{\sigma_y^\alpha b^{\beta-1}} = \varphi'\left(\frac{K_{IC}}{\sigma_y b^{1/2}}, \frac{F_P}{\sigma_y b^2}; \frac{a}{b}, \frac{t}{b}, \frac{l}{b}\right), \tag{2.19}$$

where φ' denotes the partial derivative of function φ with respect to the ratio a/b. The fracturing process of the considered system will be *stable* for positive values of the expression in equation (2.19), i.e. when the collapse load increases while the crack advances; it will be *unstable* in the opposite case. It is easy to realize then that the *stability* of a fracturing

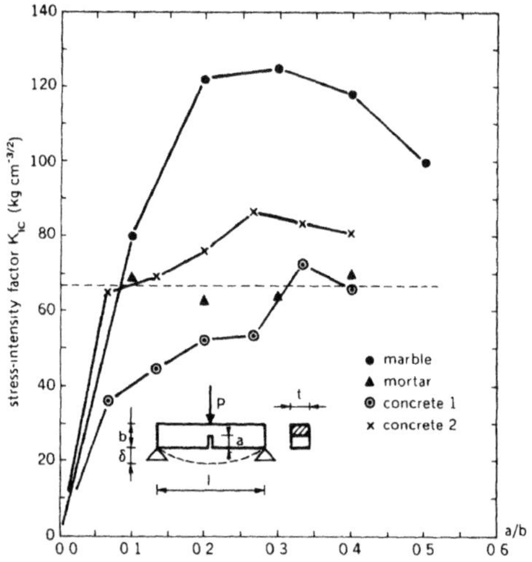

Fig. 2.8. Experimental values of fracture toughness K_{IC} against relative crack depth [30].

process follows laws of scale which are analogous to those followed by *sensitivity*.

A pertinent example will be considered in the next section.

Experimental results. The values of fracture toughness K_{IC} [29,30] for four different aggregate materials and investigated by varying the initial crack depth, are reported in Figure 2.8. It is possible to observe that, in the case of marble and concrete, K_{IC} increases for $a/b \lesssim 0.3$ and decreases for higher values of such ratio, after achieving a maximum corresponding to a/b, for which the fracture curve $s = 0.50$ is tangent to the ultimate strength curve (Figure 2.6). This is a recurring fact in fracture testing even for metallic materials when different crack depths are referred to the same specimen with crack advancing in a stable manner. Curves for $K_{IC}(a)$ or $K_{IC}(\Delta a)$ are usually called R-curves. In the author's opinion, the K_{IC} increase by increasing the crack length is not due to the expansion of the plastic or microcracked zone at the crack tip, i.e. to the development of a more and more energy-absorbing fracture process, but due to the transition of failure mode from a collapse to another one.

In Figure 2.9 the values of the ratio σ_N/σ_u of the net section rupture stress to the ultimate strength are reported. The sizes of the specimens are sufficient to induce stress concentration effects ($\sigma_N/\sigma_u < 1$). On the other hand, they are not large enough to cause a real "separation" collapse. Thus, it is possible to conclude that notch sensitivity ($\sigma_N/\sigma_u < 1$) is a necessary condition, even if not a sufficient one, to have a pure fracture

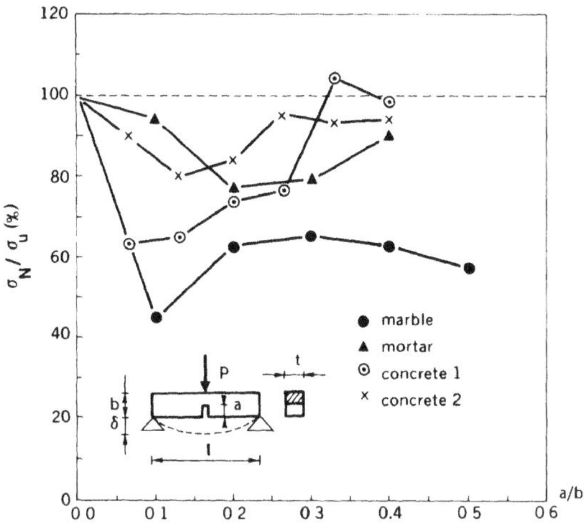

Fig. 2.9. Experimental values of notch sensitivity ratio against relative crack depth [30].

collapse ($K_I = K_{IC}$) [31]. The bell-shaped variation of K_{IC} is a necessary but not sufficient condition to have a pure ultimate strength collapse ($\sigma_N = \sigma_u$). When both conditions are satisfied, a mixed collapse occurs.

Other frequent experimental inconsistencies can also be explained. They involve the increase of K_{IC} by increasing the specimen sizes and the variability of K_{IC} by varying the test geometry. In this context, it is then possible to explain, at least in qualitative terms, the wide variability of K_{IC} and J_{IC} resistance curves and their frequent bell-shape (i.e. $\mathrm{d}J/\mathrm{d}a < 0$ for sufficiently deep cracks). Over the last few years, experimental investigations have been carried out in order to define the "mixed collapse" in terms of the J_{IC}-integral. However, in this case too, results depended sensitively on specimen and crack sizes [32]. Thus it seems plausible to assert that J_{IC}, as well as K_{IC}, can be considered only as an empirical parameter in gathering experimental data, rather than a constant property of the material. In fact, all the "fictitious" fracture parameters K_{IC}, $\mathcal{G}_{IC}$, J_{IC}, and COD do not relate merely to any unique energy-absorbing fracture process but to some collapse mechanisms different from separation.

Bažant and Cedolin [14] as well as Petersson [7] obtained numerical results analogous to those reported in the present section. The only difference is that they considered the fracture energy $\mathcal{G}_{IC}$ as a critical parameter, while previously the stress intensity factor K_{IC} was used in direct connection with external loadings. Visalvanich and Naaman have recently presented fracture test data relating to asbestos cement DCB specimens [33] (Figure 2.10). They observed that the apparent critical stress-intensity factor "increases rapidly and reaches a plateau value

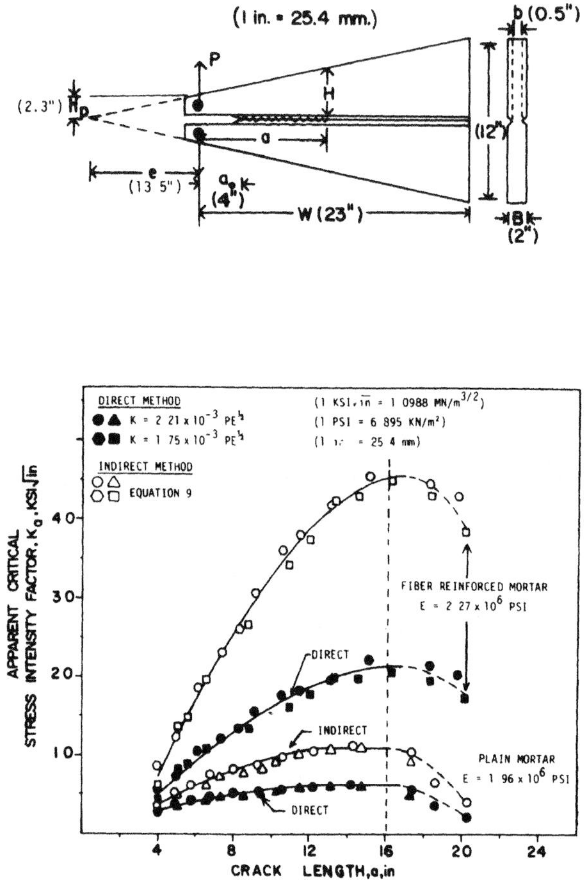

Fig. 2.10. Fracture test data relating to asbestos cement DCB specimens [33].

within a short crack extension. It decreases, however, as the crack reaches about 14 inches". Their comment is that "although this phenomenon is not yet clearly understood, it is possible that it is due to the slow response of the testing system". In the author's opinion the bell-shaped curve of Figure 2.10 is analogous to the curves of Figure 2.8. This means that for too short or too long cracks (out of the plateau) an ultimate tensile strength failure mode has occurred.

Crack propagation is the sole cause of collapse only for sufficiently large structures. This has been confirmed by the innumerable sudden failures, occurred always in large structures and due to relatively small cracks. The collapses of dams, bridges and ships have occurred in a completely brittle way, i.e. without noticeable plastic strain prior to material separation. For metallic alloys such phenomenon has been empirically explained by observing that the crack tip plastic zone at failure maintains the same extension. This explanation, if valid, applies only for sufficiently small structures. As far as aggregate materials are concerned, for which plastic effects are lacking, the explanations have been less convincing. Microcracking and the slow crack growth should be taken into consideration. The fact that a cracked specimen appears much tougher than a cracked structure many times larger and made of the same material should then depend on the fundamental laws of physical similitude rather than on effects such as the relative decrease of plastic zone, the increase of stress triaxiality or the increase of stress raiser number, etc.

Fracture sensitivity is a usual phenomenon in daily life. Adhesive tape and cloth, for example, are extremely notch sensitive materials. They need high stresses to be torn, but a very low stress to tear by introducing a small starting notch. This is due to the low ratio K_{IC}/σ_u, which makes these materials notch-sensitive, even for widths such as those of adhesive tape or cloth rolls. If materials with a much higher ratio K_{IC}/σ_u are considered, such as concretes or steel alloys, a similar situation can be produced again only with very large sizes, so as to obtain brittleness numbers s of the same magnitude. Thus a concrete dam can be made similar to an adhesive tape in the sense of the physical fracture behaviour. At the basis of such an *embrittlement for size increasing* there is perhaps only a homogeneity effect. In fact, it is important to observe that the degree of homogeneity is only a matter of scale. For example, PMMA needs to be considered as a heterogeneous material at the microscopic scale level, whereas concrete (with aggregates of $3 \div 6$ inches being regarded as a notch-insensitive material) can be considered as a homogeneous material for a dam *. Hydraulic concrete, therefore, also possesses a constant fracture toughness K_{IC} value even though it would be unpractical to measure it experimentally.

In conclusion, a first error to be avoided is that of designing and checking some structures of a particular undertaking and size without considering the most dangerous mode of failure, i.e., fracture. It would be equivalent to avoid considering the instability collapse in the case of a very slender structure. A second error to be avoided is that of believing that significant critical fracture parameters can be obtained from specimens of any geometry and size. As a general rule, it is not correct to extrapolate the results obtained from very small specimens to very large structures, even if they are made of the same material. In other words, a gradual brittleness increase occurs by increasing the structural sizes, and

* Analogously, in the dynamics of rivers, the roughness of the river-bed produces a scale effect, since a surface that is practically smooth for the prototype may be relatively rough for the model.

a transition from a plastic flow collapse (or from an ultimate strength collapse) to a *separation* takes place. In this sense, perhaps, the usual strength decrease occurring when the specimen sizes are increased may find a more convincing and even more fundamental explanation than that offered by Weibull's probabilistic argument [34].

2.3 Fracture stability in plain and reinforced concrete elements

The stability of the fracturing process in reinforced concrete beams will be taken into consideration. Attention will be focused on local phenomena relating to the cracked cross-section, while phenomena relating to the beam element to which the section belongs will be ignored. In particular, the concrete fracturing mechanism as well as the slippage and yielding of steel will be considered. The smeared damage of concrete will not be taken into account explicitly. A rigid-plastic constitutive law will be assumed for steel, while for concrete a linear elastic law, coupled with a fracturing condition according to Fracture Mechanics Theory, will be utilized. A more realistic concrete model should in fact simulate an elasto-softening behaviour with the possibility of a crushing collapse. However, the proposed model is able to predict, with sufficient accuracy, the order of magnitude of some interesting quantities, such as the dissipated energy in a loading cycle. More precisely, a reinforced concrete beam section with a through-thickness edge crack in the stretched part will be considered. The force transmitted by the reinforcement to the beam can be estimated by means of the rotation congruence condition. In the field of Linear Elastic Fracture Mechanics, such a force increases linearly by increasing the applied bending moment, until the limit force of pulling-out or yielding of steel is reached. From this point onwards a perfectly plastic behaviour of the reinforcement can be considered. In fact, it is possible to show that even the slippage can be described by a rigid-plastic law [35] and the bond stress-slip relationship for monotonic loading in tension is almost identical to that in compression [36]. Once the bending moment of slippage or yielding has been exceeded, the cracked beam section presents a linear-hardening behaviour, until the concrete fracture also occurs.

Statically undetermined reaction of reinforcement. Consider a reinforced concrete beam element of length $\Delta l \rightarrow 0$, with a rectangular cross-section of thickness t and width b, subjected to a bending moment M. Let the steel reinforcement be distant h from the external surface, while a through-thickness edge crack of depth $a \geqslant h$ is assumed to exist in the stretched part (Figure 2.11). Therefore, the cracked concrete beam segment will be in all subjected to the external bending moment M and to

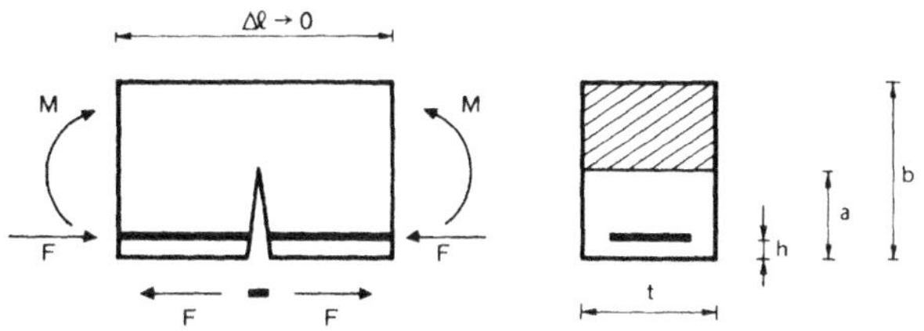

Fig. 2.11. Cracked reinforced beam element.

an eccentric axial force F, due to the statically undetermined reaction of the reinforcement. It is well-known that a bending moment M^* induces a stress-intensity factor K_I at the crack tip equal to

$$K_{\mathrm{I}} = \frac{M^*}{b^{3/2}t} Y_{\mathrm{M}}(\xi), \tag{2.20}$$

where $\xi = a/b$ is the relative crack depth and Y_{M} is the function

$$Y_{\mathrm{M}}(\xi) = 6 \times (1.99\xi^{1/2} - 2.47\xi^{3/2} + 12.97\xi^{5/2} - 23.17\xi^{7/2} + 24.80\xi^{9/2}),$$

for $\xi \leqslant 0.6$. In the same way, an axial tensile force F^* is associated with the stress intensity factor

$$K_{\mathrm{I}} = \frac{F^*}{b^{1/2}t} Y_{\mathrm{F}}(\xi), \tag{2.21}$$

in which

$$Y_{\mathrm{F}}(\xi) = 1.99\xi^{1/2} - 0.41\xi^{3/2} + 18.70\xi^{5/2} - 38.48\xi^{7/2} + 53.85\xi^{9/2},$$

for $\xi \leqslant 0.6$. On the other hand, the bending moment M^* causes a local rotation φ equal to [37,38]

$$\varphi = \lambda_{\mathrm{MM}} M^*, \tag{2.22}$$

where

$$\lambda_{\mathrm{MM}} = \frac{2}{b^2 t E} \int_0^{\xi} Y_{\mathrm{M}}^2(\xi)\, d\xi.$$

The axial tensile force F^* causes the rotation [37,38]

$$\varphi = \lambda_{\mathrm{MF}} F^*, \tag{2.23}$$

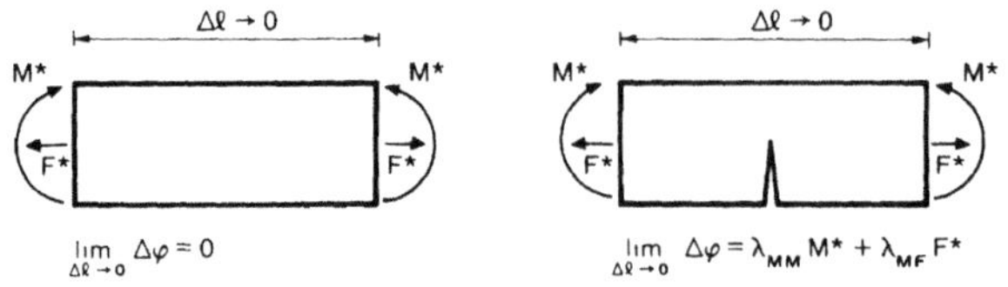

Fig. 2.12. Local rotation in (left) an uncracked and in (right) a cracked beam cross-section.

in which

$$\lambda_{\mathrm{MF}} = \frac{2}{btE} \int_0^{\xi} Y_{\mathrm{M}}(\xi) Y_{\mathrm{F}}(\xi) d\xi.$$

In other words, while an uncracked section performs an internal action of perfectly fixed joint (Figure 2.12a), a cracked section is equivalent to an elastic joint, rotating under the action of the bending moment M^* and the axial force F^* (Figure 2.12b):

$$\varphi = \lambda_{\mathrm{MM}} M^* + \lambda_{\mathrm{MF}} F^*. \tag{2.24}$$

For the statically undeterminate system, i.e. the reinforced beam element (Figure 2.11), the global moment acting on the cross-section will be

$$M^* = M - F\left(\frac{b}{2} - h\right). \tag{2.25}$$

The external moment tends to open the crack while the reinforcement reaction moment tends to close the crack. Then the axial force acting on the cross-section will be

$$F^* = -F. \tag{2.26}$$

Up to the moment of steel yielding or slippage, the global rotation, due to the bending moment M^* and to the closing force F^*, will be equal to zero:

$$\varphi = \lambda_{\mathrm{MM}} M^* + \lambda_{\mathrm{MF}} F^* = 0. \tag{2.27}$$

Equation (2.27) is the congruence condition which is able to provide the hyperstatic unknown F. Replacing expressions (2.25) and (2.26) in (2.27), the result is

$$\lambda_{\mathrm{MM}}\left[M - F\left(\frac{b}{2} - h\right)\right] - \lambda_{\mathrm{MF}} F = 0, \tag{2.28}$$

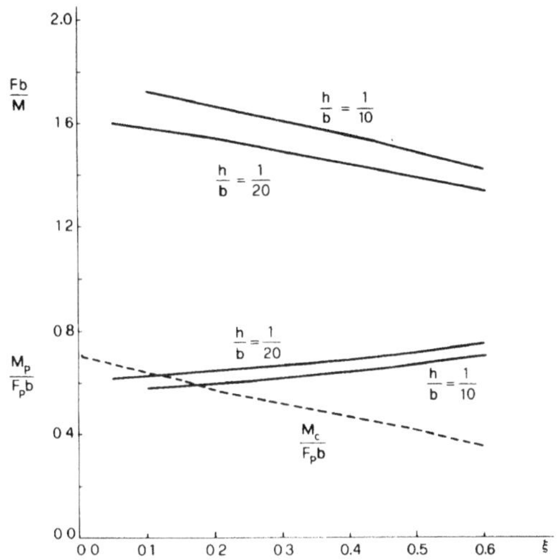

Fig. 2.13. Statically undetermined reaction of reinforcement and bending moment of reinforcement plastic flow.

and finally it is possible to obtain

$$\frac{Fb}{M} = \frac{1}{\left(\frac{1}{2} - \frac{h}{b}\right) + r(\xi)}, \tag{2.29}$$

where

$$r(\xi) = \frac{\int_0^{\xi} Y_M(\xi) Y_F(\xi) d\xi}{\int_0^{\xi} Y_M^2(\xi) d\xi}. \tag{2.30}$$

The statically undetermined reaction of the reinforcement, against the relative crack depth, for $h/b = 1/10$, $1/20$, is reported in the diagram of Figure 2.13. The decrease of the reaction by increasing the crack depth is not immediately understood, and indeed it may even surprise the reader. It can be explained, however, by observing that the compliances λ_{MM} and λ_{MF} both increase by increasing the crack length, but λ_{MF} increases more rapidly than λ_{MM}. Thus lower and lower axial forces F are needed to annul the rotation due to the external moment M.

Bending moment of reinforcement plastic flow. As equation (2.29) shows, the force F, transmitted by the reinforcement, increases linearly by

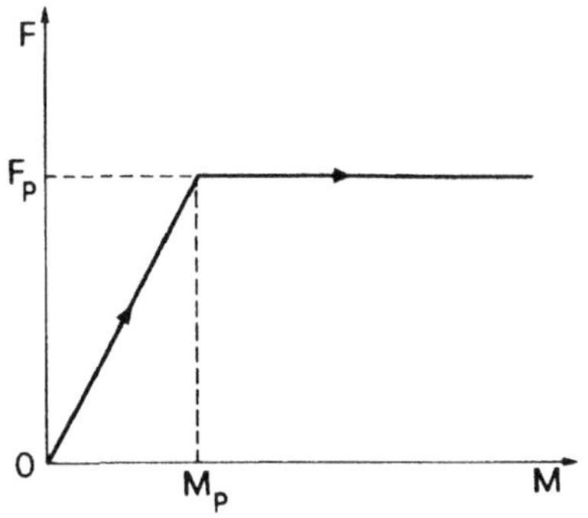

Fig. 2.14. Force transmitted by the reinforcement against the applied moment.

increasing the external moment M, until the limit force $F_P = f_y A_s$ is reached, f_y being the steel yield strength and A_s the steel area. From this point onwards a perfectly plastic behaviour of the reinforcement will be considered. This means that the infinitesimal reinforcement segment, which is uncovered, i.e. included between the two crack surfaces, will flow, always transmitting the same force F_P to the cracked concrete element (Figure 2.14). From equation (2.29) it is possible to obtain the moment of plastic flow for the reinforcement:

$$M_P = F_P b\left[\left(\frac{1}{2} - \frac{h}{b}\right) + r(\xi)\right]. \tag{2.31}$$

Such moment against the relative crack depth, for $h/b = 1/10$, $1/20$, is reported in Figure 2.13. According to the decrease of the reaction by increasing the crack depth ξ (Figure 2.13), an increase of the moment of reinforcement plastic flow M_P occurs by increasing ξ. However, it should be observed that, if concrete presents a low crushing strength f_c and steel a relatively high yield strength f_y, the concrete crushing collapse can come before the steel plastic flow. If M_c is the external moment of concrete crushing and a hypothesis of linear stress variation through the ligament holds (Figure 2.15), it follows that

$$\frac{M_c}{F_P b} = \frac{f_c/f_y}{A_s/A} \cdot \frac{(1-\xi)(2+\xi-3h/b)}{6}. \tag{2.32}$$

The dashed line of Figure 2.13 represents the diagram of the function in equation (2.32) for $f_c = 200$ kg cm^{-2}, $f_y = 3600$ kg cm^{-2}, $A_s/A = 0.024$ and $h/b = 1/10$. It can be observed that, although very favourable values to the concrete crushing collapse have been chosen, such collapse in fact

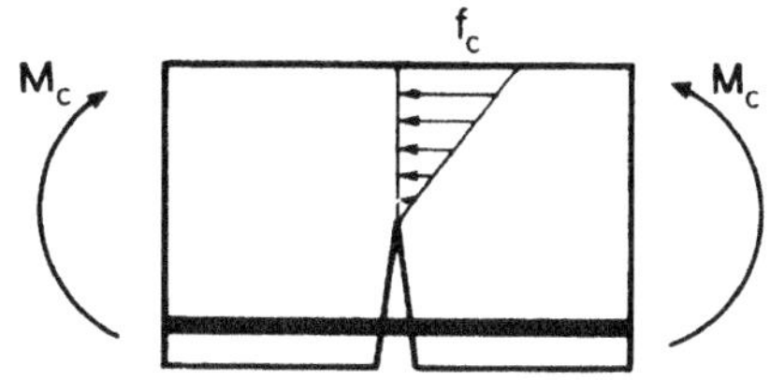

Fig. 2.15. Hypothesis of linear stress variation through the ligament.

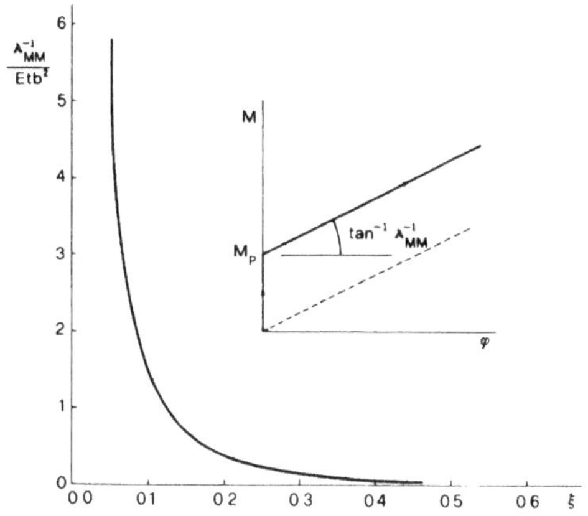

Fig. 2.16. Hardening coefficient against relative crack depth.

comes before the steel plastic flow only for sufficiently high values of the crack depth ($\xi \geqslant 0.175$).

Rigid-hardening behaviour of the cracked beam section. The purpose of the present sub-section is to describe the mechanical behaviour of the cracked reinforced concrete beam section, once the bending moment M_P of steel plastic flow has been exceeded. For

$$M \leqslant M_P, \quad \varphi = 0 \qquad \text{and for} \quad M > M_P:$$

$$\varphi = \lambda_{MM}\left[M - F_P\left(\frac{b}{2} - h\right)\right] - \lambda_{MF}F_P. \tag{2.33}$$

The $M - \varphi$ diagram for the cracked beam section is represented in Figure 2.16. This diagram expresses the equivalence of the beam section with a rigid-linear hardening spring. It is interesting to observe that the hardening line is parallel to the $M - \varphi$ diagram relating to the same cracked beam section without reinforcement (broken line). The hardening coefficient λ_{MM}^{-1} against the relative crack depth ξ is given again in Figure 2.16.

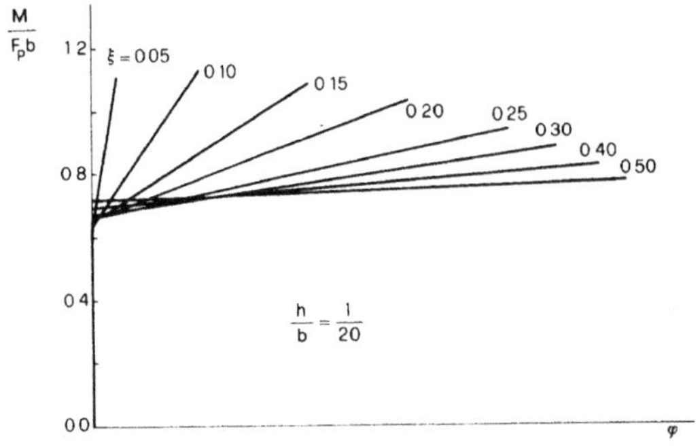

Fig. 2.17. Moment-rotation diagrams for different crack depths.

By increasing the crack depth ξ, the hardening line becomes more and more inclined, until giving rigid-perfectly plastic behaviour. On the other hand, for $\xi \to 0$, the hardening line becomes nearly vertical, until giving a rigid behaviour of the beam section simulating spring. Therefore, it is possible to conclude that, by increasing ξ, the moment of steel plastic flow increases (Figure 2.13), while the slope of the hardening line decreases (Figure 2.16). Some $M - \varphi$ diagrams, for $h/b = 1/20$, are reported in Figure 2.17, ξ varying between 0.05 and 0.50. The moment of steel plastic flow increases very little by increasing ξ; on the other hand, the slope of the hardening line decreases sharply.

Bending moment of concrete fracture. In the present sub-section the concrete fracture collapse will be examined, which is consequent to the steel plastic flow collapse, i.e. it occurs for a bending moment $M_F \geqslant M_P$. After the steel plastic flow, the stress-intensity factor acting at the crack tip will be equal to the algebraical addition of the factors in equations (2.20) and (2.21), and the actual loadings will be

$$M^* = M - F_P\left(\frac{b}{2} - h\right), \tag{2.34}$$

$$F^* = -F_P. \tag{2.35}$$

Thus, it will be

$$K_I = \frac{1}{b^{3/2}t} Y_M(\xi)\left[M - F_P\left(\frac{b}{2} - h\right)\right] - \frac{F_P}{b^{1/2}t} Y_F(\xi). \tag{2.36}$$

Presuming that equation (2.36) is equal to the concrete fracture toughness

K_{IC}, it is possible to obtain the fracture moment M_F:

$$M_F = \frac{K_{IC} b^{3/2} t}{Y_M(\xi)} + \frac{F_P b}{Y_M(\xi)} \left[Y_F(\xi) + Y_M(\xi) \left(\frac{1}{2} - \frac{h}{b} \right) \right]. \tag{2.37}$$

In non-dimensional form

$$\frac{M_F}{K_{IC} b^{3/2} t} = \frac{1}{Y_M(\xi)} + N_P \left[\frac{Y_F(\xi)}{Y_M(\xi)} + \frac{1}{2} - \frac{h}{b} \right], \tag{2.38}$$

where

$$N_P = \frac{f_y b^{1/2}}{K_{IC}} \cdot \frac{A_s}{A}$$

is a non-dimensional number, analogous to the brittleness number defined in the preceding section. The concrete fracture moment M_F against the relative crack depth ξ is given in Figure 2.18, varying the non-dimensional number $N_P (h/b = 1/20)$. For N_P values close to zero, that is for low reinforced beams or for very small cross-sections, the fracture moment decreases while the crack extends, and a typical phenomenon of *unstable fracture* occurs. For higher N_P values, a stable branch follows the unstable one of the curve, which describes the crack extension against the applied load. Already for $N_P = 1$ the minimum of the curve is evident and takes place for $\xi \cong 0.35$. For higher N_P values, the ξ value, for which the minimum occurs, is lower, while the stable branch becomes steeper. For $N_P \gtrsim 8.5$ the unstable branch disappears completely and only the stable branch remains. An analogous behaviour will be later underlined in the case of a cracked masonry wall subjected to an eccentric axial compression force. In that case, however, the unstable branch and the consequent stable one appear steeper and the existence of the minimum is then more evident. The locus of minima is represented by a dashed line in Figure 2.18. This line divides the quadrant of the diagram into two zones: the upper zone is where the fracturing process is stable, while the lower one is where the process is unstable. It is therefore possible to assert that the fracturing process in reinforced concrete becomes stable only when the beam is sufficiently reinforced or when the cross-section is sufficiently large and the crack is sufficiently deep. Such a fracture stability increase, due to the increase of beam width and steel area, appears substantially in accordance with the numerical results by Bažant [15], for which the more spaced the reinforcement in a panel (i.e. the larger the single recurrent concrete element and the larger the area of the single reinforcement), the more stable the fracturing process. If the curve

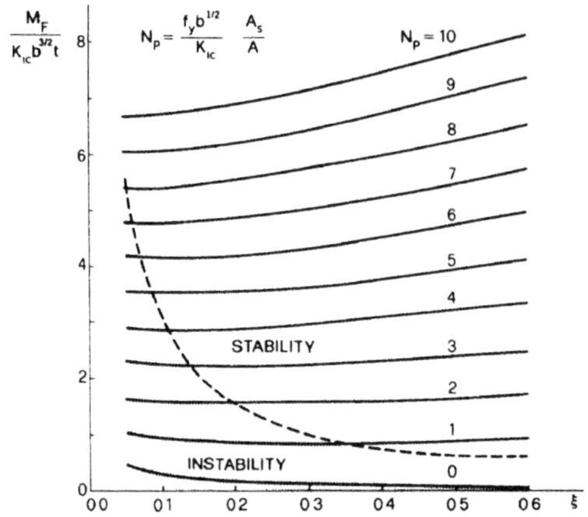

Fig. 2.18. Bending moment of concrete fracture against relative crack depth ($h/b = 1/20$).

N_P = constant were perfectly horizontal, a condition of indifferent equilibrium would occur. In fact none of the curves shows such a regularity. However, it is important to observe that the curve relating to $N_P = 1$, which could represent the fracturing phenomenon for very common reinforced concrete beams, is only slightly deflected downwards. In fact the minimum is only about 15% lower than the value of the function for $\xi = h/b = 0.05$. For $h/b = 1/10$ curves very similar to those of Figure 2–18 are obtained. Only two slight differences are present:

(1) the curves go down, i.e. fracture collapse occurs for lower moments, since the reinforcement, being internal, resists to a lesser extent than in the preceding case and

(2) the dashed line goes up, i.e. the stable zone of the diagram shrinks.

Finally, it may be interesting to compute the non-dimensional number N_P for three different reinforced concrete beams. As a first example, let us consider the following set of values

$$f_y = 2400 \text{ kg cm}^{-2}, \quad K_{IC} = 80 \text{ kg cm}^{-3/2},$$
$$b = 30 \text{ cm}, \quad A_s/A = 0.01,$$

from which $N_P = 1.64$. Thus, it is possible to verify in the diagram of Figure 2.18 that, for this very common reinforced concrete beam, the fracturing process is very close to a condition of indifferent equilibrium. Let us now examine a low reinforced beam with small cross-section:

$$f_y = 2400 \text{ kg cm}^{-2}, \quad K_{IC} = 100 \text{ kg cm}^{-3/2},$$
$$b = 20 \text{ cm}, \quad A_s/A = 0.0024,$$

from which it follows that $N_P = 0.26$. In this beam the fracturing process occurs in an unstable manner (Figure 2.18). As a third and last case, a high reinforced beam with a large cross-section is examined:

$$f_y = 3600 \text{ kg cm}^{-2}, \quad K_{IC} = 50 \text{ kg cm}^{-3/2},$$
$$b = 150 \text{ cm}, \quad A_s/A = 0.0240,$$

from which it gives: $N_P = 21.16$. In this beam the fracturing process occurs in a stable manner (Figure 2.18).

Stability of the process of concrete fracture and steel plastic flow. The stability of reinforced concrete fracturing process has been previously described on the basis of the curve representing the crack depth against the applied bending moment. In the present sub-section such stability will be studied using energetic considerations. The stress-intensity factor acting on the crack is

$$K_I = \frac{1}{b^{3/2}t} Y_M(\xi)\left[M - F\left(\frac{b}{2} - h\right)\right] - \frac{1}{b^{1/2}t} Y_F(\xi) F, \quad \text{for } M \leqslant M_P, \tag{2.39}$$

$$K_I = \frac{1}{b^{3/2}t} Y_M(\xi)\left[M - F_P\left(\frac{b}{2} - h\right)\right] - \frac{1}{b^{1/2}t} Y_F(\xi) F_P, \quad \text{for } M > M_P. \tag{2.40}$$

Inserting equation (2.29) into (2.39), it gives

$$K_I = \frac{1}{b^{3/2}t} Y_M(\xi)\left[M - \frac{M/b}{\left(\frac{1}{2} - \frac{h}{b}\right) + r(\xi)}\left(\frac{b}{2} - h\right)\right] - \frac{1}{b^{1/2}t} Y_F(\xi) \frac{M/b}{\left(\frac{1}{2} - \frac{h}{b}\right) + r(\xi)}, \quad \text{for } M \leqslant M_P. \tag{2.41}$$

Equations (2.41) and (2.40) in non-dimensional form appear as

$$\frac{K_I b^{1/2} t}{F_P} = Y_M(\xi) \frac{M}{F_P b}\left[1 - \frac{1}{1 + \frac{r(\xi)}{\left(\frac{1}{2} - \frac{h}{b}\right)}}\right] -$$

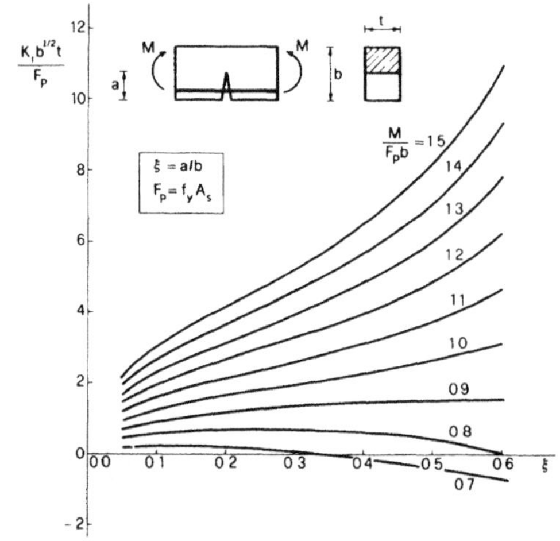

Fig. 2.19. Stress-intensity factor against relative crack depth, varying the applied moment $(h/b = 1/20)$.

$$-\frac{M}{F_P b} Y_F(\xi) \frac{1}{\left(\frac{1}{2} - \frac{h}{b}\right) + r(\xi)}, \quad \text{for } M \leqslant M_P, \tag{2.42}$$

$$\frac{K_I b^{1/2} t}{F_P} = Y_M(\xi)\left[\frac{M}{F_P b} - \left(\frac{1}{2} - \frac{h}{b}\right)\right] - Y_F(\xi), \quad \text{for } M > M_P. \tag{2.43}$$

The stress-intensity factor K_I against the crack depth ξ is reported in Figure 2.19, varying the loading parameter $M/F_P b$ $(h/b = 1/20)$. For $M/F_P b$ values lower than about 0.7, the stress-intensity factor is very low for every considered depth $h/b \leqslant \xi \leqslant 0.6$. As the curve $M/F_P b = 0.7$ clearly shows, the K_I value is positive for small depths ξ, while it becomes negative for larger depths. This means that, for $0 \leqslant M/F_P b \leqslant 0.8$ and for sufficiently deep cracks, the assumed model predicts closing of the crack as well as the non-plastic state of steel. The plastic limit, as the diagram of Figure 2.13 suggests, is very near the curve $M/F_P b = 0.7$ reported in Figure 2.19. More precisely, it is included between the two curves for $M/F_P b = 0.60$ and $M/F_P b = 0.75$. For $M/F_P b > 0.8$, the K_I factor is positive for every investigated depth ξ. For $M/F_P b \lesssim 0.9$ the function $K_I(\xi)$ presents a positive maximum. This means that for sufficiently low bending moments and sufficiently deep cracks the fracturing process is stable. In fact, from an energetic point of view, it can be asserted that the generalized crack extension force $\mathcal{G}_I = K_I^2/E$ has the same course as K_I, for $M/F_P b > 0.8$. Thus for $0.8 \lesssim M/F_P b \lesssim 0.9$ and for sufficiently high

ξ, it is found that

$$\frac{\partial \mathscr{G}_\mathrm{I}}{\partial \xi} = -\frac{\partial^2 V}{\partial \xi^2} < 0, \tag{2.44}$$

where V is the total potential energy of the concrete-steel system. In other words, for those particular values of the bending moment and the crack depth, the total potential energy V can present, as a stationary point ($K_\mathrm{I} = K_\mathrm{IC}$), only a minimum and therefore a stable equilibrium condition.

Synthesis. Up to now the only fact that has been clarified is that the concrete fracture collapse follows the reinforcement plastic collapse and that, between the two mentioned collapses, the mechanical behaviour of the cracked beam section is linear hardening. However, no indication of how much the fracture moment M_F is higher than the plastic flow moment M_P has been given yet. From equations (2.31) and (2.37) it follows that

$$\frac{M_\mathrm{P}}{M_\mathrm{F}} = \frac{\left[\frac{1}{2} - \frac{h}{b} + r(\xi)\right] Y_\mathrm{M}(\xi)}{\frac{1}{N_\mathrm{P}} + Y_\mathrm{F}(\xi) + Y_\mathrm{M}(\xi)\left(\frac{1}{2} - \frac{h}{b}\right)}. \tag{2.45}$$

In Figure 2.20 the ratio $M_\mathrm{P}/M_\mathrm{F}$ against the non-dimensional number N_P is represented, varying the crack depth $\xi (h/b = 1/20)$. From this diagram it may be deduced that, the higher the number N_P and the deeper the crack, the closer the fracture collapse is to the plastic one. In Figure 2.21 the moment-rotation diagrams $M(\varphi)$ are reported for $h/b = 1/20$, $\xi = 0.1$ and for five different values of number N_P: 0.0, 0.1, 0.3, 0.7, 3.0. Once the cross-section sizes and the mechanical properties of the material have been defined, they represent five different steel areas. Rigid behaviour ($0 \leqslant M \leqslant M_\mathrm{P}$) is followed by linear hardening behaviour ($M_\mathrm{P} < M \leqslant M_\mathrm{F}$). The latter stops when the concrete fracture occurs. If the fracture phenomenon is unstable, function $M(\varphi)$ presents a discontinuity and drops from the value M_F to the value $F_\mathrm{P} b$ with a negative jump (Figures 2.21a, b, c and d). In fact in this case a complete and instantaneous disconnection of the concrete cross-section occurs. While the rotation φ is constant, the new moment $F_\mathrm{P} b$ can be estimated according to the scheme of Figure 2.22, where each beam segment is subjected to the traction F_P of the reinforcement and to the contact compression F_P, i.e. altogether, to the moment $F_\mathrm{P}(b-h) \simeq F_\mathrm{P} b$. Increasing then the rotation φ and ignoring any phenomenon of instability, the bending

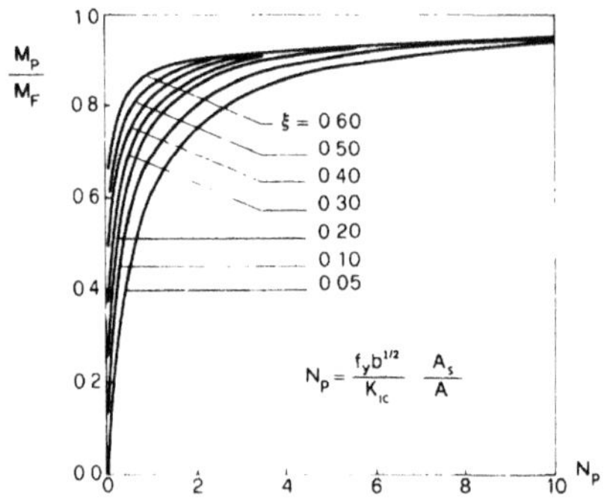

Fig. 2.20. Ratio between the moment of steel plastic flow and the moment of crack extension against the non-dimensional number N_P, varying the relative crack depth ξ $(h/b = 1/20)$.

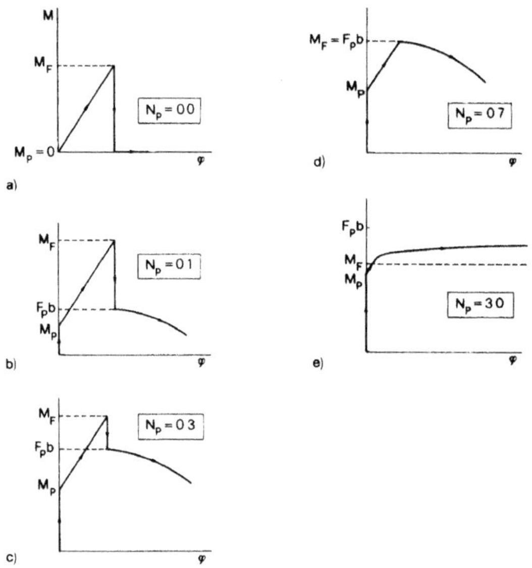

Fig. 2.21. Mechanical behaviour of the cracked reinforced beam section, for different non-dimensional numbers N_P $(h/b = 0.05;\ \xi = 0.10)$.

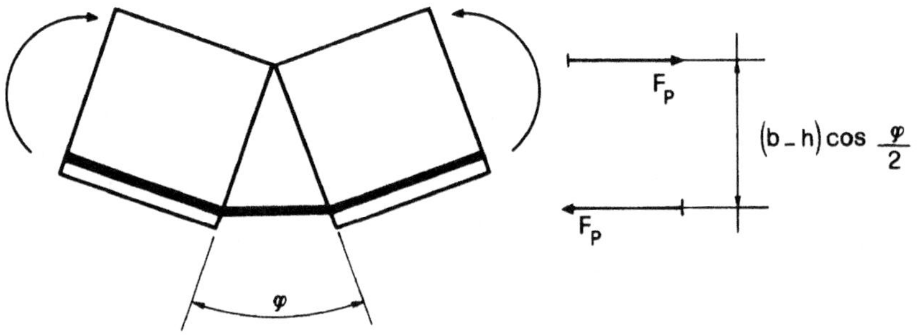

Fig. 2.22. Statical scheme after the complete disconnection of concrete.

moment decreases with a non-linear law (Figure 2.22):

$$M = F_P b \cos \frac{\varphi}{2}. \tag{2.46}$$

On the other hand, if the fracture phenomenon is stable, the function $M(\varphi)$ does not present any discontinuity and describes hardening behaviour (Figure 2.21e) analogous to that of Figure 2.17. In Figure 2.21a the case $N_P = 0$ is considered, i.e. the beam without reinforcement. The plastic flow moment M_P is naturally equal to zero, as well as the moment $F_P b$, which occurs immediately after the complete disconnection of concrete. In Figure 2.21b the case $N_P = 0.1$ is described, i.e. a low reinforced beam. By the diagram of Figure 2.13 it is possible to obtain the ratio $M_P/F_P b$, while by the diagram of Figure 2.20 the ratio M_P/M_F. The slope of the hardening line does not vary with respect to the preceding case, since it depends only on the crack length, besides the concrete elastic modulus and the cross-section sizes (Figure 2.16). In Figure 2.21c the case $N_P = 0.3$ is considered, which is analogous to the previous one, except for the fact that the ratio M_P/M_F is higher. On the other hand, the ratio $M_P/F_P b$, which is independent of N_P (Figure 2.13), remains unchanged. In Figure 2.21d the case $N_P = 0.7$ is reported. For this value $M_F = F_P b$, and then the discontinuity vanishes. Finally, in Figure 2.21e the case $N_P = 3$ is described. In this case the fracture moment M_F is only slightly higher than the plastic moment M_P, and the moment $F_P b$ would be obtained only with a positive jump of the function. On the other hand, from Figure 2.18 it is known that the fracturing process, for $N_P = 3$ and $\xi \geqslant 0.14$, is stable and thus a complete and instantaneous disconnection of concrete cannot occur (Figure 2.22). It is observed that, as for $N_P < 0.7$ it is $F_P b < M_F$, and then a discontinuity appears in the diagram $M(\varphi)$ (Figs. 2.21a, b and c), so for $N_P < 0.7$ the curves of Figure 2.18 lie completely in the unstable zone. Therefore, it is possible to conclude that, by increasing the steel percentage A_s/A, or, in the same way, by increasing the beam size b, the concrete fracturing process becomes stable.

Limit bearing capacity of cracked masonry walls. When the axial force F is a compression and the bending moment M tends to open the crack, as happens at the cross-sections of walls or arch-structures, the total stress-intensity factor can be obtained by applying the superposition principle:

$$K_I = K_I^{(M)} - K_I^{(F)} = \frac{F}{b^{1/2} t} \left[\frac{e}{b} Y_M(\xi) - Y_F(\xi) \right], \tag{2.47}$$

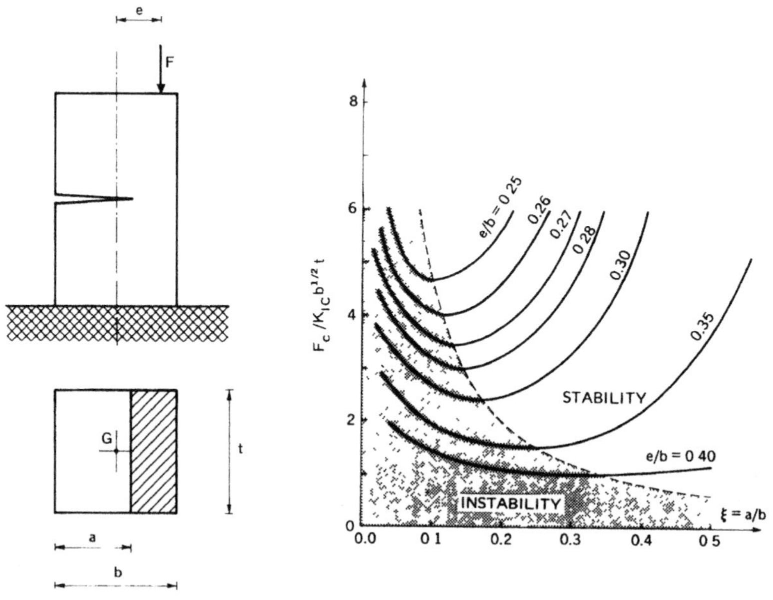

Fig. 2.23. Stability of fracturing process in a wall loaded by a compressive eccentric axial force.

where e indicates the eccentricity of the equivalent eccentric axial force (Figure 2.23). From the *critical condition* $K_I = K_{IC}$, the dimensionless axial force of crack extension can be obtained as a function of the crack depth ξ, varying the relative eccentricity e/b of the load:

$$\frac{F_c}{b^{1/2}tK_{IC}} = \frac{1}{\frac{e}{b}Y_M(\xi) - Y_F(\xi)}. \tag{2.48}$$

The curves of Figure 2.23 give a graphical representation of equation (2.48) and show that for a fixed eccentricity e/b fracturing achieves a stable stage only after presenting an unstable one. If the load F has not the possibility of following the unstable descending branch of the curve $e/b =$ constant, fracturing will have a catastrophical behaviour and the representative point will advance horizontally in the diagram of Figure 2.23 to meet again the curve $e/b =$ constant on its stable branch. On the other hand, the load relaxation possibility, and then the more or less catastrophical fracturing behaviour, depends on the geometrical and mechanical structural features, and especially on the degree of redundancy [39].

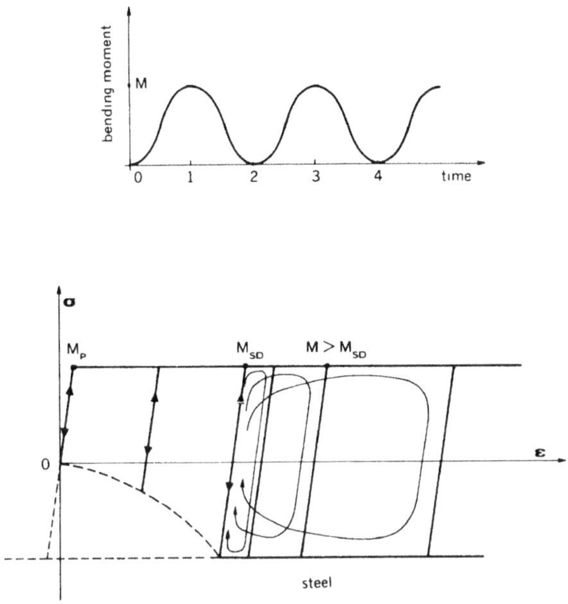

Fig. 2.24. Hysteretic loops in the steel stress-strain curve.

2.4 Hysteretic behaviour of reinforced concrete elements

When a reinforced concrete beam is subjected to seismic and generally repeated loadings, it deteriorates in a progressive manner and its stiffness and loading capacity decrease sensibly. Such effects are the result of different damage phenomena, like crushing and fracturing of concrete or pulling-out and yielding of steel bars. The Fracture Mechanics model introduced in the preceding section will be assumed again and loading and unloading processes will be considered. The bending moment range is lower than that of crack extension. Therefore, the phenomenon of shake-down due to slippage or plastic deformation of the steel bars will be studied. Up to a certain value of the bending moment an elastic shake-down occurs; above this value the shake-down becomes elastic-plastic and an hysteresis loop is described by the stress-strain diagram of steel. Thus the energy absorbed in such a dissipative phenomenon will be computed for each loading cycle.

Elastic-plastic shake-down under repeated loadings. If the cracked section is assumed to be cyclically loaded and unloaded and the crack extension possibility is ignored for the moment, the three following fields of steel

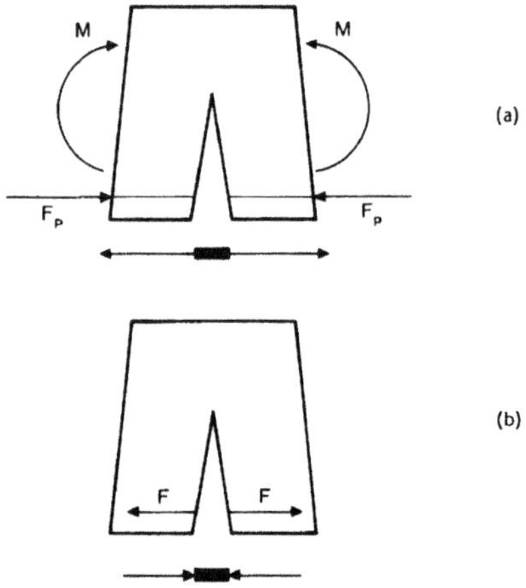

Fig. 2.25. Statical schemes under loading (a) and after unloading (b) for a reinforced beam cross-section.

behaviour (Figure 2.24) are possible:

1) $0 \leqslant M \leqslant M_P$: elastic behaviour;
2) $M_P < M \leqslant M_{SD}$: elastic shake-down;
3) $M_{SD} < M$: plastic shake-down.

The unknown element of the problem is the moment M_{SD}, above which the shake-down becomes plastic and the steel stress-strain curve presents hysteretic cycles and energy dissipation (Figure 2.24). If the cracked section is loaded with a moment M, a little higher than M_P, and then it is unloaded, a residual rotation remains, which, in the limit-case of rigid-perfectly plastic reinforcement, coincides with the under loading rotation $\varphi(M)$. In this case constraint between steel and concrete occurs, i.e. the concrete element compresses the steel segment, which works as a strut (Figure 2.25b). Therefore, it can be assumed that the unknown compression F produces the rotation $\varphi(F)$ in the cracked section. Thus, when steel is rigid-perfectly plastic, the following condition holds

$$\varphi(M) = \varphi(F), \tag{2.49}$$

from which the unknown F can be extracted, that is the steel compression, when the beam-section has been unloaded. The moment of plastic shake-down M_{SD} is defined as the lowest moment for which $F = F_P$, i.e. for which the steel yields even in compression after unloading. Now

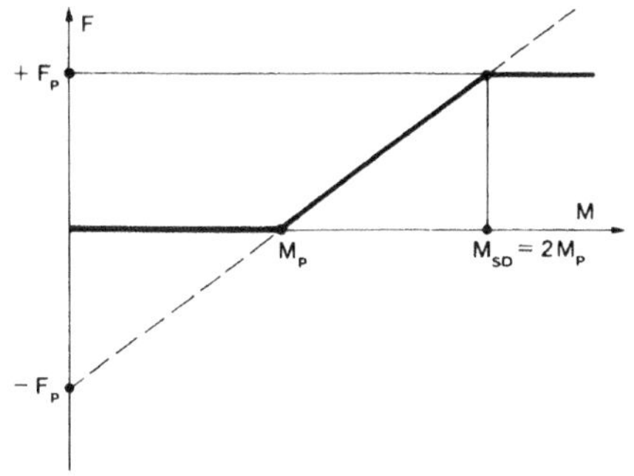

Fig. 2.26. Compression force after unloading against maximum applied bending moment.

equation (2.49) will be made explicit. Rotation $\varphi(M)$ due to the bending moment M, $M > M_P$ (Figure 2.25a), is given by equation (2.33), while rotation $\varphi(F)$ produced by the force F after unloading is (Figure 2.25b):

$$\varphi(F) = \lambda_{MM} F\left(\frac{b}{2} - h\right) + \lambda_{MF} F. \tag{2.50}$$

Applying equation (2.49) one obtains

$$F = M \frac{\lambda_{MM}}{\lambda_{MM}\left(\frac{b}{2} - h\right) + \lambda_{MF}} - F_P. \tag{2.51}$$

If both sides of equation (2.51) are divided by F_P and equation (2.31) is recalled, the following linear relationship results, connecting the external bending moment M with the compression F after unloading:

$$\frac{F}{F_P} = \frac{M}{M_P} - 1. \tag{2.52}$$

A graphic representation of such a connection is reported in Figure 2.26. For $M \leqslant M_P$, of course, steel compression after unloading is equal to zero, while for $M = M_{SD} = 2M_P$ the reverse steel yielding after unloading begins to happen. It should be observed that ratio 2, between the moment of plastic shake-down M_{SD} and the moment of direct plastic flow M_P, is the same as that which results in the case of the well-known simple model with in parallel elements of Figure 2.27. The arguments developed up to now are valid only if the crack extension does not precede the plastic shake-down:

$$M_{SD} = 2M_P < M_F, \tag{2.53}$$

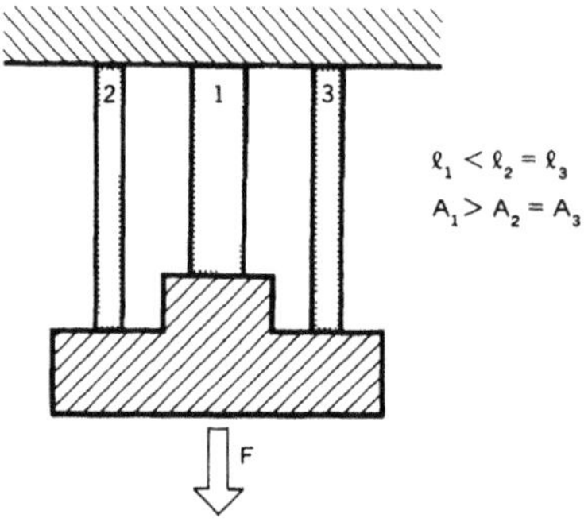

Fig. 2.27. Simplified shake-down model with in parallel-elements.

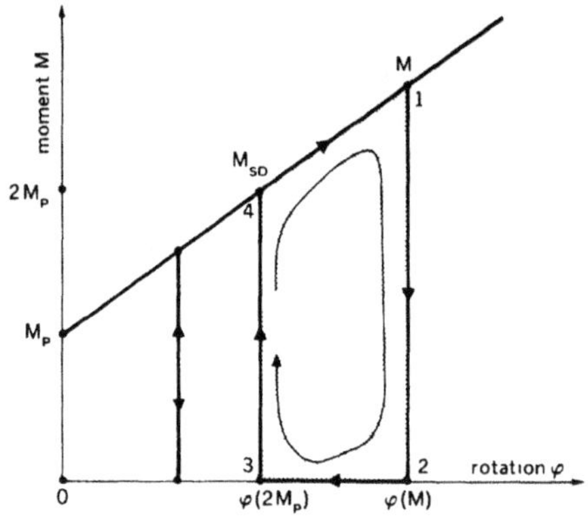

Fig. 2.28. Hysteretic loop in the moment-rotation diagram, under unidirectional cyclic loading.

with M_F being the moment of crack extension. In the opposite case, the elastic shake-down will occur for $M_P < M < M_F$, while the plastic shake-down is impossible. Recalling Figure 2.20, it is possible to specify that the plastic fatigue can theoretically occur only for sufficiently low values of the number N_P. In the case of high N_P numbers, the model then predicts the possibility of a non-linear combined effect of stable progressive fracture in concrete and shake-down in steel.

Consider now the case $M > M_{SD}$ (Figure 2.28). Once the beam-section has been unloaded, the residual rotation is reduced to that related to the moment M_{SD}. More precisely, unloading with constant rotation occurs at first (line 1-2 of Figure 2.28) and then the plastic flow of the compressed steel up to the equilibrium situation (line 2-3). When the moment is increased again, the moment M_{SD} is reached at first, the rotation being constant (line 3-4), and then the final moment M, the rotation increasing linearly in this second stage (line 4-1). As Figure 2.28 shows, the point representative of the system on the plane $M-\varphi$ describes a closed curve, which, when $M < M_{SD}$, degenerates into a segment. It is therefore quite easy to compute the plastic energy dissipated in each cycle; it is equal to the area of the rectangular trapezium of Figure 2.28:

$$\frac{\text{work}}{\text{cycle}} = \tfrac{1}{2}(M + 2M_P)[\varphi(M) - \varphi(2M_P)]. \tag{2.54}$$

Recalling from equations (2.31) and (2.33) then:

$$\frac{\text{work}}{\text{cycle}} = \frac{1}{2}\frac{2}{b^2tE}\int_0^{\xi} Y_M^2(\xi)\,d\xi \cdot \left[M^2 - 4F_P^2b^2\left(\frac{1}{2} - \frac{h}{b} + r(\xi)\right)^2\right], \tag{2.55}$$

which in non-dimensional form appears as follows:

$$\frac{\text{work/cycle}}{b^2tE} = \int_0^{\xi} Y_M^2(\xi)\,d\xi \cdot \left[\left(\frac{M}{b^2tE}\right)^2 - 4\left(\frac{F_Pb}{b^2tE}\right)^2\left(\frac{1}{2} - \frac{h}{b} + r(\xi)\right)^2\right], \tag{2.56}$$

which may be put into a more compact form:

$$\frac{\text{work/cycle}}{b^2tE} = \int_0^{\xi} Y_M^2(\xi)\,d\xi \cdot \left[\left(\frac{M}{b^2tE}\right)^2 - \left(\frac{M_{SD}}{b^2tE}\right)^2\right]. \tag{2.57}$$

In Figure 2.29 the dissipated energy per cycle is reported as a function of the maximum bending moment, varying the parameter F_P/btE. Obviously, it increases quadratically by increasing the moment and decreases by increasing F_P/btE. The points, where the curves intersect the abscissae axis, represent the moment M_{SD}, as equation (2.57) explicitly shows. In Figure 2.30 the dissipated energy per cycle is reported as a function of the crack depth ξ, varying the parameter F_P/btE. It increases with ξ in a monotonic way. In the case of doubly reinforced beam-section subjected to cyclic loadings, the considered model predicts an hysteretic behaviour as that of Figure 2.31a, when $M_P \leqslant M \leqslant M_{SD}$, as well as that of Figure 2.32a, when $M > M_{SD}$. It should be observed that the theoretical behaviour of Figure 2.31a is qualitatively similar to the experimental one of Figure 2.31b [40]. In the latter, however, further degradation phenomena are present, which the proposed model is not able to predict. On the other hand, equation (2.31) gives the following value of the steel yielding moment, in relation to the case of Figure 2.31b [40] and for $\xi \to 0$

$M_P = 0.6F_Pb = 0.6f_yA_sb = 7{,}122{,}634$ kg mm $= 618$ kips in.

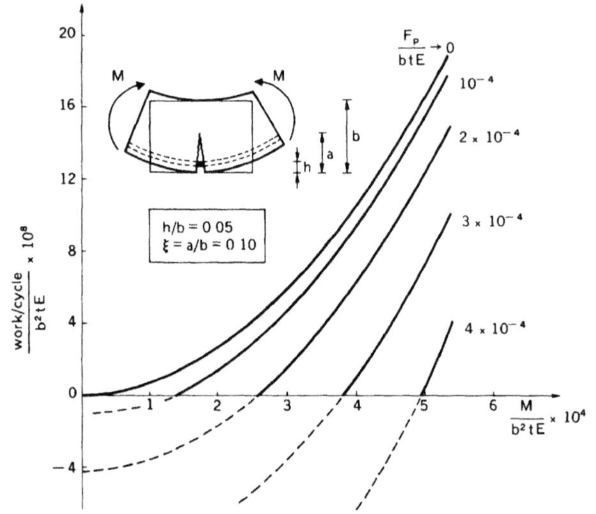

Fig. 2.29. Dissipated energy per cycle against maximum bending moment, varying the parameter F_P/btE.

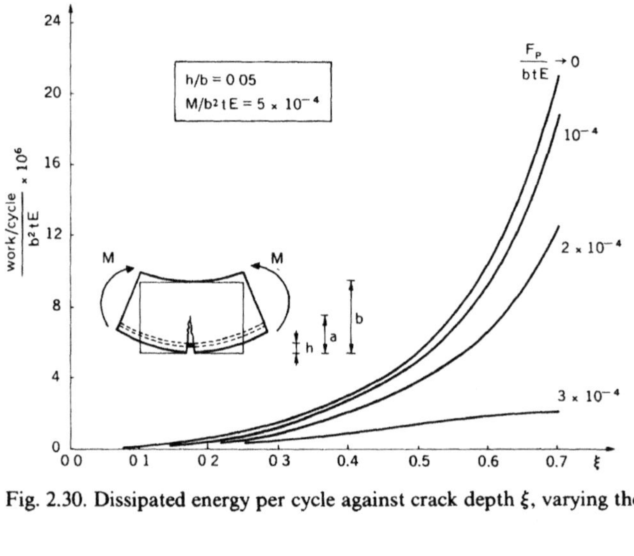

Fig. 2.30. Dissipated energy per cycle against crack depth ξ, varying the parameter F_P/btE.

Such a value is very close to the experimental result (Figure 2.31b). The theoretical behaviour of Figure 2.32a is then analogous to the experimental one of Figure 2.32b [41], which shows considerable concavities of the cycle in the second and fourth quadrants.

Observations. The following observations can be made:

(1) The considered Fracture Mechanics model does not claim to

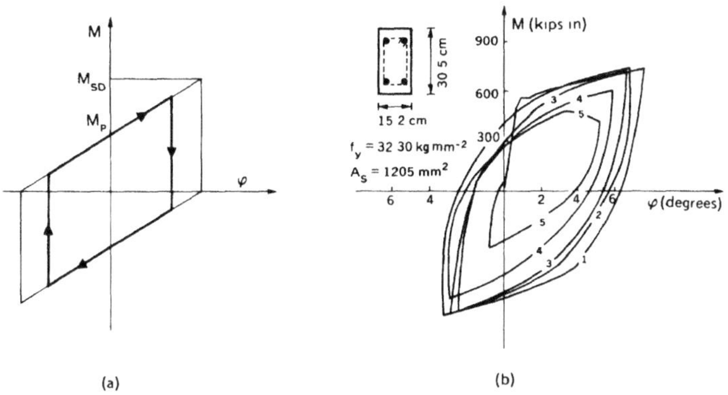

Fig. 2.31. Hysteretic behaviour of the moment-rotation diagram, under reversed cyclic loading: a) theoretical prediction when $M_P \leqslant M \leqslant M_{SD}$; b) experimental confirmation [40].

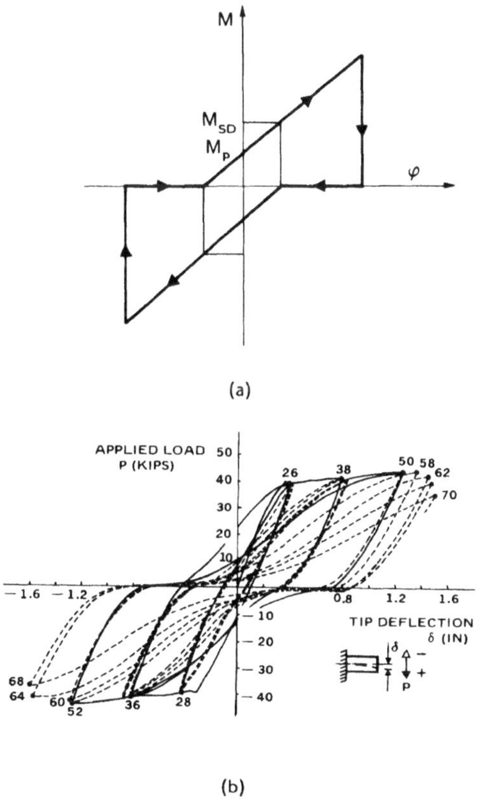

Fig. 2.32. Hysteretic behaviour of the moment-rotation diagram, under reversed cyclic loading: a) theoretical prediction when $M > M_{SD}$; b) experimental confirmation [41].

describe every aspect of reality; it only means to explain those which cannot be explained by means of the traditional concepts of Solid Mechanics (stress, strain, displacements,...). On the other hand, this is not the first attempt to consider crack contribution in reinforced concrete [42].

(2) Cracks grow in reinforced concrete beams before steel yielding only when slippage is allowed between concrete and steel. Even if slippage is not explicitly considered in the model, it can be simulated by a fictitious yield strength $\bar{f}_y$ of steel, lower than the real f_y (e.g. $\bar{f}_y \sim 0.1\, f_y$ in [35]).

(3) It is coherent to consider the reinforcement as a rigid-perfectly plastic material rather than an elastic-perfectly plastic one, since the proposed model represents only the behaviour of a cracked section or, at the most, of a cracked beam element of infinitesimal length. On the other hand, several efforts were made in order to derive plastic rotation by integration of curvatures, but the tests showed that it is caused by a concentrated rupture mechanism.

(4) The fracturing process has been analyzed up to the relative crack depth 0.6. On the other hand, the presented diagrams make clear the trends of the model even for higher depths.

(5) Creep and shrinkage of concrete could be included in the preceding fracture mechanics model on the basis of the general material model proposed by Roelfstra and Wittmann in [43].

2.5 Appendix: Dimensional independence

The fundamental mechanical quantities are: length $[L]$, force $[F]$ and time $[T]$. Consider a certain quantity $[Q] = [L^{\alpha} F^{\beta} T^{\gamma}]$. It is known that, if the length unit of measure is multiplied by λ, the force unit of measure by φ and the time unit of measure by τ, the unit measuring the quantity Q results multiplied by $\lambda^{\alpha}\varphi^{\beta}\tau^{\gamma}$. If q is the measure of Q in the units (L, F, T), in the new units $(\lambda L, \varphi F, \tau T)$ the measure results divided by $\lambda^{\alpha}\varphi^{\beta}\tau^{\gamma}$. Consider now three mechanical quantities Q_1, Q_2, Q_3:

$$[Q_1] = [L^{\alpha_1} F^{\beta_1} T^{\gamma_1}]; \tag{2.58a}$$

$$[Q_2] = [L^{\alpha_2} F^{\beta_2} T^{\gamma_2}]; \tag{2.58b}$$

$$[Q_3] = [L^{\alpha_3} F^{\beta_3} T^{\gamma_3}]. \tag{2.58c}$$

If the units of measure (L, F, T) are multiplied by λ, φ, τ, the units measuring Q_1, Q_2, Q_3 are multiplied by χ_1, χ_2, χ_3, so that:

$$\chi_1 = \lambda^{\alpha_1}\varphi^{\beta_1}\tau^{\gamma_1}; \tag{2.59a}$$

$$\chi_2 = \lambda^{\alpha_2}\varphi^{\beta_2}\tau^{\gamma_2}; \tag{2.59b}$$

$$\chi_3 = \lambda^{\alpha_3}\varphi^{\beta_3}\tau^{\gamma_3}. \tag{2.59c}$$

The three quantities Q_1, Q_2, Q_3 are said to be *fundamental* if, given χ_1, χ_2, χ_3, it is possible to univocally determine λ, φ, τ. From equation (2.59) it follows that:

$$\lg \chi_1 = \alpha_1 \lg \lambda + \beta_1 \lg \varphi + \gamma_1 \lg \tau; \tag{2.60a}$$

$$\lg \chi_2 = \alpha_2 \lg \lambda + \beta_2 \lg \varphi + \gamma_2 \lg \tau; \tag{2.60b}$$

$$\lg \chi_3 = \alpha_3 \lg \lambda + \beta_3 \lg \varphi + \gamma_3 \lg \tau. \tag{2.60c}$$

This linear system has the unknowns $\lg \lambda$, $\lg \varphi$, $\lg \tau$, i.e. λ, φ, τ, the coefficients α_i, β_i, γ_i and the known terms $\lg \chi_i$, for $i = 1, 2, 3$. Thus it admits one and only one solution, if and only if the coefficients' determinant is different from zero:

$$D = \begin{bmatrix} \alpha_1 & \beta_1 & \gamma_1 \\ \alpha_2 & \beta_2 & \gamma_2 \\ \alpha_3 & \beta_3 & \gamma_3 \end{bmatrix} \neq 0. \tag{2.61}$$

The previous condition being satisfied, the three quantities Q_1, Q_2, Q_3 are said to be *dimensionally independent* and therefore they can be taken as fundamental quantities.

Another completely equivalent definition of dimensionally independent quantities is the following one: three quantities Q_1, Q_2, Q_3 are dimensionally independent, when any quantity $[Q_0] = [L^{\alpha_0}F^{\beta_0}T^{\gamma_0}]$ can have the same physical dimensions as the product $Q_1^{\alpha_{10}}Q_2^{\alpha_{20}}Q_3^{\alpha_{30}}$, for suitable and univocally determinable values of α_{10}, α_{20}, α_{30}. In fact, from equation (2.58), if follows that

$$\alpha_0 = \alpha_1\alpha_{10} + \alpha_2\alpha_{20} + \alpha_3\alpha_{30}; \tag{2.62a}$$

$$\beta_0 = \beta_1\alpha_{10} + \beta_2\alpha_{20} + \beta_3\alpha_{30}; \tag{2.62b}$$

$$\gamma_0 = \gamma_1\alpha_{10} + \gamma_2\alpha_{20} + \gamma_3\alpha_{30}, \tag{2.62c}$$

which is a linear system with unknowns α_{10}, α_{20}, α_{30}, having as coefficient matrix the transposed matrix D^{T} of the system in equations (2.60). Equations (2.60) and (2.62) admit one and only one solution, if and only if condition in equation (2.61) holds. There are only two fundamental quantities in the statical field: length $[L]$ and force $[F]$, since time is missing. On the basis of the above reported considerations, it is possible, for example, to assert that stress $[\sigma]=[L^{-2}F]$, length $[L]$ and time $[T]$ are dimensionally independent:

$$D=\begin{bmatrix} -2 & 1 & 0 \\ 1 & 0 & 0 \\ 0 & 0 & 1 \end{bmatrix}=-1\neq 0. \tag{2.63}$$

References

1. Bažant, Z.P., Instability, ductility and size effect in strain-softening concrete, Journal of the Engineering Mechanics Division (ASCE), 102, pp. 331–344 (1976).
2. Irwin, G.R., Analysis of stresses and strains near the end of a crack traversing a plate, Journal of Applied Mechanics, 79, pp. 361–364 (1957).
3. Dougill, J.W., On stable progressively fracturing solids, Zeitschrift für Angewandte Mathematik und Physik (ZAMP), 27, pp. 423–437 (1976).
4. Bažant, Z.P., Work inequalities for plastic fracturing materials, International Journal of Solids and Structures, 16, pp. 873–901 (1980).
5. Capurso, M., Extremum theorems for the solution of the rate-problem in elastic-plastic-fracturing structures, Journal of Structural Mechanics, 7, pp. 411–434 (1979).
6. Løland, K.E., Continuous damage model for load-response estimation of concrete, Cement and Concrete Research, 10, pp. 395–402 (1980).
7. Petersson, P.E., Crack growth and development of fracture zones in plain concrete and similar materials, Report TVBM-1006, Lund Institute of Technology, Division of Building Materials (1981).
8. Hillerborg, A., Modeer, M. and Petersson, P.E., Analysis of crack formation and crack growth in concrete by means of Fracture Mechanics and Finite Elements, Cement and Concrete Research, 6, pp. 773–782 (1976).
9. Beltrami, E., Sulle condizioni di resistenza dei corpi elastici, Rendiconti del Reale Istituto Lombardo, serie II, tomo XVIII, pp. 704–714 (1885).
10. Von Mises, R., Mechanik der plastischen Formänderung von Kristallen, Zeitschrift für Angewandte Mathematik und Mechanik, 8, pp. 161–165 (1928).
11. Griffith, A.A., The phenomena of rupture and flow in solids, Phil. Trans. Royal Society, 221, 163–198 (1921).
12. Rice, J.R., A path independent integral and the approximate analysis of strain concentration by notches and cracks, Journal of Applied Mechanics, 35, pp. 379–386 (1968).
13. Sih, G.C., Some basic problems in fracture mechanics and new concepts, Engineering Fracture Mechanics, 5, 365–377 (1973).
14. Bažant, Z.P. and Cedolin, L., Blunt crack band propagation in finite element analysis, Journal of the Engineering Mechanics Division (ASCE), 105, pp. 297–315 (1979).
15. Bažant, Z.P. and Cedolin, L., Fracture mechanics of reinforced concrete, Journal of the Engineering Mechanics Division (ASCE), 106, pp. 1287–1306 (1980).
16. Carpinteri, A., Size effect in fracture toughness testing: a Dimensional Analysis

approach, Analytical and Experimental Fracture Mechanics, Edited by Sih, G.C. and Mirabile, M., Sijthoff & Noordhoff, pp. 785–797 (1981).
17. Carpinteri, A., Notch sensitivity in fracture testing of aggregative materials, Engineering Fracture Mechanics, 16, pp. 467–481 (1982).
18. Carpinteri, A., A Fracture Mechanics model for reinforced concrete collapse, Advanced Mechanics of Reinforced Concrete, IABSE-Colloquium Final Report, pp. 17–30 (1981).
19. Carpinteri, Al. and Carpinteri, An., Elastic-plastic shake-down in reinforced concrete beams subjected to repeated loadings, Proceedings 6th National Congress of the Italian Association of Theoretical and Applied Mechanics, pp. 179–190 (1982).
20. Petersson, P.E. and Gustavsson, P.J., A model for calculation of crack growth in concrete-like materials, Numerical Methods in Fracture Mechanics, Edited by Owen, D.R.J. and Luxmoore, A.R., Pineridge Press, pp. 707–719 (1980).
21. Saouma, V.E. and Ingraffea, A.R., Fracture Mechanics analysis of discrete cracking, Advanced Mechanics of Reinforced Concrete, IABSE-Colloquium Final Report, pp. 413–436 (1981).
22. Chhuy, S., Benkirane, M.E., Baron, J. and Francois, D., Crack propagation in prestressed concrete-interaction with reinforcement, Advances in Fracture Research, Edited by D. Francois, Pergamon Press, pp. 1507–1514 (1981).
23. Buckingham, E., On physically similar systems: illustrations of the use of dimensional equations, Physics Reviews, IV, p. 345 (1914).
24. Peterson, R.E., Model testing as applied to strength of materials, Journal of Applied Mechanics, 1, pp. 79–85 (1933).
25. Langhaar, H.L., Dimensional Analysis and Theory of Models, Huntington, Robert E. Krieger Publ. Co. (1980).
26. Carpinteri, A., Application of Fracture Mechanics to Concrete Structures, Journal of the Structural Division (ASCE), 108, pp. 833–848 (1982).
27. Carpinteri, A., Plastic flow collapse vs. separation collapse (fracture) in elastic-plastic strain-hardening structures, Materials & Structures, 16, pp. 85–96 (1983).
28. Carpinteri, A., Sensitivity and stability of progressive cracking in plain and reinforced cement composites, International Journal of Cement Composites and Lightweight Concrete, 4, pp. 47–56 (1982).
29. Carpinteri, A., Experimental determination of fracture toughness parameters K_{IC} and J_{IC} for aggregative materials, Advances in Fracture Research, Edited by D. Francois, Pergamon Press, pp. 1491–1498 (1981).
30. Carpinteri, A., Static and energetic fracture parameters for rocks and concretes, Materials & Structures, 14, pp. 151–162 (1981).
31. Ziegeldorf, S., Müller, H.S. and Hilsdorf, H.K., Effect of aggregate particle size on mechanical properties of concrete, Advances in Fracture Research, Edited by D. Francois, Pergamon Press, pp. 2243–2251 (1981).
32. Schmidt, R.A. and Lutz, T.J., K_{IC} and J_{IC} of Westerly Granite-effects of thickness and in-plane dimensions, American Society for Testing and Materials, STP-678, pp. 166–182 (1979).
33. Visalvanich, K. and Naaman, A.E., Fracture methods in cement composites, Journal of the Engineering Mechanics Division (ASCE), 107, pp. 1155–1171 (1981).
34. Leicester, R.H., Effect of size on the strength of structures, Paper no. 71, Forest Products Laboratory, Division of Building Research, Australia (1973).
35. Giuriani, E., Experimental investigation on the bond-slip law of deformed bars in concrete, Advanced Mechanics of Reinforced Concrete, IABSE-Colloquium Final Report, pp. 121–142 (1981).
36. Ciampi, V., Eligehausen, R., Bertero, V. and Popov, E., Analytical model for deformed bar bond under generalized excitations, Advanced Mechanics of Reinforced Concrete, IABSE-Colloquium Final Report, pp. 53–67 (1981).

37. Okamura, H., Watanabe, K. and Takano, T., Applications of the compliance concept in Fracture Mechanics, American Society for Testing and Materials, STP-536, pp. 423–438 (1973).
38. Okamura, H., Watanabe, K. and Takano, T., Deformation and strength of cracked member under bending moment and axial force, Engineering Fracture Mechanics, 7, pp. 531–539 (1975).
39. Carpinteri, Al. and Carpinteri, An., Softening and fracturing process in masonry arches, Proceedings Sixth International Brick Masonry Conference, pp. 502–510 (1982).
40. Brown, R.H. and Jirsa, J.O., Reinforced concrete beams under load reversal, Journal of the American Concrete Institute, pp. 380–390 (1971).
41. Popov, E.P., Seismic behavior of structural subassemblages, Journal of the Structural Division (ASCE), 106, pp. 1451–1474 (1980).
42. Westergaard, H.M., Stresses at a crack, size of the crack, and the bending of reinforced concrete, Journal of the American Concrete Institute, 30, pp. 93–102 (1934).
43. Roelfstra, P.E. and Wittmann, F.H., Stress analysis of structural members taking complex materials behaviour into consideration, Advanced Mechanics of Reinforced Concrete, IABSE-Colloquium Final Report, pp. 185–195 (1981).

A. HILLERBORG

3

Numerical methods to simulate softening and fracture of concrete

3.1 Introduction

The tensile fracture of concrete is as a rule regarded as brittle. Concrete has no yield behaviour of the type found in metals. Its tensile stress-strain diagram is nearly linear up to the maximum point, whereupon it immediately starts to descend. In spite of this concrete however can be said to have a considerable toughness. This toughness causes the fracture process zone in front of a growing crack to be of the order of 100–200 mm or even longer [1], i.e. much longer than what is normally found for metals. Because of these long fracture process zones linear elastic fracture mechanics (LEFM) can as a rule not be applied to concrete. On the other hand those methods which have been developed to take into account yielding within the non-linear zone for metals cannot be applied directly to concrete, as concrete does not yield in the way metals do.

The toughness of concrete has to do with the softening, i.e. the existence of a descending branch in the stress-deformation diagram.

This chapter describes the possibility of analysing the tensile fracture and fracture mechanics of concrete by means of methods based on the softening behaviour. The starting point will therefore be the softening properties of concrete in a simple tension test.

3.2 The behaviour of concrete in a tension test

Suppose that a tension test on a bar of concrete is performed. The concrete is assumed to be homogeneous on the macroscale and the test is assumed to be displacement-controlled and stable. By measuring the total elongation Δl_A of the specimen length l_A, a curve of the type shown in the upper diagram (Figure 3.1) is obtained. Before final rupture a fracture zone will form somewhere along the bar, as indicated. The increasing damage within the fracture zone when the deformation is further increased causes a decrease in strength within that zone and thus a decreasing stress in the specimen.

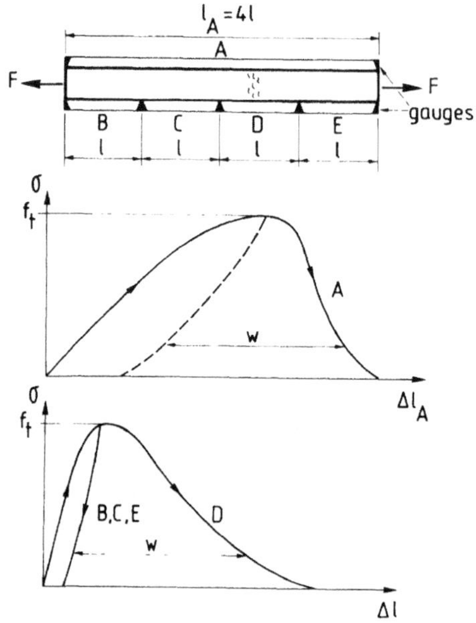

Fig. 3.1. A tension test with elongation measurements.

As the stress decreases the parts outside the fracture zone are unloaded. Thus the fracture zone does not spread along the bar, but it is limited to the position where it first started. This phenomenon is sometimes called "strain localization" [2–3]. In a completely homogeneous material the length (in the stress direction) of the fracture zone should theoretically approach zero, but for concrete the length can be estimated to be of the same order as the maximum aggregate size.

A closer examination of the deformation of the bar by means of 4 gages B-E, each of length $l = l_A/4$ is indicated in the figure. The fracture zone is situated within gage length D. The result is shown in the lower diagram of Figure 3.1.

As soon as the fracture zone starts developing at D, the stress starts to decrease. Thus the material at B, C and E is unloaded and the deformation within these parts follow an unloading branch with decreasing deformations at the same time as the deformation is increasing at D.

As the material is assumed to be homogeneous, the curves B, C and E will be equal and curve D will have the same ascending branch. What differs D from the others is the additional deformation due to the fracture zone. We introduce the notation w for this *additional deformation within the fracture zone*.

From Figure 3.1 it can be seen how w can be measured. The material at B, C and E is still homogeneous (on the macroscale) as it contains no

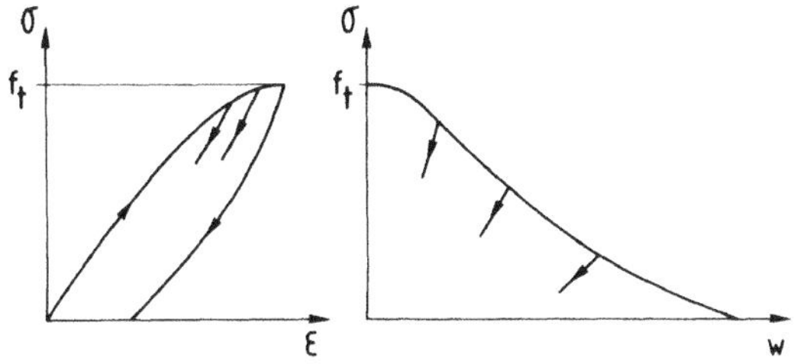

Fig. 3.2. The two relations $\sigma - \varepsilon$ and $\sigma - w$ for the general description of the tensile deformation.

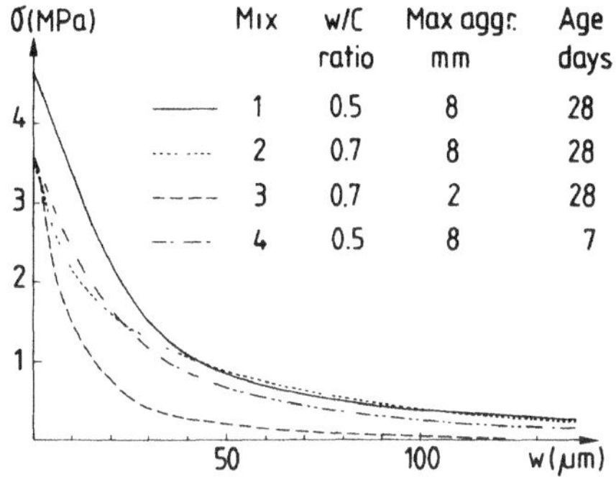

Fig. 3.3. Examples of $\sigma - w$ curves for mortar and concrete [4].

fracture zone. The stress-deformation properties of such a homogeneous material can be expressed by means of a stress-strain diagram, including unloading branches. The total deformation Δl of a gage length l which does not contain any fracture zone can thus be expressed as

$$\Delta l = \varepsilon l \tag{3.1}$$

where ε is the strain (taking into account the stress history).

On a gage length l, which contains a fracture zone (D in Figure 3.1) we also get the additional deformation w, thus

$$\Delta l = \varepsilon l + w \tag{3.2}$$

It is thus possible to calculate the stress-deformation curve for any gage length if the two curves relating σ and ε and σ and w, Figure 3.2 are known with reference to the site of fracture zone. Each of the two curves can also contain unloading branches, as indicated.

It must be noted that the $\sigma - \varepsilon$ diagram has a relative elongation (strain, e.g. expressed as percent or microstrain) on its horizontal axis, whereas the $\sigma - w$ diagram has a deformation (e.g. expressed in mm).

It is assumed that the whole fracture zone is outside, equation (3.1), or inside the gage length l, equation (3.2). As long as this assumption is fulfilled, the length of the fracture zone is unimportant.

The two curves of Figure 3.2 are necessary in order to describe the stress-deformation properties. A single curve can not give the relevant information, as its shape depends on the gage length, cf. curves A and D in Figure 3.1.

Figure 3.3. shows examples of $\sigma - w$ curves for mortar and concrete [4].

3.3 A comparison between concrete and steel

The above description of the tension test is in principle valid for any type of material. The behaviour of the fracture zone is however very different for different materials. Here the difference between concrete and steel will be discussed, but each of these materials may be looked upon as representing a large group of materials.

The additional deformation w of the fracture zone depends on some kind of damage within this zone. The kind of damage is quite different for concrete and steel.

For *concrete* the damage consists of cracks mainly perpendicular to the maximum tensile stress direction. This damage is not followed by any appreciable lateral deformation, nor is it to any great extent influenced by tensile stresses in other directions. For this reason the $\sigma - w$ curve is assumed to be a material property, which is independent of the size of the specimen and it can also be assumed to be rather little dependent on stresses in other directions, with the exception of high compressive stresses. One consequence of this is that for concrete the difference between plane strain and plane stress conditions is insignificant.

For *steel* a major part of the damage in the fracture zone consists of necking due to yielding of the material. The size of the necking zone depends on the size and shape of the specimen. For this reason the $\sigma - w$ curve determined from a tension test is not a material property, but it depends on the size and shape of the specimen. It is also well-known that the yield strength of steel depends on the stresses in the lateral directions. For these reasons the behaviour of steel under plane strain and plane stress conditions is very different.

To the author's knowledge it has not been possible to measure the complete stress-deformation curve for steel, as unstability with a sudden fracture occurs at a rather high stress level. This of course does not mean that such a curve does not exist. It just means that the curve is very steep.

In fracture mechanics the energy absorption associated with fracture is important. In a tension test according to Figure 3.1 the total energy

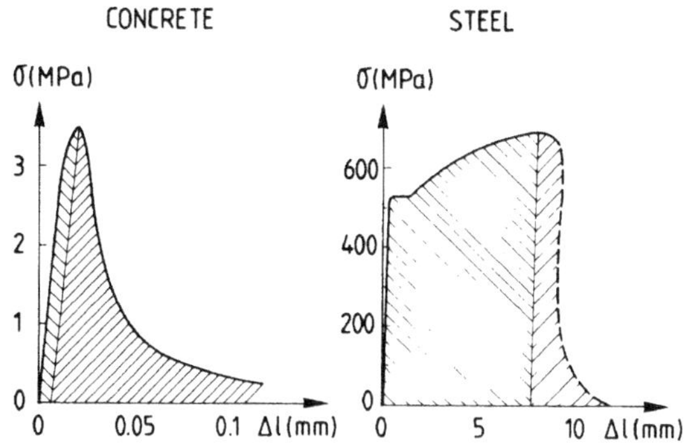

Fig. 3.4. Typical stress-deformation curves on a gage length of about 100 mm for concrete and steel, showing that the main energy absorption is associated with the descending branch ($\sigma - w$ curve) for concrete, but with the ascending branch ($\sigma - w$ curve) for steel.

absorption equals the area below the load-deformation curve for the whole specimen. With the deformation described by means of equation (3.2), the total energy W, absorbed within the gage length l, is

$$W = \int F \mathrm{d}\Delta l = A \int \sigma(l \mathrm{d}\varepsilon + \mathrm{d}w) = Al \int \sigma \mathrm{d}\varepsilon + A \int \sigma \mathrm{d}w \tag{3.3}$$

where F = load, A = cross sectional area. Thus one part of the energy is related to the area below the $\sigma - \varepsilon$ curve, another part to the area below the $\sigma - w$ curve. In the complete stress-deformation diagram these two parts are separated by the unloading-curve from the maximum point, cf. Figure 3.1.

Figure 3.4. illustrates typical stress-deformation curves for concrete and steel, with the areas below the curves divided into the two parts belonging to the $\sigma - \varepsilon$ and $\sigma - w$ curves respectively. Both curves are valid for a gage length of about 100 mm. The curve for steel is very approximate, as the specimen size and the steel grade are not taken into account and as the shape of the descending part is not known. In spite of this uncertainty the conclusion can be drawn that, for test pieces of ordinary size, in concrete the energy absorption is mainly associated with the $\sigma - w$ curve, whereas in steel most (or at least much) of the energy absorption is associated with the $\sigma - \varepsilon$ curve, i.e. with yielding of the material. Thus the energy absorption is mainly associated with the ascending branch of the stress-deformation curve for steel, but with the descending branch for concrete.

The following conclusions can thus be drawn from this comparison:

(1.) For concrete the $\sigma - w$ curve can be assumed to be a material property, which is as a rule not very sensitive to the specimen size, stress state etc. For steel this assumption can not be used. This difference is very important for the possibility of applying the analytical approach discussed below.

(2.) For concrete the difference between plane strain and plane stress conditions is insignificant for the fracture process, whereas for steel this difference is of great importance and gives rise to great problems regarding the application of fracture mechanics. From this point of view the analysis of fracture is thus much easier for concrete than for steel. The special techniques which have been developed to overcome this difficulty for steel – e.g. the R-curve analysis – can not be expected to be of the same use for concrete as for steel.

(3.) The importance of the descending branch for the energy dissipation at fracture in concrete means that the J_c approach can not be expected to be applicable, like it is for steel.

This comparison thus shows that those analytical techniques which have been developed for the fracture of steel often can not be expected to be suitable for concrete. On the other hand concrete has some properties which make it possible to apply simplifying assumptions, which are not admissible for steel. Thus it is justified to use an approach for the analysis of concrete fracture, which differs rather much from the fracture mechanics methods used for steel.

The basic difference between concrete and steel is that before fracture steel yields under practically constant volume, whereas concrete develops microcracks. Yielding of this kind is accompanied by great lateral deformations or stresses, which is not the case for cracking. In the discussion above steel represents all yielding materials, concrete represents all non-yielding materials like rock and ceramics. Concrete also may represent fiber-reinforced materials, as these also deform without giving rise to great lateral deformations or stresses. Thus the methods which will be described below can be applied to concrete, ceramics, rock, fiber-reinforced concrete and many other materials, but not, without modifications, to most metals or other materials where yielding is accompanied by great lateral stresses or deformations (the possible application to such materials is discussed in [5]).

3.4 Tensile fracture zones

One very essential conclusion from the previous discussion is that the properties of a tensile fracture zone in concrete can be assumed to be rather independent of the size and shape of the specimen and of the type of stress field as long as a tensile stress is the main cause of fracture. Therefore we can assume that a tension fracture zone behaves in the same way independently of the cause of the tensile stress field. Figure 3.5

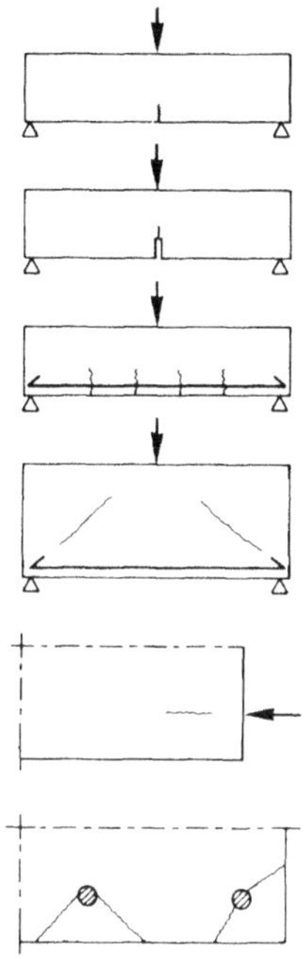

Fig. 3.5. Examples of cases where tensile fracture zones may form in concrete.

illustrates some cases where tensile fracture zones occur in concrete and where this assumption may be applied.

The assumption that the behaviour and properties of the tensile fracture zone is independent of the type of stress field can not always be regarded as justified. If the stress field is such that one of the principal stresses is a high compressive stress, this will influence the tensile strength and thus the properties at tensile fracture. Another case where the assumption has to be modified is where the principal stress direction changes appreciably during the loading process, which may happen e.g. for shear failure.

In many cases, however, it seems reasonable to assume that the behaviour of the tensile fracture zone is a material property. This property can be measured in a tension test, and it is suitably described by means of the $\sigma - w$-diagram.

The fracture zone has some length in the stress direction. The additional deformation w of the fracture zone can be looked upon as being caused by additional strains ε_w within the fracture zone according to

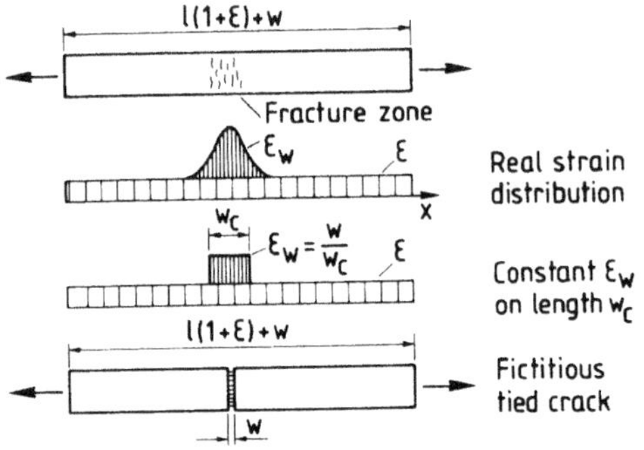

Fig. 3.6. Strain distribution during fracture, and two possible simplifying assumptions.

Figure 3.6. Thus the total additional deformation w is

$$w = \int \varepsilon_w \mathrm{d}x \tag{3.4}$$

which is the area below the ε_w curve.

The length of the fracture zone has a very limited influence upon the behaviour of a specimen as long as this length is small compared to the specimen size. The distribution of the additional strain ε_w within the fracture zone is then also of a minor importance. Thus for the introduction of the $\sigma - w$ curve into a numerical analysis the choice of the assumed ε_w distribution is unimportant.

One natural assumption is an even distribution of ε_w within a length w_c, Figure 3.6. In finite element analyses w_c is then suitably chosen so as to coincide with the mesh, normally as the length of one element. This approach has been applied by Bazant et al. [6].

A further simplification in the description can be achieved by assuming that the fracture zone consists of only a tied crack, Figure 3.6, i.e. a crack with width w and with the ability to transfer stresses σ according to the $\sigma - w$ curve. As this assumed crack is not a real crack, but just a simplified way of describing the fracture zone, the crack can be said to be fictitious and the corresponding model has been called "the fictitious crack model" [5,7–9].

Thus for the numerical calculations the fracture zone is suitably modelled either as a discrete fictitious crack or as a zone with an evenly distributed additional strain over one finite element (or sometimes more than one). The two models will normally give practically the same result. That model should be chosen, which is the simplest with regard to the available FEM-program and to the type of problem that is to be analysed. The discussion below refers to the fictitious crack model.

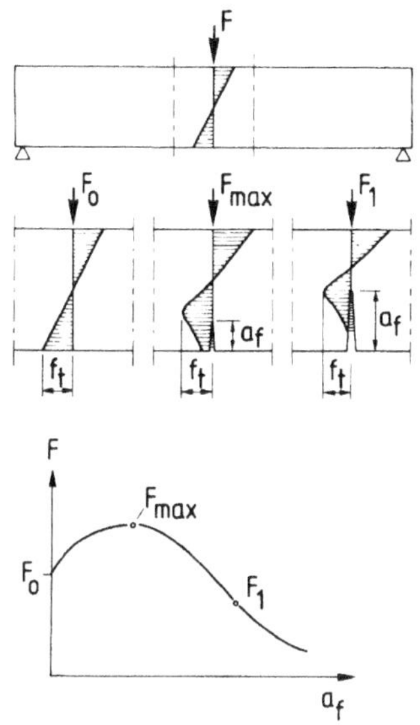

Fig. 3.7. Change in stress distribution and load with increasing length a_f of fracture zone for a bent beam.

However, the author is well aware that the approach with a distributed strain may sometimes have numerical advantages.

3.5 A general model for the tensile fracture of concrete

The discussion above can be summarized in the following model for the tensile fracture process:

A tensile fracture zone starts developing as soon as the strain corresponding to the tensile strength f_t is exceeded. The fracture zone undergoes an additional deformation w, but it can still transfer some stresses as long as w is small. Between the additional deformation w of the fracture zone and the transferred tensile stress σ a $\sigma - w$ relation is valid, which is a material property.

Figure 3.7 shows the application of this model to a bent beam, where it is assumed that only one tensile fracture zone developes and that the $\sigma - \varepsilon$ diagram is a straight line. At low loads there is no fracture zone. When the load is increased to a value F_0, the tensile strength f_t is reached and a further increase in F will start the development of a fracture zone. The fracture zone will start at the bottom of the beam and grow upwards. As it grows, the deformation at the bottom of the beam increases, and then the stress will decrease (or possibly remain constant). In the figure the development of the fracture zone and the corresponding stress

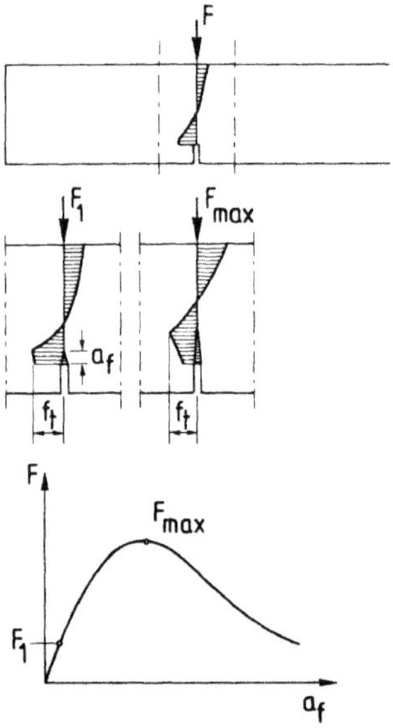

Fig. 3.8. Change in stress distribution and load with increasing length a_f of fracture zone for a notched beam.

distributions are indicated. The fracture zone is shown as a "fictitious crack", which is often a suitable way of illustration in connection with theoretical discussions and explanations.

As the fracture zone grows, the corresponding load F grows, reaches a maximum F_{max} and decreases. The fracture zone later changes to a real crack, which does not transfer stresses. All this process can be analysed by means of the model.

Figure 3.8 shows the corresponding train of events for a beam with a notch. The only essential difference is that in this case we have a theoretical stress singularity at the notch tip. Therefore the fracture zone starts to develop as soon as the load is applied. We can follow the development of this fracture zone, calculate the maximum load etc., just as in the previous case.

Thus the same model can be applied whether the beam is notched or unnotched. It can of course also be applied where we have holes or other stress concentrators, and to situations where shrinkage or thermal strains occur. The model thus has a general applicability in contrast to conventional fracture mechanics, which can only be applied where a crack already exists.

3.6 Material properties

The most essential properties for the analysis of tensile fracture are the $\sigma - \varepsilon$ and $\sigma - w$ curves. In the practical application it is however conveni-

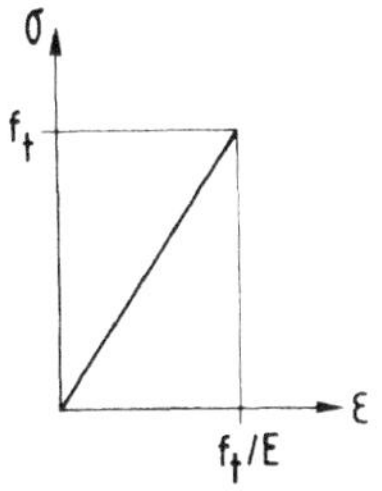

Fig. 3.9. Linear approximation of the $\sigma - \varepsilon$ curve.

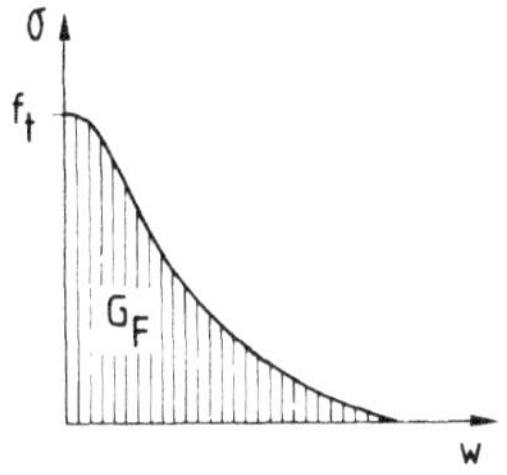

Fig. 3.10. The area below the $\sigma - w$ curve, which represents the fracture energy per unit area, is denoted by G_F.

ent to use a simplified material characterization by means of a few values instead of two complete curves.

Most numerical calculations are for practical reasons based on a linear $\sigma - \varepsilon$ curve, which is a reasonable approximation for ordinary concrete. Such a curve (Figure 3.9) is completely defined by means of the modulus of elasticity E and the tensile strength f_t.

Poisson's ratio υ is also a material property. It has however a very small influence on the numerical results, and it will not be further discussed. In the calculations the value $\upsilon = 0.2$ has been used.

Regarding the $\sigma - w$ curve one of its most essential properties is the area under the curve, as this area equals the total energy absorption per unit area in the fracture zone, cf. equation (3.3). Denote this area by G_F and use the term fracture energy for it, Figure 3.10. This leads to the material property G_F as the fracture energy.

From the shape of the $\sigma - w$ curve, the value of f_t and G_F can be completely defined. If for example the shape is a straight line, which is the simplest possible approximation, the $\sigma - w$ thus corresponds to that in Figure 3.11. This curve is often used in calculations, due to its simplicity.

For ordinary concrete tests have shown that the shape is rather constant [4], Figure 3.12. A good approximation of this shape can be achieved by means of the bilinear relation of Figure 3.13.

With the shape of the $\sigma - w$ curve given (or assumed) and with the

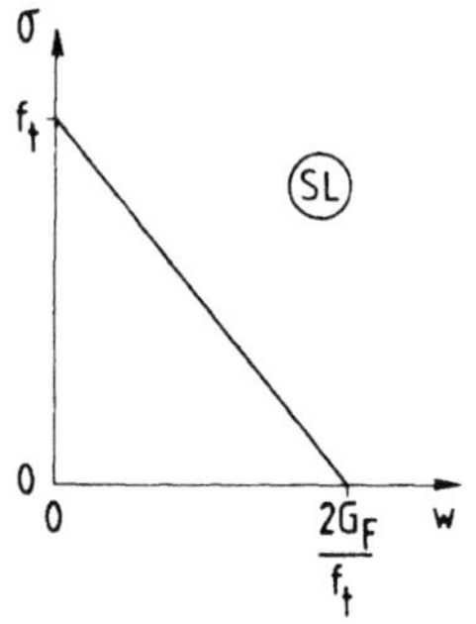

Fig. 3.11. Approximation of the $\sigma - w$ curve by means of a straight line (SL).

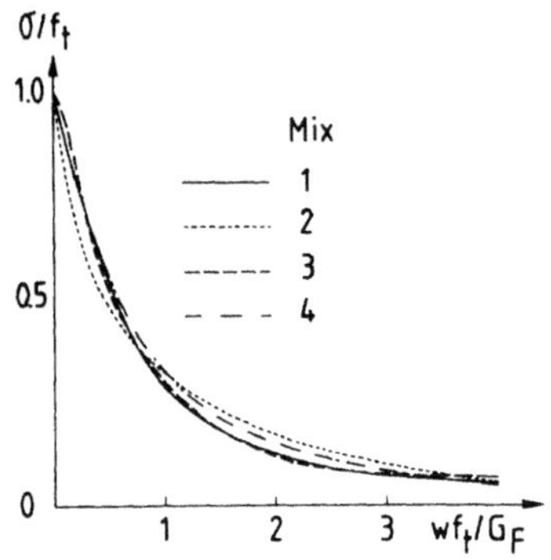

Fig. 3.12. The curves of Figure 3.3. redrawn to show that their shape is similar.

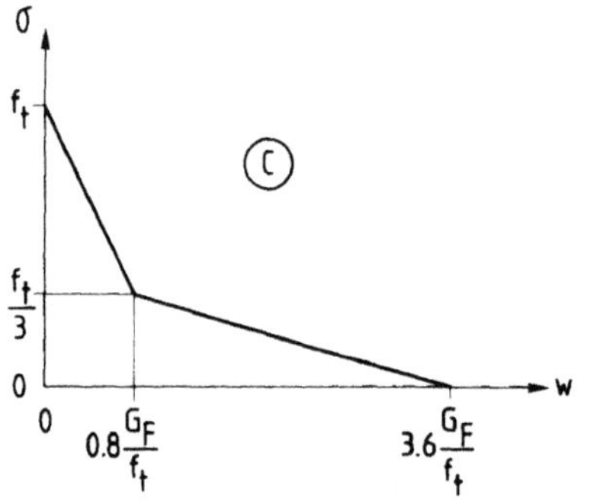

Fig. 3.13. Bilinear representation of the $\sigma - w$ curve, approximating the measured curves for concrete (C) [4].

assumption of a linear $\sigma - \varepsilon$ curve, the material properties which govern the tensile failure behaviour are fully defined by the values E, f_t and G_F; see Figures 3.9, 3.11 and 3.13.

For the presentation of analytical results it is often convenient to use a combination of these values, called the characteristic length, denoted l_{ch},

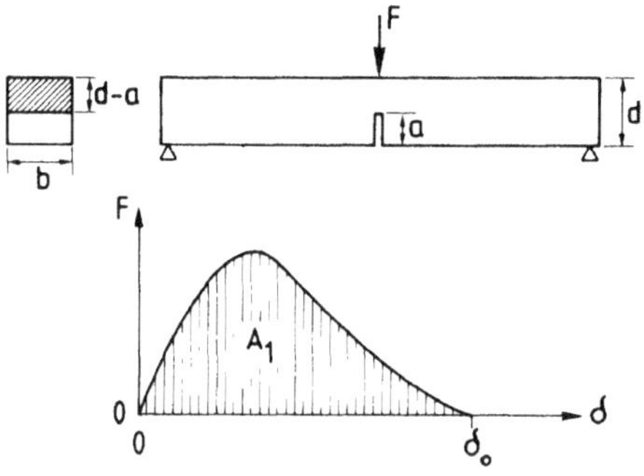

Fig. 3.14. Test for the determination of the fracture energy G_F.

and defined

$$l_{ch} = \frac{EG_F}{f_t^2} \tag{3.5}$$

The characteristic length is thus a material property, which has no direct physical interpretation, but which is calculated from the three properties E, G_F and f_t.

Typical values of l_{ch} are of the order [4,8]

cement paste	5– 15 mm
cement mortar	100–200 mm
normal concrete	200–400 mm

Lightweight concrete and other concrete materials where the strength ratio between aggregate and cement paste is lower than normal will show smaller values of l_{ch}.

If the concrete is fibre reinforced, this may lead to a strongly increased fracture energy G_F without any appreciable change in E or f_t [9]. This gives rise to very high formal values of l_{ch}, up to many meters. As the shape of the $\sigma - w$ curve in this case is quite different, these l_{ch}-values are not directly comparable to those for normal concrete.

The determination of G_F can be made by means of a stable tension test. As the stable tension test is rather difficult to perform, it is however preferable to use e.g. a three-point bend test on a notched beam. According to [4] this can be done as shown in Figure 3.14 with a registration of the deflection δ of the loading point versus the load F. The value of G_F is calculated from

$$G_F = \frac{A_1 + Mg\delta_0}{b(d-a)} \tag{3.6}$$

where M is the mass of the beam between the supports and g is the acceleration due to gravity.

3.7 FEM analysis of a fracture zone: coincident with predetermined crack path

Numerical analyses based on the fictitious crack model are suitably performed by means of the finite element method, FEM. The fictitious crack is then represented as a separation of elements with a simultaneous introduction of forces to represent the stresses transferred through the fracture zone.

The fictitious crack, which later develops into a real crack, will follow some crack path, which depends on the state of stress. Let us first assume that we can foresee this crack path and that we arrange our element mesh so that this crack path coincides with element boundaries. It is then possible to model the development of the crack directly by separating elements.

In the description below it is assumed that forces between elements are acting only in the node points. Other assumptions are of course also possible, but they do not change the general principles.

As an example let us look at the three-point bend test on a notched beam, Figure 3.15. In this case we can assume that the crack path will be vertical, starting at the notch tip, i.e. we can use a predetermined crack path. The scope of the analysis will be to follow the development of the fracture zone and the crack and to calculate the corresponding loading history, stresses, strains and deformations.

In Figure 3.15 the element mesh at the notch tip is shown and also its change as the fracture zone grows. When we start applying the load F it will cause a high stress at the notch tip, node 1. Our first step will be to calculate the load F_1, which gives a stress equal to f_t, the tensile strength, at node 1.

It must be noted that the value of F_1 will depend on the size of the element, and it will approach zero as the element size approaches zero. This size dependency of the first step is however of no importance for the results of the whole analysis.

Any further increase of F above F_1 would cause a stress higher than f_t at node point 1, which is not possible. Thus a fracture zone starts at node point 1, and we model this by separating the node into two nodes 1′ and 1″. The distance between 1′ and 1″ corresponds to the width of the fictitious crack. The nodes 1′ and 1″ are called a node pair in the following discussion. Between the node points 1′ and 1″ a tensile force is introduced, corresponding to the stress σ where σ varies with w according to the $\sigma - w$ curve for the material. For the further calculations we thus

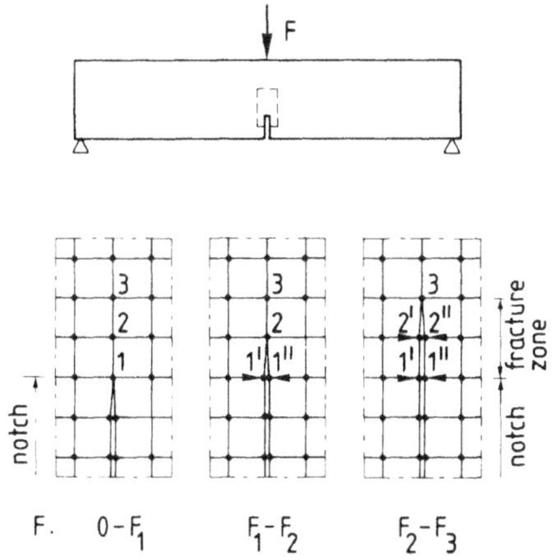

Fig. 3.15. Example of FEM representation of fracture zone development.

have to apply the $\sigma - w$ relation to the node pair $1' - 1''$ instead of the condition that $1'$ and $1''$ coincide in the original mesh.

With this revised mesh we now proceed by calculating the load F_2, which causes a stress equal to f_t at node point 2. At this load level we separate node 2 into a node pair $2'$ and $2''$, introducing forces corresponding to the $\sigma - w$-curve. In this way we can proceed and follow the variation in the load when the fracture zone develops. We can then of course also calculate the corresponding variations in stresses, strains and deformations.

In the general case, with nonlinear $\sigma - \varepsilon$ and/or $\sigma - w$ curves each step involves an iteration. If, however, the $\sigma - \varepsilon$ relation is linear and the $\sigma - w$ relation is piecewise linear, the calculation can be separated into incremental steps, where the material properties are linear within each step. Then a direct linear FEM-calculation can be made for the increments within each step and iterations can be avoided. The forces between the node pairs $1'$-$1''$, $2'$-$2''$ etc., will decrease as the distances increase, which will formally correspond to springs with negative spring constants.

In this stepwise linear technique the limit between two steps corresponds to the change of stiffness properties at some node pair along the crack path. This can happen because f_t is reached at a point or because a pair of nodes reaches a w-value where there is a change of slope in the σ-w relation.

The changes in stiffness properties occur only between the pair of node points along the crack path. It is then suitable to use a technique where the properties of the crack path are first determined. This can e.g. be made on a finite element mesh where all nodes, where a fracture zone is expected to occur, are separated, and the influences of unit forces

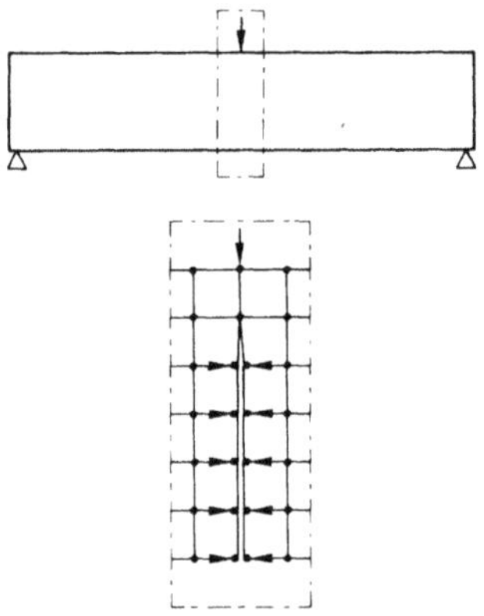

Fig. 3.16. Example of the principle of substructure technique. The node pairs shown separated coincide in the unloaded beam. Positions of unit loads are shown.

between the separated node pairs and of a unit load are calculated, Figure 3.16. Based on these values a simple system of equations can be established, relating the forces in and the separation of the node pairs along the crack path and the acting load F.

For each of the node pairs some kind of restriction is then valid during each step in the calculation of the development of the fracture zone.

(1.) For a node pair where f_t has not been reached $w = 0$.

(2.) For a node pair in the fracture zone the force and the separation are related according to the $\sigma - w$ curve.

(3.) For a node pair at a notch or at a real crack the force is zero.

The number of equations to be solved for node forces equals the number of node forces in the simple symmetrical example shown in Figure 3.16. By applying a condition for the end of the step, as described above, the load at the end of each step can be calculated, and the corresponding node forces etc. These calculations can be made on a small computer, as the number of equations is very limited.

The method sketched above is a force formulation. In practice a corresponding displacement formulation is often used, or a mixed formulation.

Petersson [4] has used this technique for an extensive study of the three-point bend test. Some of his results will be shown in order to demonstrate the practical usefulness of the fictitious crack model.

Figure 3.17 shows the theoretical development of the fracture zone, and the corresponding stress distribution, load and deflection for a notched beam in bending. The $\sigma - \varepsilon$ and $\sigma - w$ curves have both been assumed to be linear in this example. At maximum load in this case the length of the fracture zone is about 50 mm, which approximately corresponds to the length of the notch and to 1/3 of the length of the ligament.

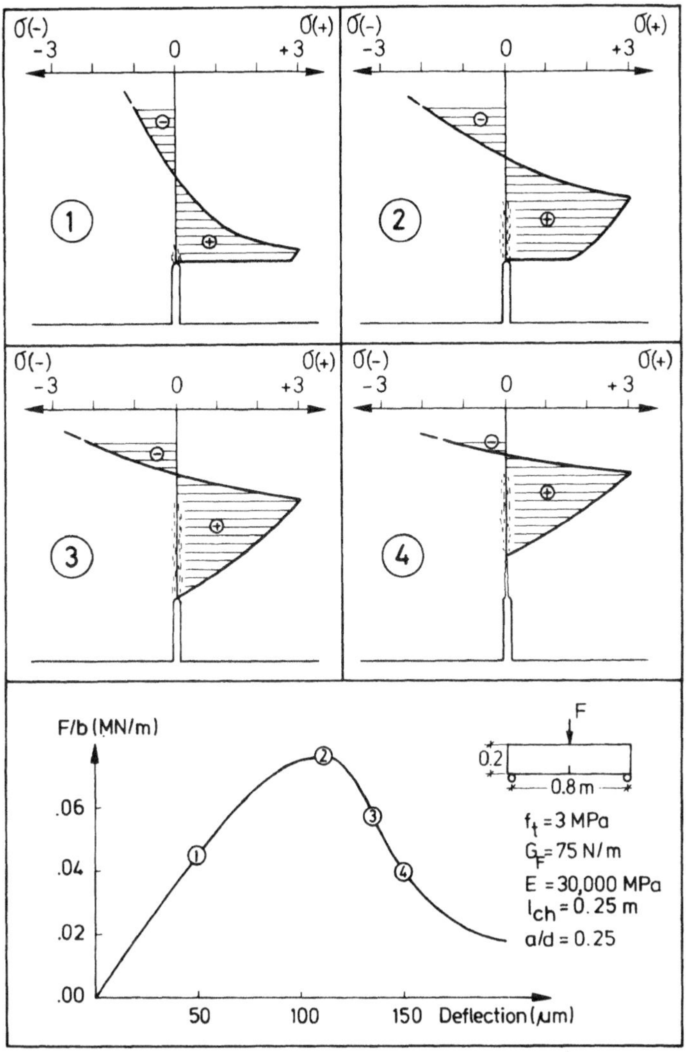

Fig. 3.17. The theoretical development of the fracture zone, and the corresponding stress distribution, load and deflection. The $\sigma - w$ relation is assumed to be a straight line [4].

Figure 3.18 shows the corresponding stress distributions at maximum load for different beam depths. It is evident that the stress distribution at maximum load very much depends on the beam depth. For small beams the distribution approaches that according to the theory of plasticity, whereas for deep beams it approaches that according to the theory of elasticity (but the stress never exceeds f_t).

The variation in the length of the fracture zone at maximum load with the beam depth is shown in Figure 3.19. For deep beams this length is of the order 0.5 l_{ch}, i.e. 0.1–0.2 m for ordinary concrete.

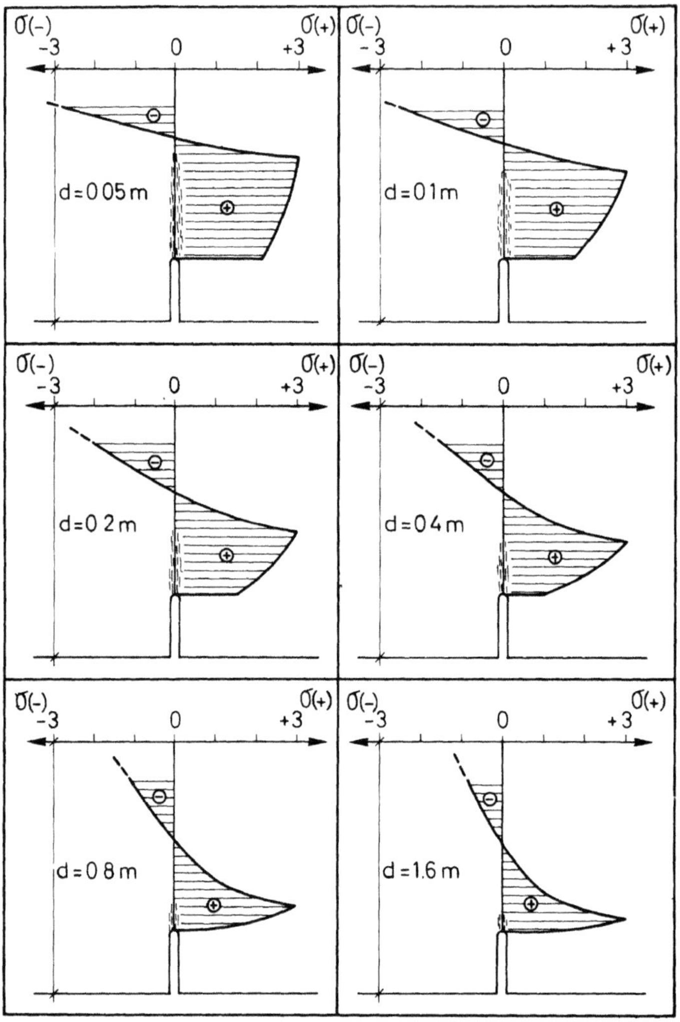

Fig. 3.18. Stress distribution in front of the notch tip at maximum load for different beam depths d. Properties according to Figure 3.17; beam length = $4d$ [4].

With these long fracture zones it is evident that linear elastic fracture mechanics (LEFM) can not be expected to yield accurate results for concrete specimens of ordinary size, used in laboratories.

In laboratory tests sometimes the width of the specimen is reduced along the expected crack path in order to force the crack to follow this path, e.g. [1]. It can be demonstrated that this leads to an increase in the length of the fracture zone. The length will then be approximately proportional to the ratio between the width of the surrounding material and the width at the crack path. If thus the width along the crack path is

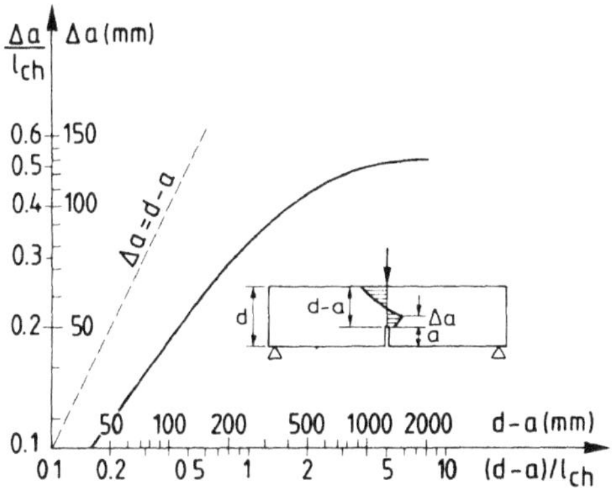

Fig. 3.19. Variation of fracture zone length Δa at maximum load with ligament depth $d - a$. The curve is approximately valid for all notch depths a. The outer scales are general. The inner scales are valid for an average concrete quality with $l_{ch} = 250$ mm [4].

1/3 of the width of the specimen the length of the fracture zone at maximum load will be about 3 times as great as the value given above.

LEFM is based on the assumption of the existence of a crack tip, in front of which a stress concentration exists, whereas there are no stresses across the crack. It is evident from the above analyses, e.g. Figures 3.17 to 3.19, that no well-defined crack tip exists for concrete, but a fracture zone with an appreciable length. As no well-defined crack tip exists, any attempt to determine its position is more or less meaningless. This is true whether the determination is based on visual or other observations or on compliance measurements. Any determination of LEFM parameters based on the determination of the crack tip position and crack growth is thus unreliable and more so the smaller the specimen.

In order that the shape of the fracture zone and the stress distribution within this zone shall be adequately represented in the FEM analysis, the distance between the nodes along the crack path must be small enough compared to the length of the fracture zone. With a normal shape of the $\sigma - w$ curve for concrete, a node distance of 0.2–0.4 of the fracture zone length at maximum load seems to give a sufficient accuracy. This corresponds to 0.1–0.2 l_{ch} for big specimens and smaller values for smaller specimens, cf. Figure 3.19. These estimations have been confirmed by test calculations, Figure 3.20.

As the demonstrated analysis gives the complete load-deflection relation, stresses, strains, etc., the analysis can be used to predict the corresponding test results.

From the maximum load the values G_c and K_c can be calculated according to ordinary formulas based on LEFM. Figure 3.21 shows which values of G_c and K_c that can be expected from this type of tests.

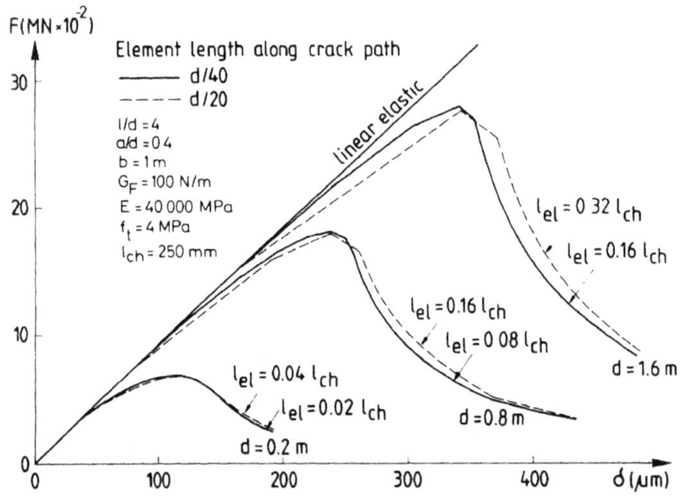

Fig. 3.20. Calculated load-deflection curves for different beam depths and different finite element lengths l_{el} along the crack path. Linear $\sigma - \varepsilon$ and $\sigma - w$ relations [4].

The three curves refer to different assumptions regarding the shape of the $\sigma - w$ curve.

According to all the curves G_c approaches G_F and K_c approaches $\sqrt{EG_F}$

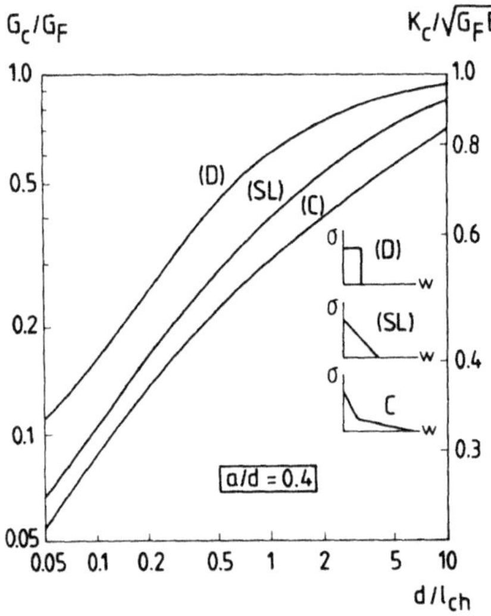

Fig. 3.21. Theoretical values of linear elastic fracture mechanics parameters G_c and K_c, which can be expected from three-point bend tests on notched beams of different depths. The curves are valid for different shapes of the σ-w-curves, cf also Figures 3.11 and 3.13. D stands for Dugdale, i.e. a constant yield stress in the fracture zone [4].

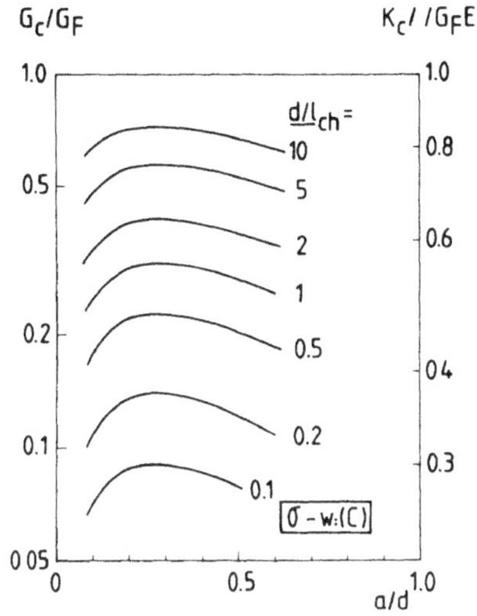

Fig. 3.22. Theoretical variation of G_c and K_c with notch depth a and beam depth d [4].

when the beam depth d increases. We know that the applicability of LEFM increases as the specimen size increases, and thus we can draw the conclusion that for big specimens

$$G_c \rightarrow G_F \tag{3.7}$$

$$K_c \rightarrow \sqrt{EG_F} \tag{3.8}$$

For big specimens thus

$$\left(\frac{K_c}{f_t}\right)^2 \rightarrow l_{ch} \tag{3.9}$$

From the diagram we can see that the beam depth must be greater than about 2–3 l_{ch} for a Dugdale material (steel) but greater than 10–20 l_{ch} for concrete in order to get a K_c-value which is not more than 10 percent to low. For ordinary concrete this corresponds to a beam depths of 2–8 m. Such beams are evidently unsuitable for handling in a laboratory.

Many tests have been performed on beams with a depth of about 0.1–0.2 m, corresponding to d/l_{ch} about 0.5. The measured values of G_c and K_c then can be expected to be less than 0.25 and 0.5 of the correct ones respectively.

Figure 3.22 shows an example of the theoretical influence of the notch

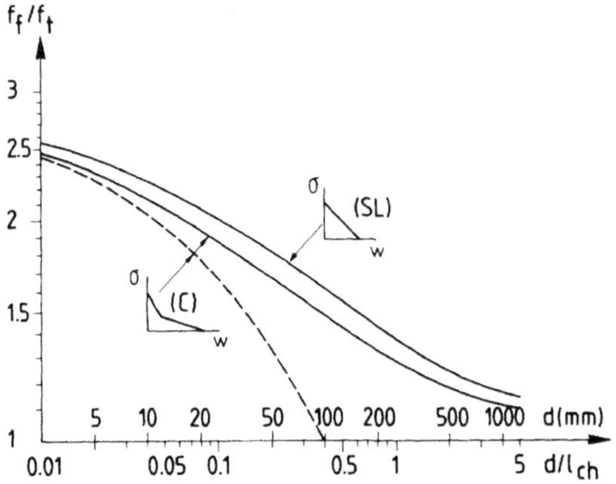

Fig. 3.23. Theoretical variation of flexural strength f_f with beam depth d. The upper scale on the horizontal axis is valid for an average concrete quality with $l_{ch} = 250$ mm. The dashed line indicates the theoretical influence of shrinkage in an indoor climate.

depth on the measured values of K_c and G_c. This influence is rather small for notch depths between 0.2 and 0.5 times the beam depth.

Figure 3.23 shows the theoretical variation of the flexural strength (modulus of rupture) with the beam depth for an un-notched beam. The curves are calculated on the assumption that only one fracture zone develops. The theoretical influence of shrinkage stresses is also indicated.

Figures 3.24 to 3.25 show the theoretical variation of the net bending strength

$$f_f^{net} = \frac{6M}{b(d-a)^2} \tag{3.10}$$

with a/d and d/l_{ch}. Test results showing the variation of f_{net} with a/d according to Figure 3.24 are sometimes used as a measure of the notch sensitivity of a material. From Figure 3.24 it is evident that the notch sensitivity is not a pure material property, but a property which depends on d/l_{ch}, i.e. on the size d of the specimen and on the material property l_{ch}. For d/l_{ch} below about 1.0 a specimen is practically not notch sensitive at all. This corresponds to a beam depth of about 0.2–0.4 m for concrete. Tests on smaller notched concrete beams thus can not be expected to give any information regarding fracture mechanics parameters, but just values of the flexural strength. This was pointed out already in 1962 by Blakey et al [10] with reference to tests by Kaplan [11].

In the above simple examples it has been possible to foresee the crack path due to symmetry. In more complicated cases it is often impossible to

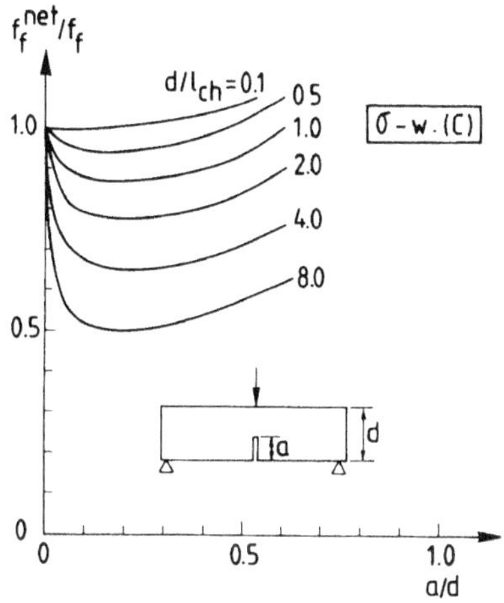

Fig. 3.24. Theoretical variation in net bending strength with notch depth a and beam depth d [4].

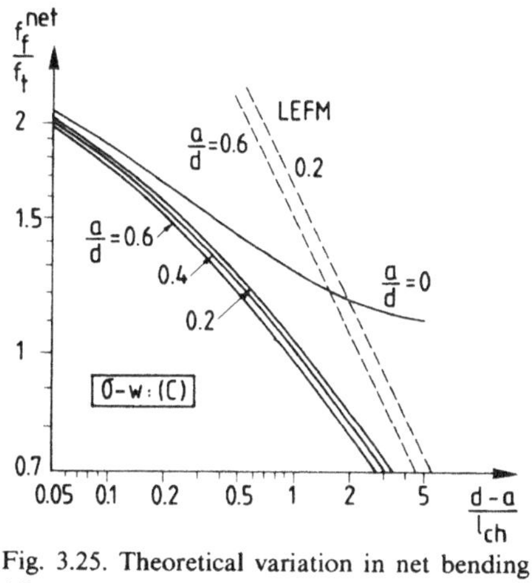

Fig. 3.25. Theoretical variation in net bending strength with notch depth a and ligament $(d-a)$.

foresee the crack path. In spite of this it may be advantageous to use a predetermined crack path for the analyses, as this may be much simpler than trying to follow the correct path. The accuracy of the chosen crack path can be checked by comparison with the principal stress direction within the fracture zone. Ideally the first principal stress should be perpendicular to the crack path. The calculation may have to be repeated with successively more accurate predetermined crack paths.

This technique has been applied by Gustafsson [12] for the analysis of the shear strength of reinforced concrete beams.

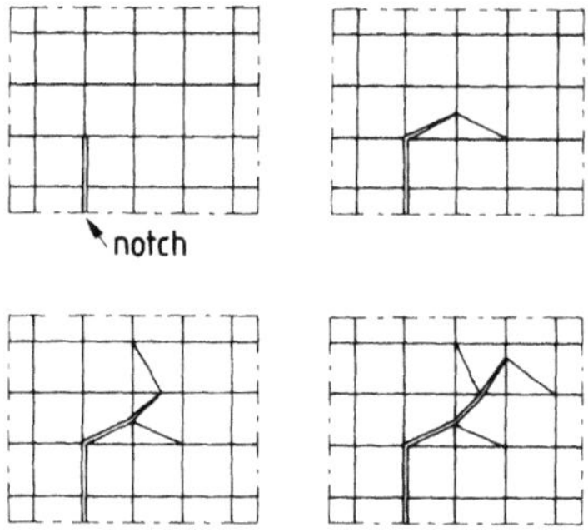

Fig. 3.26. Example of possible gradual rearrangement of a finite element mesh in order to follow the growth of a fracture zone and a crack.

3.8 FEM analysis of a fracture zone: not coincident with predetermined crack path

For concrete it is natural to assume that microcracks open perpendicular to the first principal stress. Thus it is assumed that the fictitious crack tends to grow in the direction perpendicular to the first principal stress at the tip of a real or fictitious crack.

It should be noted that in the model discussed here only finite stresses occur. For this reason, the crack growth direction cannot be determined by LEFM.

The crack growth direction in every instant can in principle be determined theoretically. In the general case, this crack path will not follow the element boundaries in the original finite element mesh. This problem can be solved in two different ways.

One method is to rearrange the mesh and successively add new boundaries, simulating the crack path according to Figure 3.26. This method has e.g. been used in [13], although in connection with a different model.

An other method is to let the crack path be represented by a path following the boundaries of the original mesh, but approximating the probable correct path as closely as possible. The general idea is outlined in Figure 3.27.

The details of a FEM program for such an analysis depends on the type of elements etc., but the following general principles can be used as a guideline:

(1.) The fictitious crack starts forming at the point of maximum tensile stress when this stress reaches f_t. If a notch or other point of stress concentration exists, this will be the starting point.

(2.) The direction of growth of a fictitious crack is perpendicular to the direction of the principle tensile stress at the crack tip.

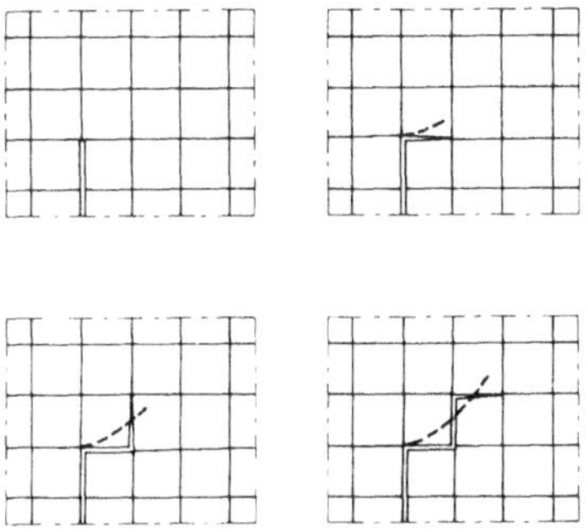

Fig. 3.27. Example of the approximation of the crack path by separating elements along original boundaries. The dashed line indicates the corresponding probably correct crack path.

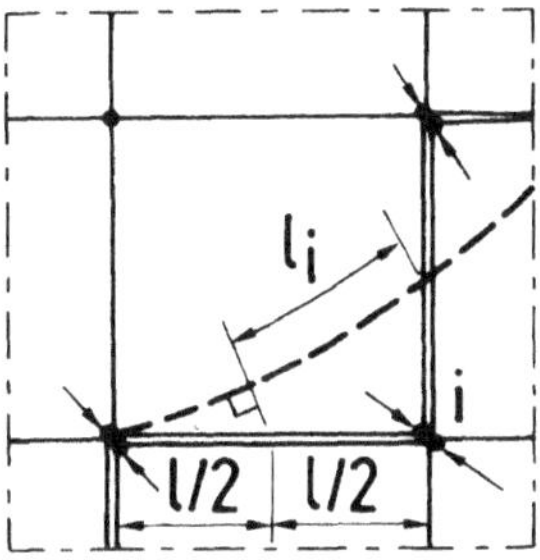

Fig. 3.28. Detail of Figure 3.27.

(3.) One pair of node points is separated at a time. The separation should take place when the stress reaches f_t at the crack tip, but if this stress is not known from the analysis, some other criterion has to be used, e.g. the highest principal tensile stress in the elements next to the crack tip.

(4.) The inter-element crack path and its properties are chosen in such a way, that the path and properties of the correct crack are represented as accurately as possible. Thus e.g. the force-deformation properties of node pair i in Figure 3.28 may represent the part l_i of the correct crack.

(5.) The force in a node pair has a certain value and direction at the moment of separation, Figure 3.28. This value and direction form the starting-point for the further calculations. The direction is kept constant in the calculation. In this way residual forces are avoided. The force-deformation relation is chosen so that it represents the correct energy absorption for the part l_i of the probable crack.

If shear stresses in the fracture zone (e.g. aggregate interlock) shall be taken into account, certain modifications are necessary. These are not treated here.

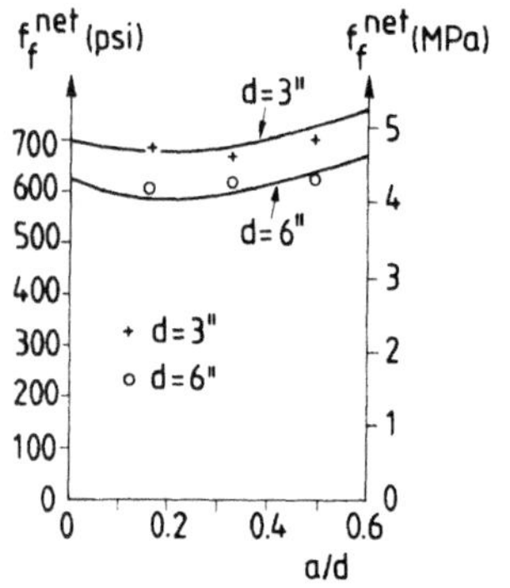

Fig. 3.29. Average values of net strength according to Kaplan [11], compared to theoretical curves from Figures 3.23 and 3.24 with assumed $l_{ch} = 12''$ (305 mm), $f_t = 335$ psi (2.3 MPa).

3.9 Some comparisons with test results

A basic idea of the model is that the necessary material properties shall be determined by means of simple specimens and that it shall be possible to calculate the behaviour of different structures by means of these properties.

For practically no tests reported in the literature the necessary material properties have been measured. Therefore a real check of the agreement between theoretical results and tests can not be performed. What can be done is to calculate which values of the unknown properties that will best explain the test results. If these properties are of a reasonable order, it supports the validity of the model. A lot of tests have been checked in this way, and as a rule it seems that the model can describe the test results rather well. Only a few examples will be shown here in order to demonstrate how comparisons with tests may be performed.

The tests of Kaplan [11] were among the first which were reported, and they are often cited. These tests were made on notched beams in three- or four-point bending, with a beam depth of 3″ (75 mm) or 6″ (150 mm) and a notch depth of 1/6, 2/6 or 3/6 of the beam depth.

Figure 3.29 shows the average net strength, (equation (3.10)), for the beams with different beam and notch depths. As a comparison the theoretical curves with reasonable values of l_{ch} and f_t are shown.

For each a/d-value, concrete mix and loading condition one test was made with $d = 3''$ and one with $d = 6''$. For each of these pairs the ratio between the net stresses has been calculated. The mean value of this ratio was 1.12 ± 0.07. With the depth ratio of 2, the value of this ratio according to LEFM should have been 1.41. This shows that LEFM is not applicable in this case. According to the theoretical curves of Figure 3.25 this ratio should be about 1.23 if $l_{ch} = 4''$ (100 mm), 1.19 if $l_{ch} = 8''$ (200

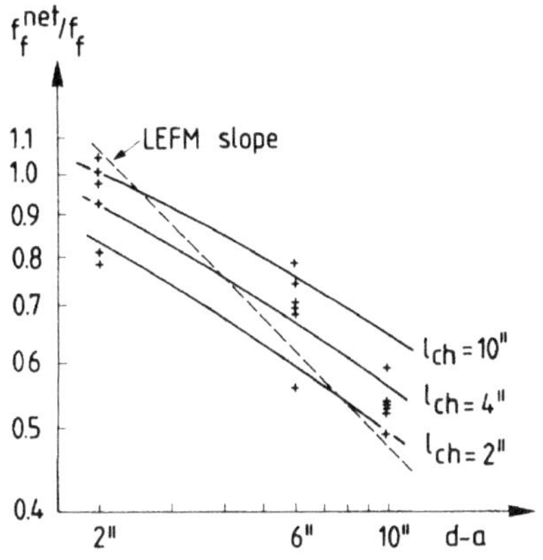

Fig. 3.30. Test results according to Walsh [14], compared to theoretical values from Figure 3.25. The value f_f in this case is the flexural strength of a beam with $d = 3''$.

mm) and 1.17 if $l_{ch} = 12''$ (300 mm). Thus $l_{ch} = 12''$ gives a reasonable agreement. A still higher l_{ch} would have given an even better agreement, but so high values are not probable for the concrete in question. If we use $l_{ch} = 12''$, we can find from Figure 3.21 that the G_c-values of the order 0.08 lb/in (14 N/m) for $d = 3''$ and 0.12 lb/in (21 N/m) for $d = 6''$, given in the paper, both correspond to a correct G_c-value ($= G_F$) of about 0.50 lb/in (90 N/m), which is in agreement with normally found values of G_F. It must however be pointed out that this type of correction is very uncertain. No certain conclusions regarding G_c can be drawn from tests on such small specimens. The only certain conclusion is that LEFM can not be applied to tests on such small specimens.

Another series of tests which is often cited is that by Walsh [14]. In his tests the a/d-ratio was constant 1/3, but d was given the values 3″, 9″ and 15″. Walsh also reported the flexural strength, measured on beams with $d = 3''$. In Figure 3.30 all Walsh's test results are given as the ratio between the net strength of the notched beam and this flexural strength for the concrete in question. In the diagram are also shown theoretical curves for different values of l_{ch}, based on Figure 3.25, as well as a straight line, the slope of which corresponds to the expected relation according to LEFM. It seems like $l_{ch} \cong 4''$ gives the best agreement, but the curves for 2″ and 10″ are within about ±10 percent from that for 4″, and it is thus not possible to determine l_{ch} with any good accuracy from these tests. With $l_{ch} = 4''$ the calculated K_c-values are about 30 percent too low for $d = 15''$ and 35 percent too low for $d = 9''$ according to Figure 3.21.

Naus [15] made tests on center-notched specimens. He determined the length of the fracture zone by means of strain gages and an assumption

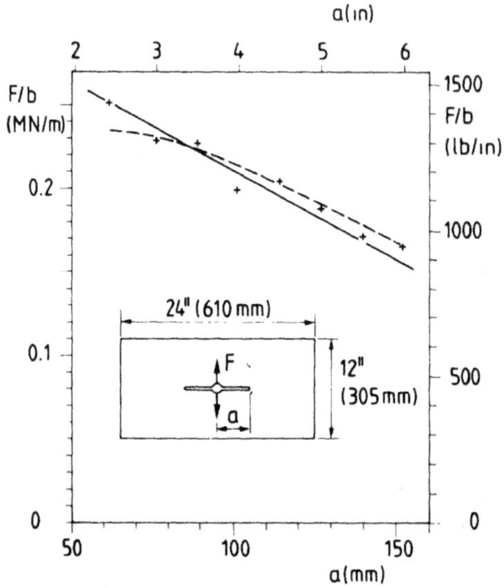

Fig. 3.31. Comparison between theoretical values and test results from Naus [15], test series M-O-AD-C, mortar specimens. ——— Fictitious crack model, $f_t = 3.04$ MPa, $l_{ch} = 110$ mm ------ LEFM $K_c = 0.77$ MNm$^{-3/2}$

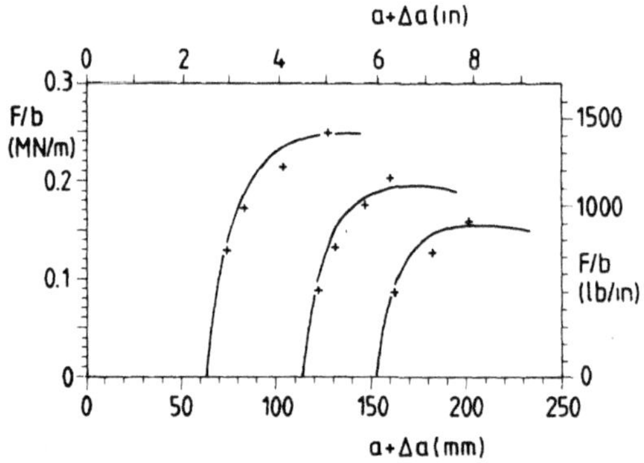

Fig. 3.32. Comparison between theoretical and measured lengths Δa of fracture zones for three of the specimens of Figure 3.31.

that fracture started when $\varepsilon = 150 \cdot 10^{-6}$. His tests have been theoretically analysed by means of the fictitious crack model and with LEFM by Gustafsson [16]. An example of his results is shown in Figures 3.31 and 3.32. The value $f_t = 3.04$ MPa is taken as the reported tensile splitting strength, whereas the values $l_{ch} = 110$ mm for the fictitious crack model and $K_c = 0.77$ MNm$^{-3/2}$ for LEFM are chosen in order to fit the test results. It can be seen that the fictitious crack model in this case gives a good agreement with the measured values and that LEFM also gives a reasonable agreement with the maximum load. From the latter fact the conclusion might be drawn that LEFM is applicable in this case. LEFM, however, is only applicable if the size of the fracture zone is small compared to the notch length and to the ligament length. It can be seen from Figure 3.32 that this condition is not fulfilled in this case, as measured as well as calculated fracture zone lengths are 50–70 mm, i.e. of the same magnitude as the notch length *a*.

The value of K_c corresponding to $f_t = 3.04$ MPa and $l_{ch} = 110$ mm is 1.01 $\text{MNm}^{-3/2}$ according to equation (3.9). Thus the value 0.77, determined from LEFM, in this case can be estimated to be approximately 25 percent to low.

One conclusion from this is that the fact that the variation of the maximum load with the notch depth is in a reasonable agreement with LEFM is no proof of the applicability of LEFM. The same conclusion can also be drawn from Figure 3.22 regarding bent notched beams with notch depths between 0.2 and 0.5 times the beam depth.

3.10 Summary and conclusions

The complete fracture behaviour of concrete can be theoretically analysed by means of a model, where the softening of the material due to the damage within the fracture zone is taken into account. The softening is described as a relation between the additional deformation w within the fracture zone, and the stress σ which can still be transferred in spite of the damage.

The relation between σ and w is assumed to be a material property, which can be determined from a tension test. This assumption seems to be valid in most cases, but not where high compressive stresses are acting.

The model can not without modifications be applied to metals like steel. The difference between concrete and steel is discussed.

The model is general in the sense that it can be applied to un-notched as well as to noched specimens and to cases where shrinkage or thermal strains are acting.

For the practical applications it is as a rule necessary to use some numerical technique, like the finite element method. As no stress singularities occur, it is not necessary to use special crack tip elements for this analysis.

By means of the analysis the complete fracture behaviour can be followed, from the onset of cracking, over the maximum load, and to the final collapse. Thus also the applicability of other approaches can be analysed, e.g. linear elastic fracture mechanics. Results of such analyses are given and it is concluded that the application of linear elastic fracture mechanics to specimens of an ordinary size will result in great errors.

Comparisons indicate a good agreement between test results and predictions from the theoretical analysis. The validity of this comparison is however limited by the lack of complete data regarding the material properties of the test specimens reported in the literature.

The model is well adapted for the analysis of practical problems where tensile fracture plays an important role for the behaviour of a structure, e.g. shear failure of reinforced beams, splitting and spalling caused by reinforcement or crack widths and spacings.

References

1. Sok, C., Baron, J. and Francois, D., Mécanique de la rupture appliquée au beton hydraulique, Cement and Concrete Research, 9, pp. 641–648 (1979).
2. Bazant, Z.P., Instability, ductility and size effect in strain-softening concrete. Journal of the Engineering Mechanics Division, ASCE, 102, pp. 331–344 (1976).
3. Heilmann, H.G., Hilsdorf, H.H. and Finsterwalder, K., Festigkeit und Verformung von Beton unter Zugspannungen. Deutscher Ausschuss für Stahlbeton, 203, W. Ernst & Sohn, Berlin (1969).
4. Petersson, P.-E., Crack growth and development of fracture zones in plain concrete and similar materials, Report TVBM-1006, University of Lund, Sweden (1981).
5. Hillerborg, A., The fictitious crack model and its use in numerical analyses, Fracture Mechanics in Engineering Application, Proc. Int. Conf. Bangalore (1979).
6. Bazant, Z.P. and Oh, B.H., Concrete fracture via stress-strain relations, Report No. 81-10/665c, Center for Concrete and Geomaterials, Northwestern University (1981).
7. Hillerborg, A., A model for fracture analysis, Report TVBM-3005, University of Lund, Sweden (1978).
8. Modéer, M., A fracture mechanics approach to failure analysis of concrete materials, Report TVBM-1001, University of Lund, Sweden (1979).
9. Hillerborg, A., Analysis of fracture by means of the fictitious crack model, particularly for fibre reinforced concrete, The International Journal of Cement Composites, 2, pp. 177–184 (1980).
10. Blakey, F.A. and Beresford, F.D., Discussion of a paper by Kaplan, Journal of the American Concrete Institute, 58, pp. 919–923 (1962).
11. Kaplan, M.F., Crack propagation and the fracture of concrete, Journal of the American Concrete Institute, 58, pp. 591–610 (1961).
12. Gustafsson, P.J., Analysis of the shear strength of r/c beams, Report, Division of Building Materials, University of Lund, Sweden (1982).
13. Saouma, V.E., Interactive finite element analysis of reinforced concrete: a fracture mechanics approach, Report 81-5, Department of Structural Engineering, Cornell University (1981).
14. Walsh, P.F., Fracture of plain concrete, Indian Concrete Journal, Nov. 1972, pp. 469–476 (1972).
15. Naus, D.J., Applicability of linear elastic fracture mechanics to Portland cement concretes, thesis, University of Illinois, Urbana (1971).
16. Gustafsson, P.J., Private communication (1982).

A.R. INGRAFFEA AND V. SAOUMA

4

Numerical modeling of discrete crack propagation in reinforced and plain concrete

4.1 Introduction

The earliest applications of the finite element method to concrete structures included crack modeling [1–3]. Such modeling, then and now, can be completely characterized by two essential elements. These are the manner of inclusion of a crack in the finite element mesh, and the criterion for crack instability and direction of growth.

In the early efforts cracks were modeled discretely. That is, a crack was formed by separation of previously common element edges. The resultant change in nodal connectivity was made at each stage of propagation, as depicted in Figure 4.1. If the crack trajectory was assumed a priori, double-nodes could be provided along the element edges that would become the crack path [2]. Alternatively, all nodes could be doubled initially and only those close to a predicted path would be separated [4].

The early discrete crack approaches all suffered two major drawbacks:

(1) Lack of physical fidelity in that crack trajectory was constrained to follow predefined element boundaries.
(2) Increasing cost and decreasing efficiency in that additional degrees-of-freedom had to be added with each crack increment. This means that the size of the analysis grew, as did the bandwidth of the structure stiffness matrix.

The "smeared crack" approach, introduced by Rashid [5] and amply described in other chapters of this book, overcame these drawbacks and quickly replaced the discrete approach.

In the early efforts the criterion for instability and direction of growth, the second essential modeling element, did not derive from any fracture mechanics notion. Rather, a simple tensile strength approach was employed. When the principal tensile stress computed in the elements ahead of the discrete or smeared crack tip reached some measure of strength, those elements were "cracked." Direction of propagation was assumed to be normal to that of the principal stress.

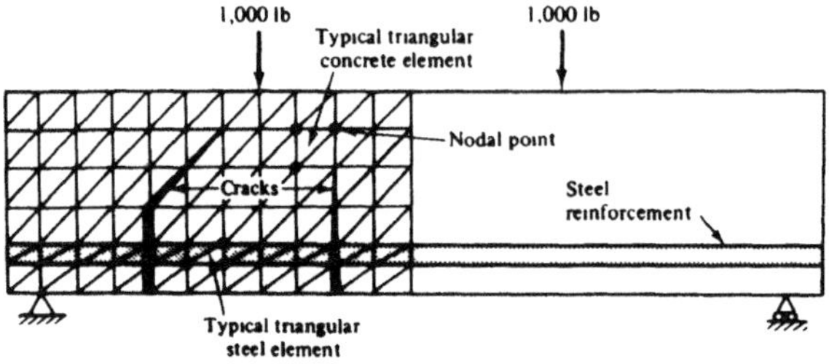

Fig. 4.1. Early example of discrete crack modeling [2].

Given this starting point, one must ask what has changed in the ensuing fifteen years of usage of the finite element method for cracking analysis of concrete structures. In the authors' opinion, the answer is: not much. A smeared crack, tensile strength fracture model is the state-of-the-practice for research and production oriented codes. The capabilities for aggregate interlock in shear and the intersecting of cracks have been added. However, this model has remained largely unaffected by ongoing developments in finite elements, equation solving, user-computer interaction, and fracture mechanics. It is essential to the developments of the present chapter to recognize that, in the formative years of this model, the constant strain triangle, simple equation solvers, key-punching of data, batch analysis mode, and elementary constitutive models were the order of the day.

In the present chapter, crack modeling approaches will be described which markedly depart from today's state-of-the-practice. It will be shown that these methods integrate many of the innovations in the aforementioned areas of development. As crack propagation models, these methods are characterized by:

(1) a return to the discrete crack,
(2) linear and nonlinear fracture mechanics, and, as a new essential modeling element,
(3) the use of interactive-adaptive computer graphics techniques as an integral part of the analysis process.

Before proceeding to the new models themselves, additional background material related to the evolution of these essential elements needs to be detailed.

The discrete crack vs. the finite element mesh. It is not surprising that the initial usage of the discrete crack approach was short lived. To overcome the first of the drawbacks cited above, manual remeshing after each crack increment would have been required. It was well known that bandwidth minimization was essential to analysis efficiency, but, to overcome the second drawback, manual renumbering for minimal bandwidth of an increasingly complex mesh also would have been required. Clearly, given the mesh generation and bandwidth minimization capability of the mid-1960's, an adversary relationship existed between the discrete crack and the finite element method. In the authors' opinion, it was the encumbrances of the above mentioned drawbacks, and not a new view of the physics of the problem, which led to the early and nearly ubiquitous adoption of the smeared crack approach. There have been arguments proposed, ex post facto, that the smeared crack is more realistic; however, given the scale differences between the finite element width typically used in analysis and usually predicated by other considerations [7] and the physical width of a crack and associated process zone as observed in experiments [8], the authors find these arguments difficult to accept.

In trying to return to the more physically palatable discrete crack idea there have been sporadic efforts towards eliminating its drawbacks, such as the novel approach of Ngo [6]. However, if a discrete model were to become acceptable at any level of use, a method of automatic regeneration of the mesh to accommodate arbitrary and incrementally changing crack trajectories would have to be found. This method would properly introduce the displacement and traction variations across and along the crack, properly mesh in the crack and around its tip, allow cracks to cross reinforcement, and minimally increase the bandedness of the system stiffness matrix. Let us review the various advances over the last decade that have, in fact, led to the development of such a method.

First, take the capability for automatic mesh generation. Few would, today, consider the complete manual generation of even a two-dimensional mesh. Automatic mesh generators of various degrees of sophistication are now in widespread use. At their highest level they make use of generalized lofting, splining, and other elegant topological manipulations to produce complex two- and three-dimensional meshes from minimal user input [9–11]. But, in fact, the problem at hand is simpler in that discrete crack modeling requires only local modifications to an existing mesh.

A solution to this problem for two-dimensional and axisymmetric meshes was found by the second author [12], and implemented in a code called the Finite Element Fracture Analysis Program (FEFAP). Overall, the algorithm developed is based on the local remeshing requirements which result from:

(1) A discrete crack crossing an element.
(2) On emerging from an element side or a corner node, ascertaining which element is the next to be entered.
(3) Checking if the crack will stop in the next element entered.

A crack may cross a triangular or a quadrilateral element in many different ways. The 25 cases programmed in FEFAP are shown schematically in Figure 4.2.

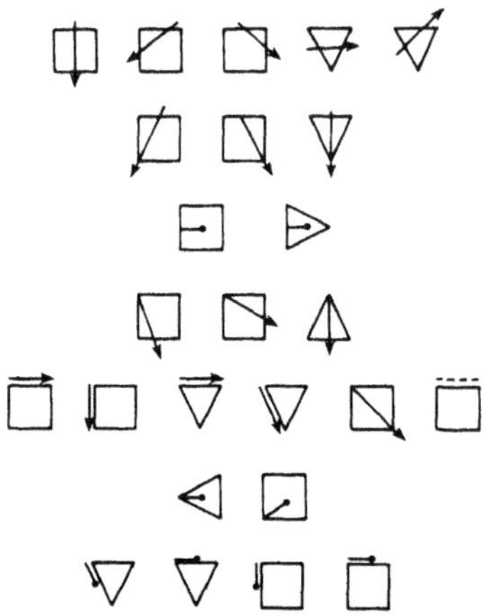

Fig. 4.2. Possible crack paths around, into, and through quadrilateral and triangular elements programmed into FEFAP [12].

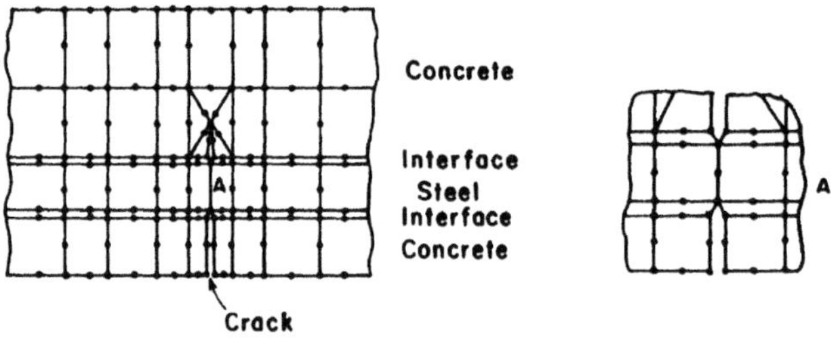

Fig. 4.3. Detail of discrete crack crossing reinforcement bordered by bond elements [12].

The algorithm, although still under modification, has proven to be versatile and robust. For example, a number of common pathological situations may arise in remeshing. These are automatically diagnosed and corrected. Examples include:

(1) A crack about to cross interface and reinforcement elements. Since the latter should not be broken by the crack, modifications such as those shown in Figure 4.3 are automatically performed.

(2) When a crack breaks an element in two, and if one of the new elements would have a large aspect ratio, then one of the sides of the original element will be shifted to coincide with the crack. Consequently, the crack now propagates between two elements of acceptable aspect ratio.

Particular examples of these and other capabilities of the remeshing algorithm in FEFAP will be seen in example problems later in this chapter.

To summarize, an algorithm has been developed which automatically remeshes in the vicinity of a crack to accommodate arbitrary, predicted trajectory. This eliminates the first drawback to the original discrete crack approach.

Next, let's take a modern view of the bandwidth problem. Two potential solutions to this problem have arisen over the last decade. These are the use of either the frontal solution technique or automatic bandwidth minimizers. The latter approach has been adopted by the authors in the form of an algorithm developed by Gibbs, Poole, and Stockmeyer [13]. After each crack increment has created new nodes all nodes are automatically renumbered to approach minimum bandwidth and all data keyed to node numbers is relabeled. The entire process is transparent to the user and typically adds only a few percent additional CPU time. Consequently, the second drawback originally ascribed to the discrete crack approach has also been eliminated. It is true that the size of the problem still increases with crack propagation as compared to a smeared crack approach, but, in the authors' opinion, this is as it should be. Clearly, the deformation field of a structure becomes more complex with cracking, and mesh refinement is justified to maintain a reasonable level of analysis accuracy.

Had automatic remeshing and bandwidth minimization capability existed in the mid-1960's, it is likely that the discrete model would have undergone continued development. It is also likely that the original criterion for crack propagation in the presence of a sharp crack tip would then have come under review. Consequently, the fracture mechanics of concrete could well have become part of a discrete cracking model at a much earlier date. Those elements of concrete fracture mechanics which have been integrated into the authors' new discrete models are described next.

The fracture mechanics of concrete. A recent survey [14] indicates over 400 research papers and reports on the subject of concrete fracture mechanics since the mid-1960's. However, as manifested by the continued use of a simple stress criterion for crack propagation, this research has yet to make its influence felt in practice. Reasons for this somewhat startling observation are the following.

First, fracture mechanics has traditionally been associated with a mathematically sharp crack. Since the cracks modeled in the finite element programs of the last fifteen years have been mathematically smeared, there has been no imperative to connect the two.

Second, the geometrical restrictions on cracked concrete structures that would define the domain of valid application of the simplest fracture mechanics model, classical, linear elastic fracture mechanics, have yet to be determined to the satisfaction of most researchers in the field. A

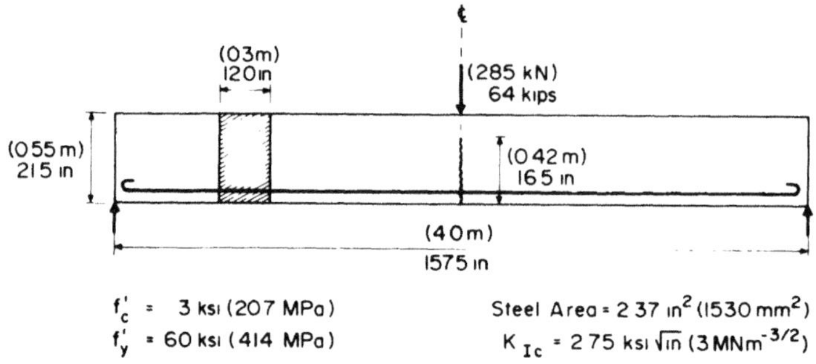

Fig. 4.4. Example problem for Equation (4.1).

reluctance on the part of practitioners to even consider a change of fracture criterion naturally follows.

Third, there continues to be a confusion concerning the mesh dependence of both fracture-mechanics-based and the traditional tensile strength approaches. Again, in the face of such inconclusiveness, the practice conservatively sticks to the tried and true methods of the past.

Indeed, the apparent continued success of the tensile strength criterion is its own most compelling reason for its continued use. But is it possible that this apparent success has been circumstantial, that, in fact, the strength, serviceability, and crack trajectory predictions of the past fifteen years have all been virtually insensitive to the fracture criterion employed? Bazant and Cedolin [15] have raised this possibility in the context of the mesh dependency issue. The authors would like to raise it again in the form of a simple heuristic proof of the following statement:

> The strength, deflection, and crack trajectory predictions for many reinforced concrete structures are insensitive to concrete fracture toughness.

That this is so for beams designed by ACI Code provisions can be easily seen from the following example. Assume that the under-reinforced concrete beam shown in Figure 4.4 has developed a single flexural crack at mid-span. The conditions shown in Figure 4.4 correspond to nominal ultimate load at which point the reinforcement has just reached yield and the crack is assumed to have propagated to the bottom of the corresponding concrete compressive stress block.

According to linear elastic fracture mechanics, the beam will not fracture as long as

$$K_{\mathrm{I}}^{\mathrm{P}} < K_{\mathrm{I}}^{\mathrm{R}} + K_{\mathrm{Ic}}^{\mathrm{c}} \tag{4.1}$$

where

$K_{\mathrm{I}}^{\mathrm{P}}$ = applied stress intensity factor due to load P

K_I^R = crack arresting stress intensity factor due to the load in the reinforcement

K_{Ic}^c = fracture toughness of the concrete

For the condition shown in Figure 4.4, it can be shown that,

$$K_I^P \simeq 82 \text{ ksi} \sqrt{\text{in}} \, (90 \text{ MNm}^{-3/2})$$

For a reasonable value of K_{Ic}^c, say about 2.75 ksi $\sqrt{\text{in}}$ (3 $\text{MNm}^{-3/2}$), it can be seen that K_{Ic}^c is a negligible term in Equation (4.1). Consequently, the crack is stable as long as

$$K_I^P < K_I^R \tag{4.2}$$

Assuming perfect bond, K_I^R depends only on the yield force in the steel and the distance from the steel level to the centroid of the compressive stress block. Since, at failure, the latter ultimately depends on the concrete compressive strength, the beam strength is seen to be independent of its fracture toughness. Since maximum deflection would occur under the condition of Equation (4.2), this quantity also is independent of toughness. Finally, it can be shown that a crack length prediction for the structure shown would be sensitive to toughness only for relatively short crack length.

Indeed, one can further show with the above example that if tensile stress and strength terms are substituted for the analogous stress intensity and toughness terms, the analogous result is obtained: strength and deflection predictions are insensitive to tensile strength. Recently, this has been conclusively proven for a variety of beam configurations by Dodds et al [7].

These conclusions may be obvious physically; they are, after all, implicit in the ACI Code, but they seem to have been overlooked in the debate over valid fracture criteria and mesh sensitivity. The above example clearly shows that a finite element prediction of strength of such a structure will be quite insensitive to *any* cracking criterion. Consequently, a dependence of any criterion on any mesh characteristic will not influence strength prediction to any great degree. In summary, the continued use of the tensile strength fracture criterion for certain classes of concrete structures is correct, but only because, in effect, it can't be wrong.

Two issues remain to be clarified here concerning the fracture mechanics of concrete before proceeding to details and examples of the authors' cracking models. Obviously, it is now essential to know for which concrete structures response to load is sensitive to cracking model. Then, it becomes equally important to identify the proper fracture-mechanics-based model for their analysis.

Clearly, the crack arresting effect of reinforcement is the key element in the first issue. It should be obvious that predictions of response of plain concrete structures will be fracture-model and mesh sensitive. However, given the enormous range of shapes, reinforcement patterns, and loadings of prototype reinforced concrete structures, it is highly likely that among them there are also cases of fracture model and mesh sensitivity. In the authors' opinion, these cases have yet to be conclusively identified.

With respect to the second issue, for the plain concrete case, recent experimental research [16,17,18] clearly indicates that both linear and nonlinear fracture mechanics approaches will be required. Descriptions of the approaches and their implementation in the authors' finite element codes are presented in subsequent sections. Surprisingly, current research [19,20] shows that even for very long cracks in large reinforced concrete structures which appear to be candidates for fracture model sensitivity, linear elastic fracture mechanics is not applicable.

To summarize, the authors have incorporated both linear and nonlinear fracture mechanics approaches into their new discrete cracking programs. The usage of these programs is being directed towards practical analysis of plain concrete structures, such as dams and abutments, and research applications in plain and some reinforced structures. Examples of both types of usage are presented later in this chapter.

To complete the scenario of evolutionary development of the authors' programs, it is necessary to investigate the influences of the change in user-computer interface in finite element analysis. This is done in the next section.

Interactive graphics: Luxury or necessity? Perhaps the most exciting development in computer analysis in the last fifteen years is the area of computer graphics. The essential element of all CAD/CAM systems, interactive graphics has begun to put the structural engineer back in control of his sophisticated analyses. Despite the widespread adoption of computers for analysis of concrete structures, there remain at least two serious obstacles to realization of their full potential. These obstacles involve engineering productivity and control by the engineer of decision-making in the design-analysis cycle.

Productivity in analysis involves more than just the actual analytical operations. In the input or preprocessing part of the analysis extremely voluminous and complex descriptions of problems must be prepared and fed into the computer. This is true even for the initial database of a smeared-crack-based model. At the output or postprocessing end of the analysis vast quantities of results must be sifted and evaluated to make design decisions or to discern the pattern of behavior. These two aspects – preprocessing and postprocessing – not only consume the preponderance of time and manpower, but also, by their nature, are subject to human error. The need in any computer analysis of a concrete structure is for greater efficiency and reliability, and one way to achieve these is through better man-machine communication, an auspicious advantage of interactive computer graphics. An example of this advantage is depicted in Figure 4.5. Here a three-dimensional boundary element preprocessor developed by Perucchio and the first author [11] is used to generate a mesh for the analysis of a concrete dam buttress. Generation of the mesh required only four man-hours. In addition to speed and immediate visual checking, the images of Figure 4.5 show the additional capabilities of partial image refreshing, zooming, and rotation afforded by high-level interactive graphics.

Control of decision-making in analysis has undergone drastic change. Not much control needed to be exercised over the analysis process in the simple linear analyses of the mid-1960's; with the initial database specified there were no more experience or judgment decisions left to be made by the analyst. However, with nonlinear analyses encouraged by the complex constitutive models of the 1970's, processing of the initial database often necessitated revisions such as load step size, error tolerances, mesh gradation, or the solution algorithm itself. Such analyses are far from being "black box" automated, and this is as it should be. The engineer must be able to exercise judgment effectively, and, ideally, in real time, in making key analysis decisions. The ability to keep machine and engineer in proper balance during the analysis process is becoming recognized as another characteristic advantage offered by interactive graphics. In fact, were it not for the productivity and control advantages offered by interactive graphics, the means for efficient implementation of and for control over the remeshing algorithm in FEFAP and the consequent development of new discrete crack models would not have been available. Interactive graphics are used in FEFAP to display in real time:

(1) all stages of overall mesh development,
(2) "zoomed" details of meshes,
(3) deflected shapes,
(4) principal stress fields,
(5) crack trajectories,
(6) energy release rate curves,
(7) predicted crack increment directions, and, finally, and most importantly
(8) numerical information essential to the analysis decision making process.

The last item implies that the codes developed by the authors are not only interactive-graphic but that they are also "interactive-adaptive". This term is used to describe methods, essential to nonlinear analysis, in which parameters, algorithms, meshes, and problem descriptions are

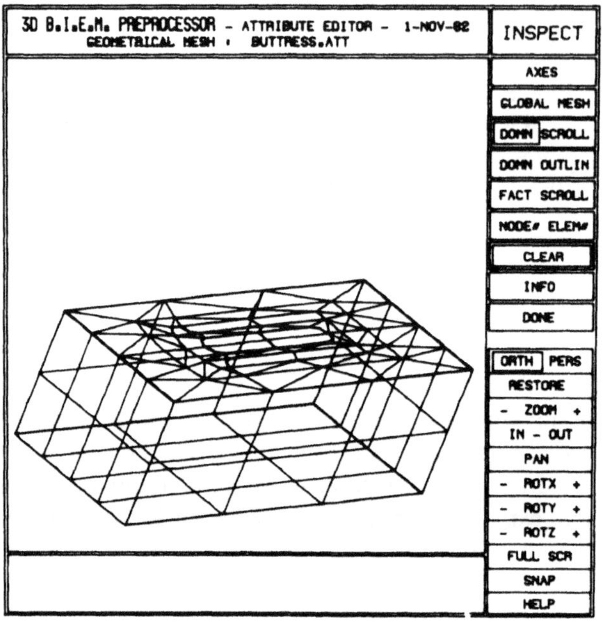

a

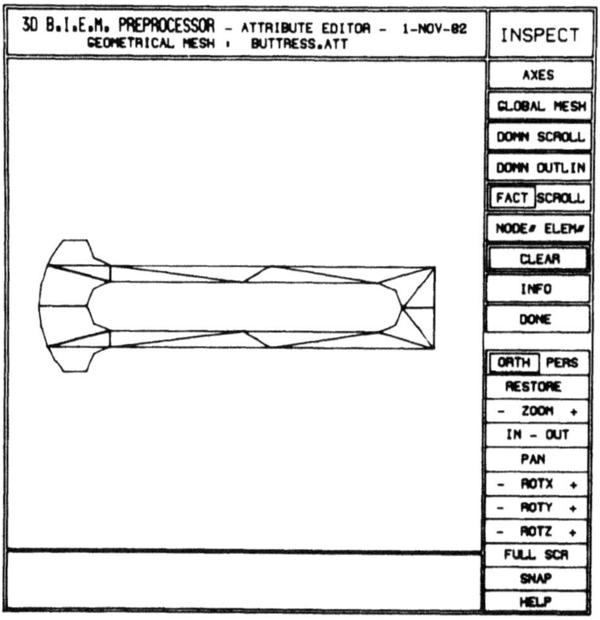

b

selected or changed by the user during the analysis itself.

With an interactive-adaptive computer-graphics system, both the machine and the person do what they do best: The machine takes on the

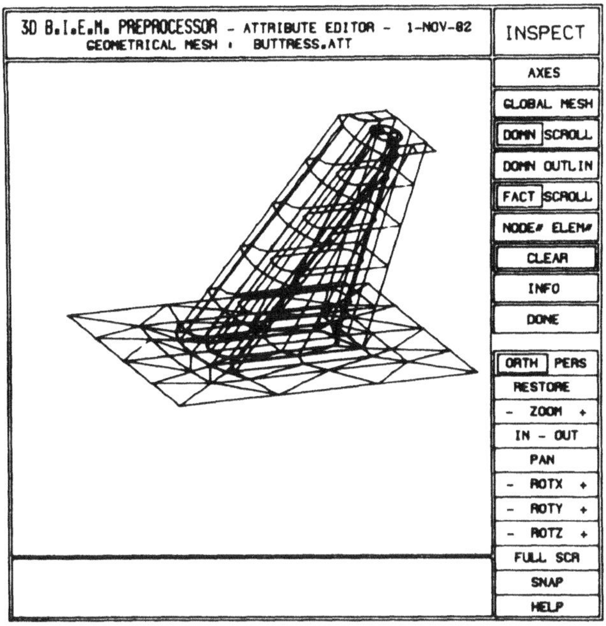

c

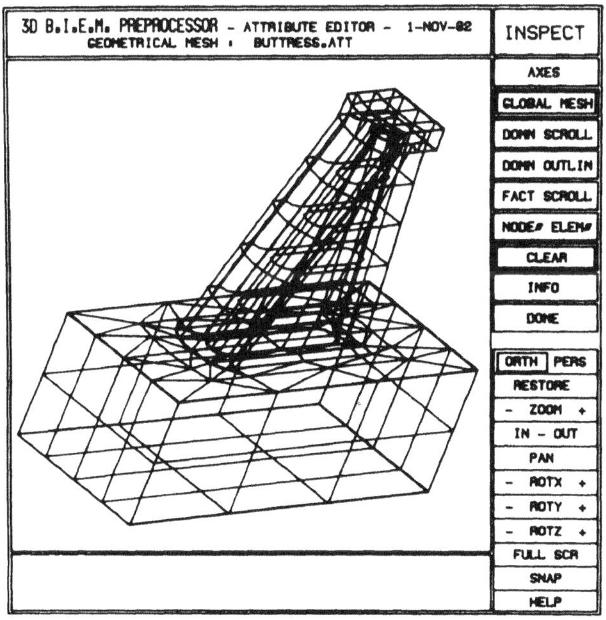

d

Fig. 4.5. Three-dimensional boundary element mesh generation for concrete dam buttress using high-level interactive graphics. a) Portion of foundation; b) Typical cross-section through buttress; c) Placement of cross-sections in space; d) Final mesh including buttress cap.

tedious calculations, the data manipulation, and the figure-drawing, while the person visually integrates and evaluates patterns of behavior. Effectively, the engineer and the computer become active partners in the analysis process.

Background summary. Fifteen years of finite element, computer, and fracture mechanics developments have created a backdrop for the introduction of new, discrete cracking models for the analysis of concrete structures. The smeared crack, tensile strength approach was adopted for reasons of convenience, not for realism and rigor. It is still useful and accurate for macro-analysis of many types of reinforced concrete structures. However, for cracking analysis of plain concrete structures and for micro-analysis of reinforced concrete elements, a fracture mechanics approach is necessary and a discrete model will be shown to be more useful and realistic.

4.2 Discrete crack models for concrete

Two types of fracture models have been implemented into the new discrete cracking approach [12,16,17,19]. The first model is based on classical, linear elastic fracture mechanics. Governing parameters are the stress-intensity factors and fracture toughness. The latter quantity is still designated K_{Ic}, despite sparsity of information regarding the difference, if any, between plane strain and other fracture toughness measures in concrete. As usual the assumption is made that the process zone associated with the crack is much smaller than the crack length and ligament (again, information concerning thickness effects on process zone size is inconclusive to date). At the scale of the crack length concrete is thus assumed to act as an homogeneous solid.

Whereas the linear model assigns a critical stress intensity to the crack tip, the second, nonlinear model, similar to the Dugdale model, assumes that stress intensity is always zero. This model is based on the assumption that a narrow process zone exists ahead of the traction free crack. The process zone itself is idealized as a discontinuity in displacement, but not in stress; in effect it is a "fictitious crack" [21], a concept whose origin is thoroughly discussed in another chapter of this book. The process zone constitutive law is one of traction, both normal and tangential, versus COD.

These models are related by scale. The small crack-tip process zone in the linear model can be viewed as the nonlinear process zone at a much smaller scale. A very short crack would be described by the nonlinear model, a very long crack *with sufficient COD over most of its length* by the linear approach. In pursuing the matter of scale further, it is obvious that "large aggregate" and "matrix material" can be re-interpreted in scale for mortar, with sand and hardened cement paste occupying these roles in this case.

Details concerning formulation of these models and their manner of implementation into a discrete crack idealization are presented next.

4.3 The linear model

In the linear model stress intensity factors control crack stability and trajectory. The sequence of events in the implementation of this model into a two-dimensional mixed-mode crack propagation code is:

(1) Compute stress intensity factors for present crack tip location and loading.

(2) Substitute K_{I} and K_{II} into mixed-mode interaction formula. Compute new crack direction and assess stability. If crack is unstable, continue. If stable, go to step 4.

(3) Remesh for a selected increment of propagation. Repeat steps 1 through 3 until crack is stable or fracture occurs.

(4) If crack is stable, raise load level until instability is predicted by interaction formula. Continue with step 3.

The accurate prediction of load level, angle change, and length corresponding to each increment of crack propagation requires accurate computation of mixed-mode stress intensity factors. Their efficient computation is also desirable since many analyses may need to be performed in a single problem. The next section presents background and formulation for a technique which is accurate and efficient. Moreover, it will be shown to be comparatively trivial algorithmically.

Stress intensity factor computation. The displacement correlation technique has been adopted for use with the discrete crack propagation programs developed by the authors. In this method displacement computed at nodes near the crack tip are correlated with the theoretical values. For example, in pure Mode I, near tip displacements relative to the crack tip are,

$$u = \frac{K_{\mathrm{I}}}{4G}\sqrt{\frac{r}{2\pi}}\left[(2\kappa - 1)\cos\frac{\theta}{2} - \cos\frac{3\theta}{2}\right] + \ldots$$

$$v = \frac{K_{\mathrm{I}}}{4G}\sqrt{\frac{r}{2\pi}}\left[(2\kappa + 1)\sin\frac{\theta}{2} - \sin\frac{3\theta}{2}\right] + \ldots \qquad (4.3)$$

where,

u = displacement parallel to crack axis

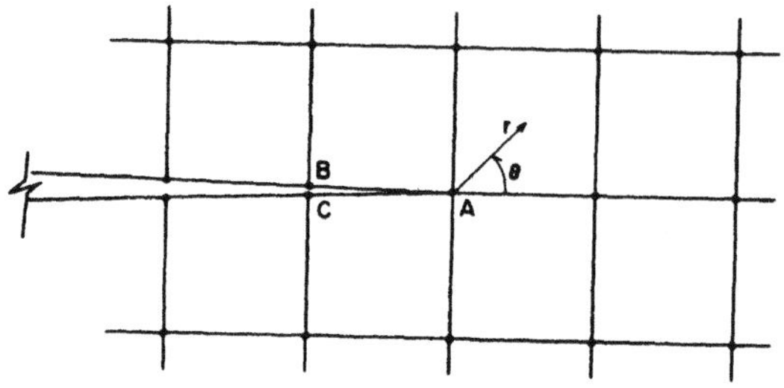

Fig. 4.6. Crack-tip nodal lettering for Equation (4.4), (4.6), and (4.8).

u = displacement parallel to crack axis
v = displacement normal to crack axis
ν = Poisson's ratio
G = shear modulus
κ = $(3-4\nu)$ for plane strain
κ = $(3-\nu)/(1+\nu)$ for plane stress

Taking $\theta = 180°$, and $r = r_{AB}$, in Figure 4.6, the K_I computed by the finite element method is expressed by,

$$K_I = \sqrt{\frac{2\pi}{r_{AB}}}\frac{2G}{(\kappa+1)}(v_B - v_A) \tag{4.4}$$

For pure Mode II and plane stress,

$$u = \frac{K_{II}}{4G}\sqrt{\frac{r}{2\pi}}\left[(2\kappa+3)\sin\frac{\theta}{2} + \sin\frac{3\theta}{2}\right] + \ldots \tag{4.5}$$

$$v = \frac{-K_{II}}{4G}\sqrt{\frac{r}{2\pi}}\left[(2\kappa-3)\cos\frac{\theta}{2} + \cos\frac{3\theta}{2}\right] + \ldots$$

And an expression for K_{II} is, for $\theta = 180°$, $r = r_{AB}$,

$$K_{II} = \sqrt{\frac{2\pi}{r_{AB}}}\frac{2G}{(\kappa+1)}(u_B - u_A) \tag{4.6}$$

For mixed-mode loading, the near-tip displacements are given by linear combinations of Equation (4.3) and Equation (4.5). However, it can be seen that the combined expressions for u and v uncouple at $\theta = \pm 180°$. Consequently, Equations (4.4) and (4.6) can be used for a crack on a symmetry or anti-symmetry line. In the general, asymmetrical case, the stress intensity factors must be computed from crack-opening and crack-sliding displacements, COD and CSD, respectively, where, for $r = r_{AB} = r_{AC}$,

$$\mathrm{COD} = v_B - v_C$$

$$\mathrm{CSD} = u_B - u_C \tag{4.7}$$

Substituting Equations (4.3) and (4.5) into Equation (4.7), and solving for K_I and K_{II} yields,

$$K_I = \sqrt{\frac{2\pi}{r_{AB}}}\frac{G}{(\kappa+1)}(v_B - v_C)$$

$$K_{II} = \sqrt{\frac{2\pi}{r_{AB}}}\frac{G}{(\kappa+1)}(u_B - u_C) \tag{4.8}$$

Equations (4.8) are the most general form for computing stress intensity factors using the displacement correlation technique.

Obviously, the success of this technique depends on an accurate modeling of the theoretical $r^{1/2}$ displacement variation near the crack tip. Many workers have applied a number of finite element method techniques to this end. Earliest efforts employed brute force. Constant strain triangles (CST) were used with extremely fine meshes near the crack tip. The obvious major drawback to use of the CST in fracture problems is the very large number of degrees of freedom required for engineering accuracy even for simple geometry, loading, and a single crack.

The next step in finite element method development for the displacement correlation technique was to take a hint from the elasticity solution for the crack tip displacements and develop an element containing an $r^{1/2}$ interpolation function. In practice, this has been done by a number of workers using elements of various shapes. The most convenient of these, the quarter-point isoparametric triangle [22,23], is the singular element employed by the authors. Combination of this singular element with the extraction technique outlined below makes programming for stress intensity factor computation almost trivial compared to the efforts of less than a decade ago.

The application of the displacement correlation technique to quarter-point singular element was first proposed by Shih et al [24] for Mode I problems. The generalization of mixed-mode is straightforward [25] and goes as follows. First, let a crack tip be surrounded by quarter-point singular elements and the crack face nodes be lettered as shown in Figure 4.7. The only meshing constraint at this point is that the lengths of the two elements containing the lettered nodes be the same, i.e. $\overline{AC} = \overline{AE}$. Next, expand the lettered nodal displacements in terms of the element

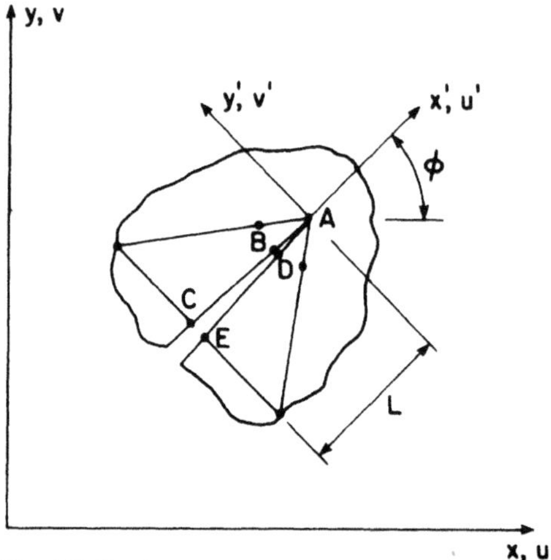

Fig. 4.7. Crack-tip nodal lettering for quarter-point singular elements, Equation (4.10).

shape functions and with respect to the crack tip coordinate system shown in Figure 4.7. This leads to expressions of the following form:

$$v' = v'_A + (-3v'_A + 4v'_B - v'_C)\sqrt{r/L} + (2v'_A - 4v'_B + 2v'_C)r/L$$

$$u' = u'_A + (-3u'_A + 4u'_B - u'_C)\sqrt{r/L} + (2u'_A - 4u'_B + 2u'_C)r/L \tag{4.9}$$

Similar expressions are obtained along the $\overline{\mathrm{ADE}}$ ray. Taking the difference between the expressions along the two rays yields a computed crack opening or crack sliding profile.

Now the displacement expansions from the asymptotic analytical solution, Equation (4.3), are evaluated at $\theta = \pm 180°$ to compute a theoretical crack opening or crack sliding profile. Equating like powers of r in the computed and theoretical profiles leads directly to the simple expressions,

$$K_I = \sqrt{\frac{2\pi}{L}}\,\frac{G}{\kappa + 1}\left[4(v'_B - v'_D) + v'_E - v'_C\right]$$

$$K_{II} = \sqrt{\frac{2\pi}{L}}\,\frac{G}{\kappa + 1}\left[4(u'_B - u'_D) + u'_E - u'_C\right] \tag{4.10}$$

in which

L = length of singularity element side along the ray,
v' = crack-opening nodal displacements,
u' = crack-sliding nodal displacements.

The primes indicate that the global coordinate nodal displacements have

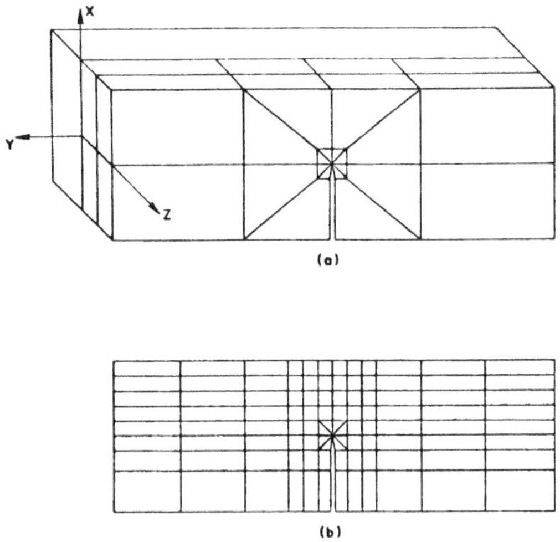

Fig. 4.8. Two- and three-dimensional finite element meshes for ASTM 3PB specimen. With symmetry invoked, three-dimensional model has 139 nodes and 20 elements; 175 nodes and 50 elements in two-dimensional model. From [25].

been transformed to the crack-tip coordinate system defined in Figure 4.7. The above procedure has been generalized to the three-dimensional case by Ingraffea and Manu [25].

Algorithmically, the displacements and coordinates of the crack face nodes belonging to the quarter-point elements need to be flagged and retrieved for each crack increment solution. These are then transferred to a simple subroutine which codes Equations (4.10). With the efficiency of the method obvious, its accuracy will be discussed with reference to the meshes shown in Figure 4.8. These are for the ASTM standard three-point bend specimen.

The accuracy of Equations (4.10) depends on a number of mesh characteristics. For example, Figure 4.9 shows two- and three-dimensional finite element results compared to the ASTM solution, known to be accurate within 1 percent, with L/a as a parameter. The differences in the two-dimensional results vary from -8 percent at $L/a = 0.20$ to -1 percent at $L/a = 0.03$. The three-dimensional results bracket the ASTM value, and centerline values are within ± 5 percent difference over the entire L/a range tested. The conclusion here is that there is an optimum L/a ratio; however, it is problem and mesh dependent. From the authors' experience on a large number of two- and three-dimensional analyses, it is safe to say that, other conditions being met as discussed below, engineering accuracy is assured with L/a ratios in the 0.05 to 0.15 range. Although the automatic remeshing algorithm in FEFAP does not include the L/a ratio as a constraint parameter, the user can interactively modify the size of crack tip elements after remeshing by dragging nodes

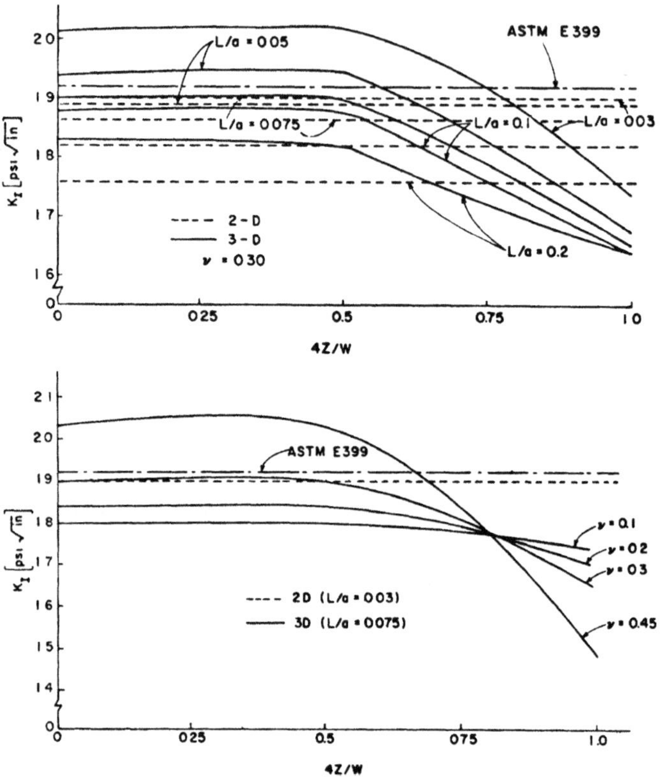

Fig. 4.9. Results from meshes shown in Figure 4.8 compared to ASTM standard calibration [25].

and changing common diagonal orientation of adjacent triangular elements.

A second mesh characteristic which can strongly influence accuracy is the number of singular elements positioned around the crack tip. More specifically, the maximum circumferential angle subtended by any of these elements must be limited to less than about 60 degrees, with 45 degrees seeming to be an optimal choice. The reason for this is that the theoretical circumferential variation in displacement is trigonometric and has a number of inflection points; the quarter-point elements are piecewise quadratic in their approximation to this variation. Take the three-point bend problem of Figure 4.8 and the results shown in Figure 4.9, for example. If the number of singular elements around the crack tip were reduced from 8 to 6 (4 to 3 using the symmetry of the problem), the error for any given value of L/a would double, approximately. Although the result might still be acceptable for small L/a, one would, of course, rather use the largest L/a practicable so that the mesh is minimally disturbed by the crack tip. The remeshing algorithm in FEFAP automati-

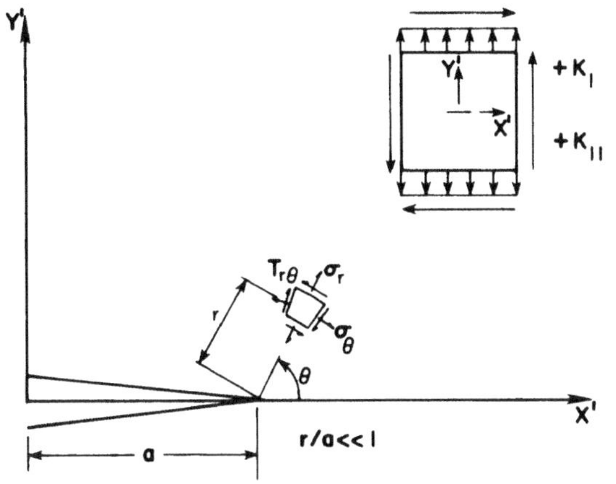

Fig. 4.10. Crack-tip stress components.

cally varies the number of elements around the crack tip as a function of a user specified subtended angle.

With proper care in near-tip meshing with quarter-point modified elements, Equations (4.10) and their three-dimensional generalizations can easily produce engineering accuracy in stress intensity factor calculations. No special program or element is required.

Mixed-mode interaction formulae. As outlined above, the next step after calculation of stress intensity factors is the assessment of crack stability and the angle of incipient propagation. This is done using a mixed-mode fracture initiation theory. Details of and comparisons among a number of such theories are given in references [26–28]. Only final formulations for two theories will be discussed here.

The first of these was formulated by Erdogan and Sih [29]. The parameter governing fracture initiation in their theory is the maximum circumferential tensile stress, $(\sigma_\theta)_{max}$, near the crack tip.

Given a crack under mixed-mode conditions, the stress state near its tip can be expressed in polar coordinates as,

$$\sigma_r = \frac{1}{\sqrt{2\pi r}} \cos\frac{\theta}{2}\left[K_I\left(1+\sin^2\frac{\theta}{2}\right) + \tfrac{3}{2}K_{II}\sin\theta - 2K_{II}\tan\frac{\theta}{2}\right] + \ldots$$

$$\sigma_\theta = \frac{1}{\sqrt{2\pi r}} \cos\frac{\theta}{2}\left[K_I\cos^2\frac{\theta}{2} - \tfrac{3}{2}K_{II}\sin\theta\right] + \ldots \tag{4.11}$$

$$\tau_{r\theta} = \frac{1}{\sqrt{2\pi r}} \cos\frac{\theta}{2}\left[K_I\sin\theta + K_{II}(3\cos\theta - 1)\right] + \ldots$$

These stress components are shown in Figure 4.10. The $(\sigma_\theta)_{max}$ theory states that:

(1) Crack extension starts at the crack tip and in a radial direction.
(2) Crack extension starts in a plane normal to the direction of greatest tension, i.e., at θ_0 such that $\tau_{r\theta} = 0$.
(3) Crack extension begins when $(\sigma_\theta)_{max}$ reaches a critical, material constant value.

The theory is stated mathematically using Equations (4.11).

$$\sigma_\theta\sqrt{2\pi r} = \text{constant} = \cos\frac{\theta_0}{2}\left[K_I\cos^2\frac{\theta_0}{2} - \tfrac{3}{2}K_{II}\sin\theta_0\right] = K_{Ic} \tag{4.12}$$

or

$$1 = \cos\frac{\theta_0}{2}\left[\frac{K_I}{K_{Ic}}\cos^2\frac{\theta_0}{2} - \tfrac{3}{2}\frac{K_{II}}{K_{Ic}}\sin\theta_0\right] \tag{4.13}$$

and,

$$\tau_{r\theta} = 0 \Rightarrow \cos\frac{\theta_0}{2}\left[K_I\sin\theta_0 + K_{II}(3\cos\theta_0 - 1)\right] = 0. \tag{4.14}$$

Equations (4.13) and (4.14) are the parametric equations of a general fracture initiation locus in the $K_I - K_{II}$ plane, shown in Figure 4.11. Also, the direction of the initial fracture increment, θ_0, can be found from Equation (4.14) which gives,

$$\theta_0 = \pm\pi \text{ (trivial)}$$

$$K_I\sin\theta_0 + K_{II}(3\cos\theta_0 - 1) = 0. \tag{4.15}$$

In summary, the governing equations of the $(\sigma_\theta)_{max}$ theory are (4.13) and (4.15). Algorithmically, the stress intensity factors for a given crack tip location and loading are first substituted into Equation (4.15) to obtain the new angle of propagation, θ_0. The stress intensity factors and the angle θ_0 are then substituted into Equation (4.13). If it is not satisfied, the stress intensity factor pair plots either within or outside the fracture locus shown in Figure 4.11. If within, then that crack cannot propagate without a sufficient increase in stress intensity factors. If outside, then the crack is unstable and can continue to propagate until it reaches a free surface or until the stress intensity factor pair returns to within the locus.

The second of the mixed-mode fracture initiation theories to be discussed was formulated by Sih [30–33]. The parameter governing fracture initiation in his theory is the strain energy density near the point of initiation.

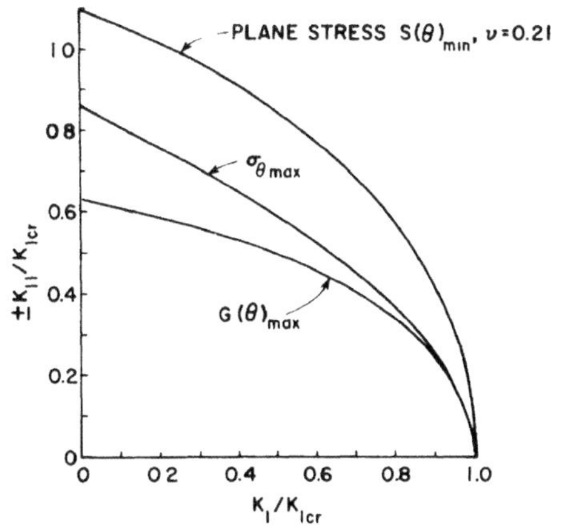

Fig. 4.11. Mixed-mode fracture initiation loci.

Sih has shown [31] that the strain energy density variation at a distance r from a crack tip is,

$$\frac{\partial U}{\partial V} = \frac{1}{r}\left(\frac{a_{11}K_I^2 + 2a_{12}K_IK_{II} + a_{22}K_{II}^2}{\pi}\right) \tag{4.16}$$

where,

$$a_{11} = \frac{1}{16G}[(1+\cos\theta)(\kappa - \cos\theta)]$$

$$a_{12} = \frac{\sin\theta}{16G}[2\cos\theta - (\kappa - 1)] \tag{4.17}$$

$$a_{22} = \frac{1}{16G}[(\kappa+1)(1-\cos\theta) + (1+\cos\theta)(3\cos\theta - 1)].$$

If the quantity in parentheses in Equation (4.16) is called S, i.e.,

$$\frac{\partial U}{\partial V} = \frac{S}{r_0}, \tag{4.18}$$

it can be seen that, at a constant value of r, S is the varying intensity of the strain energy density around the crack tip.

The $[S(\theta)]_{min}$ theory proposes that:

(1) Crack extension occurs in the direction along which $\partial U/\partial V$ pos-

sesses a minimum value, i.e., θ_0 such that,

$$\frac{\partial S}{\partial \theta} = 0 \qquad \frac{\partial^2 S}{\partial \theta^2} \geqq 0. \tag{4.19}$$

(2) Crack extension occurs when $S(\theta_0)$ reaches a critical, material value, S_c.

(3) $S(\theta)$ is evaluated along a contour $r = r_0$, where r_0 is a material constant. By combining the condition statistics (2) and (3) it can be seen that,

$$\left(\frac{\partial U}{\partial V}\right)_c = \frac{S_c}{r_0}. \tag{4.20}$$

That is, specifying S_c and r_0 to be material constants is equivalent to specifying a material critical strain energy density.

A relationship between S_c and fracture toughness can be obtained in the following manner. For $K_{II} = 0$, Equation (4.19) predicts that $\theta_0 = 0°$. Then, for $K_I = K_{Ic}$, equating Equation (4.20) to Equation (4.16) yields,

$$S_c = \frac{(\kappa - 1) K_{Ic}^2}{8\pi G} \tag{4.21}$$

A fracture initiation locus in the K_I v.s. K_{II} plane can then be obtained from,

$$\frac{S_c}{r_0} = \frac{1}{\pi r_0}\left(a_{11} K_I^2 + 2a_{12} K_I K_{II} + a_{22} K_{II}^2\right)$$

or

$$1 = \frac{8G}{(\kappa - 1)}\left[a_{11}\left(\frac{K_I}{K_{Ic}}\right)^2 + 2a_{12}\frac{K_I K_{II}}{K_{Ic}} + a_{22}\left(\frac{K_{II}}{K_{Ic}}\right)^2\right] \tag{4.22}$$

where θ_0 is obtained from the conditions Equation (4.19).

Although the direction $\theta_0 = 0°$ for fracture initiation in pure Mode I is not elasticity constant dependent, θ_0 for all other cases is a function of Poisson's ratio in the $[S(\theta)]_{min}$ theory.

Again, to summarize, Equations (4.19) and (4.22) govern the $[S(\theta)]_{min}$ theory. The algorithm begins with satisfaction of Equation (4.19) to obtain θ_0. It and stress intensity factors are then substituted into Equation (4.22). The crack stability implications are then the same as outlined in the section describing the $(\sigma_\theta)_{max}$ theory.

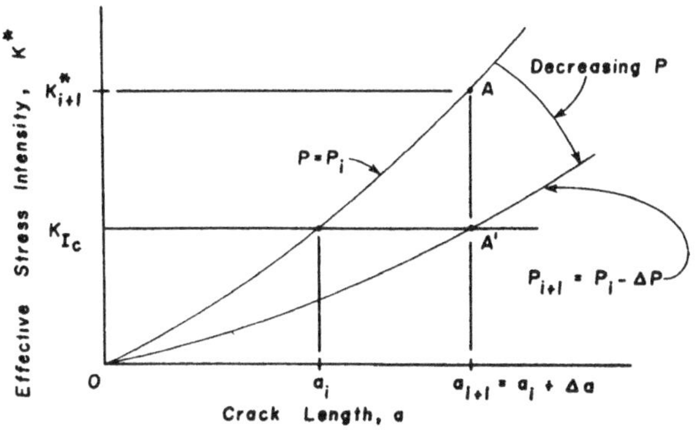

Fig. 4.12. Stress intensity factor variation for Case 1.

Of course, in a quasi-static crack propagation analysis the governing equations for one of the theories would be applied at the end of each growth step or load step for each crack tip. It may not be necessary to increase loads to bring the stress intensity factors of a previously stable crack tip onto the fracture locus. The propagation of another crack may cause the same effect. Algorithmically, this implies that the interaction factor for each crack tip, the right-hand-side of Equation (4.13) or (4.22), be updated in memory after each crack or load increment. Depending on the mode of interaction between the program and the user, the former or the latter will use the interaction factors to decide which one or more of the crack tips should be propagated in a given fracture step.

Predicting crack increment length. The third and last step in the crack propagation algorithm according to the linear model is prediction of the load change required to drive a crack a specified distance. Alternatively, a load change may be specified and the corresponding crack increment sought.

The fundamental principle here is that a crack, once initiated, will continue to propagate as long as there is sufficient energy or, equivalently, effective stress intensity, available. Effective stress intensity, K^*, here refers to a mixed-mode case and is the combination of Mode I, II, and III stress intensity factors required by the particular mixed-mode theory in use. The right-hand sides of Equations (4.13) and (4.22) can, therefore, be viewed as normalized effective stress intensity factors.

A number of possible stability cases must be considered in creating an algorithm for predicting crack increment length. Some examples will be considered here with reference to Figures 4.12 and 4.13. Assume first that propagation along some predicted direction θ_0 is being investigated.

Case 1. Effective stress intensity increases monotonically with crack length, curve OA in Figure 4.12.

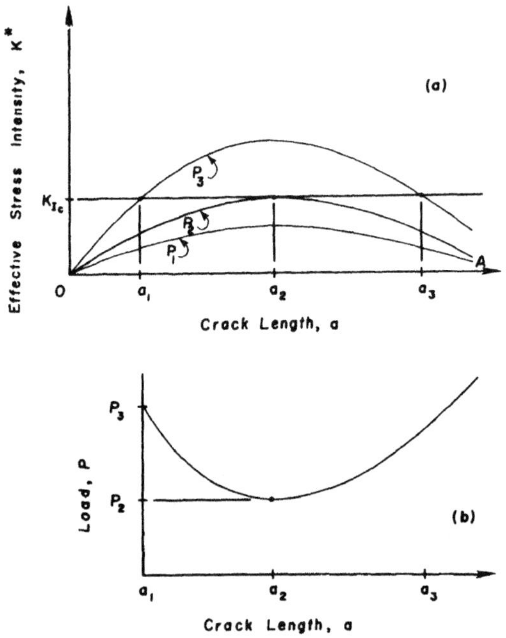

Fig. 4.13. Case 2. (a) Stress intensity factor variation; (b) Load variation.

If the initial crack length is less than a_i, no propagation occurs. For $a = a_i$, propagation can occur and it will continue at $P = P_i$; that is, a condition for local instability has been met. Of course, an algorithm could be written which would place such a scenario in displacement or crack-length control: A crack increment, Δa, could be specified and the load decrement required to just bring the crack tip to $a_i + \Delta a$ could be computed. This situation is depicted by curve OA′ in Figure 4.13. To compute P_{i+1} recall that LEFM specifies that at instability,

$$K_{\mathrm{Ic}} = \alpha_i P_i \sqrt{a_i} = \alpha_{i+1} P_{i+1} \sqrt{a_{i+1}} \tag{4.23}$$

where the factor α depends on geometry and interaction theory. Therefore,

$$P_{i+1} = \frac{\alpha_i}{\alpha_{i+1}} P_i \sqrt{\frac{a_i}{a_{i+1}}} \tag{4.24}$$

Equation (4.24) is only directly useful, however, if the α_{i+1} coefficient is known at Step i. For arbitrary problems, this is certainly not the case. An alternative is to propagate the fracture an amount Δa (in the direction θ_0)

and compute K^*_{i+1} at load level P_i. The new load level is then,

$$P_{i+1} = \left(\frac{K_{\text{Ic}}}{K^*_{i+1}}\right) P_i \tag{4.25}$$

as can be seen in Figure 4.13.

Case 2: Effective stress intensity increases, reaches a maximum value, and then decreases with increasing crack length, curve OA in Figure 13a. For the value of K_{Ic} shown and at load level P_1, no crack propagation is possible. At load level P_2, propagation is possible only at crack length $a = a_2$, but the corresponding, theoretical crack increment length is $\Delta a = 0$. At load level P_3, propagation can occur for a crack of length a_1, and it would be unstable in load control. Again, as in Case 1, above, by using a crack length or displacement control algorithm the crack of initial length a_1 could be propagated stably to length a_2 by decreasing the load incrementally from level P_3 to level P_2, as shown in Figure 4.13b.

For crack lengths longer than a_2, crack propagation is stable in the load control sense. An effective stress intensity monotonically decreasing with increasing crack length implies that a monotonically increasing load is required for continued propagation. In Figure 4.13, it can be seen that if the load is again increased to P_3 propagation to crack length a_3 is possible.

If one is starting with crack length a_1, and load level P_3 (Figure 4.13), the prediction technique is the same as described under Case 1: Propagate the crack an amount Δa in the direction θ_0 at load P_3, compute the effective stress intensity for the new crack length, and apply Equation (4.25).

Suppose, however, that one is at load level P_2 and crack length a_2. The same algorithm can still be used: the only difference is that the quantity in parentheses in Equation (4.26) will now always be less than one.

The reverse of Case 2 is also possible: Effective stress intensity can at first decrease and then begin to increase with increasing crack length. (See, for example, References [34] and [35].) This implies nothing new algorithmically, however, since the implications of this situation are handled by the techniques described in Cases 1 and 2.

A number of alternative numerical techniques for crack increment length prediction are available [12,28,36]. Some are based on energy balance, some are more approximate than others. The simple technique described here is theoretically exact for pure Mode I, colinear propagation. However, any numerical technique which employs finite, straight crack increments to model curvilinear propagation will be approximate. One is updating effective stress intensity incrementally, rather than continuously. Stress intensity factors and angle changes will be somewhat

Fig. 4.14. View of downstream face of Fontana Dam. Crack seen as discontinuity in water flow in right foreground.

in error. The error depends on the specified length of the fracture increment.

The analogy here is with dynamic analysis where the time step controls accuracy and stability of the solution. It is the authors' experience with their codes that predicted trajectories sometimes oscillate about an average path. This is a manifestation of error in K_{II} simulation which is a result of "kinking" the crack path rather than allowing it to continuously curve. Spuriously high K_{II} values are computed which, alternating in sign with each increments, zig-zag the crack. However, it is quite possible that if too large an increment is used divergence of predicted trajectory could occur.

All the theoretical ingredients for crack propagation modeling under mixed-mode, LEFM assumptions have now been presented. These are combined in the following example problem.

Linear model example problems. In 1972 a large crack was discovered on the downstream face of the Tennessee Valley Authority's (TVA) Fontana Dam [37], Figure 4.14. The crack was observed to intersect a drainage gallery and continue beyond the gallery for an unknown distance, Figure 4.15. A case study [38] was performed on Fontana Dam to confirm the supposed crack mechanism, to compute the approximate time of occur-

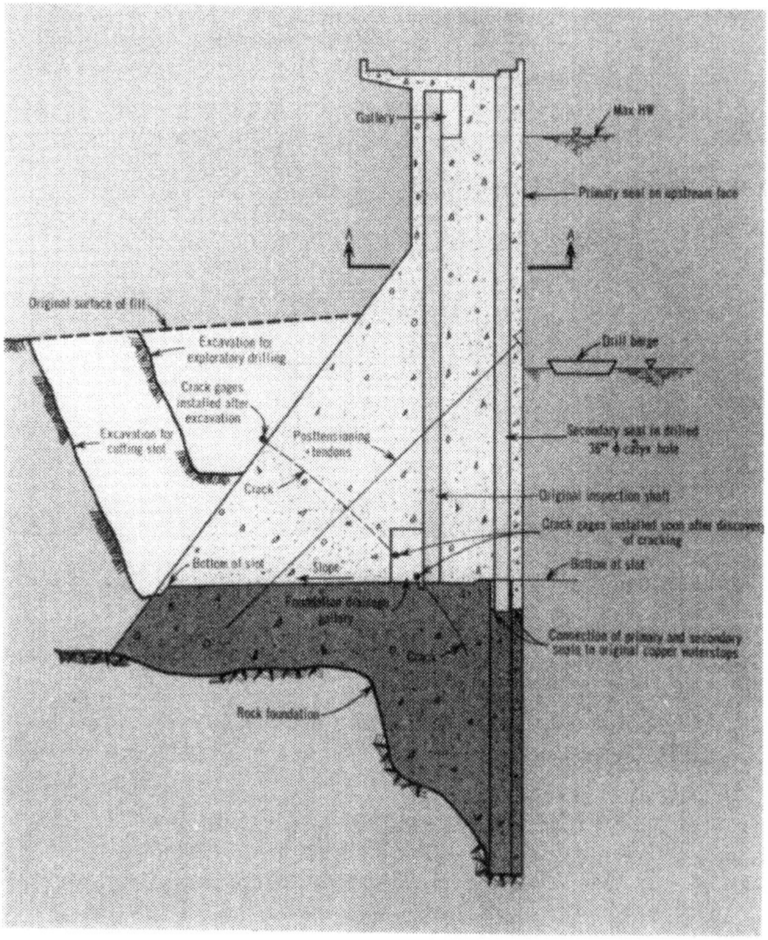

Fig. 4.15. Cross-section through typical cracked block [37].

rence of initiation, to evaluate remedial measures performed by the TVA, and to predict the current extent and location of the crack.

A three-dimensional finite element analysis of the entire dam was first performed. The mesh for this analysis, a part of which is shown in Figure 4.16, was generated in a few hours using a preprocessor developed at the Cornell University Program for Computer Graphics [10]. This analysis confirmed that the crack mechanism was a combination of thermal expansion and thermally induced concrete growth.

The three-dimensional study generated boundary conditions for the two-dimensional finite element mesh shown in Figure 4.17. This mesh idealized a representative cross-section through the cracked portion of

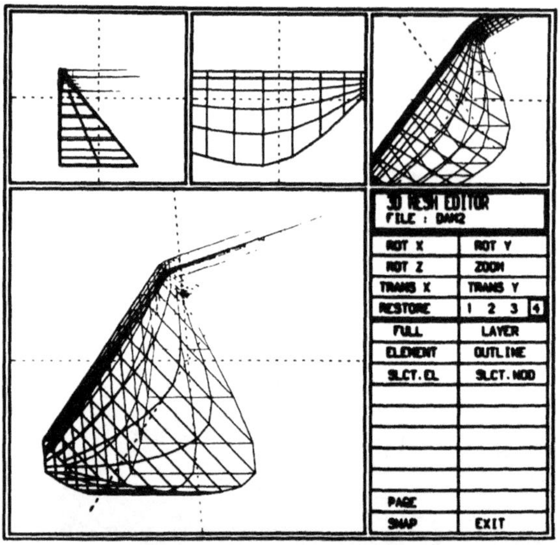

Fig. 4.16. Three-dimensional idealization of Fontana Dam generated with high-level interactive graphics [10].

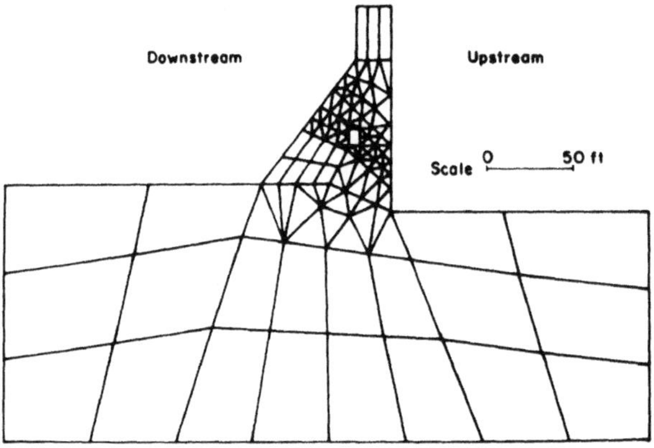

Fig. 4.17. Initial plane strain idealization of block in region of crack [38].

the dam. The three-dimensional analysis had predicted substantial tensile stresses over the middle third of the height of the dam's downstream face, indicating that a crack could have initiated anywhere in this region. Consequently, a crack was initiated in the two-dimensional mesh at a point in this region where initiation was observed to have occurred. A propagation analysis was then performed using the algorithms described above, with the following qualification. Since no information concerning

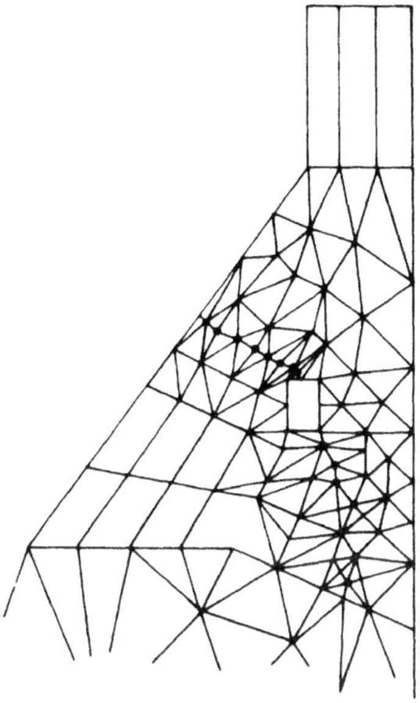

Fig. 4.18. Detail of final mesh after 15 increments of cracking [38].

the K_{Ic} of the dam's concrete was available, it was decided not to apply an interaction formula. Rather, the trajectory of the crack would be predicted and the corresponding stress-intensity factor history computed.

The crack was propagated in 15 increments. A detail of the final mesh is shown in Figure 4.18. A comparison between the predicted trajectory and known crack locations is shown in Figure 4.19, while Figure 4.20 shows the corresponding computed stress intensity factor variations. Table 4.1 presents a comparison of angle-change predictions for three mixed-mode interaction theories. The trajectory shown in Figure 4.18 was obtained using the angle-change predictions of a maximum energy release rate, $G(\theta)_{max}$, formulation [39]. Table 4.1 shows that trajectories predicted by the two theories discussed in a previous section would not differ appreciably from that shown in Figure 4.18.

Figure 4.20 clearly shows the K_{II} oscillations previously mentioned. Note also that the average oscillation amplitude grew when crack increment length increased from about 3 feet (0.92 m) at Step 11 to about 5 feet (1.53 m) thereafter.

Three of the four original objectives of this case study were met successfully. The three-dimensional analyses confirmed TVA consultants' suspicions that cyclic, reversible thermal expansion could not in itself have caused crack initiation. Additional irreversible deformation induced by temperature dependent volume expansion of the concrete was identified by the case study as the most likely cause of crack initiation. The

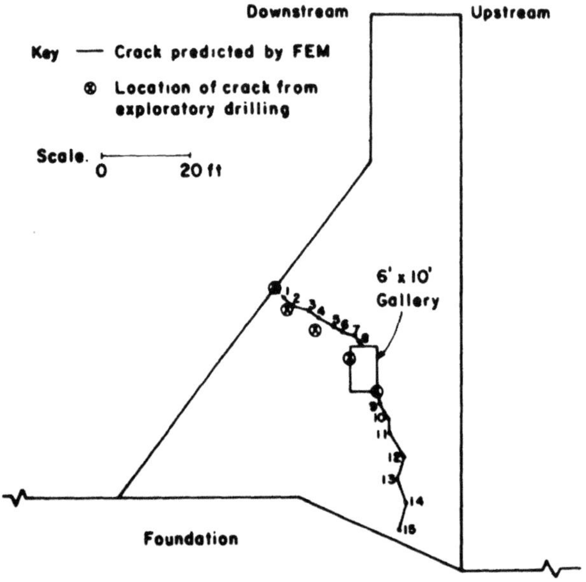

Fig. 4.19. Comparison of predicted and observed crack trajectories [38].

TABLE 4.1.
Comparison of angle change predictions.

Crack increment	Δa (ft.)	$[G(\theta)]_{max}$ (deg.)	$[S(\theta)]_{min}$ (deg.)	$(\sigma_\theta)_{max}$ (deg.)
1	3.25	−8	−8	−8
2	3.	+9	+9	+9
3	3.	−11	−11	−11
4	3.	+12	+12	+12
5	3.	−12	−12	−12
6	3.	+20	+18	+20
7	3.	−28	−25	−29
8	3.	+2	+2	+2
9	3.19	+12	+12	+12
10	3.	−28	−25	−32
11	3.	+35	+29	+37
12	6.	−18	−18	−18
13	5.	+30	+24	+32
14	5.	−30	−25	−31
15	5.	+9	+9	+9

three-dimensional analyses also produced a correlation between observed dam displacement and computed tensile stresses in the region of crack initiation. It was found that one inch of displacement at the crest of the

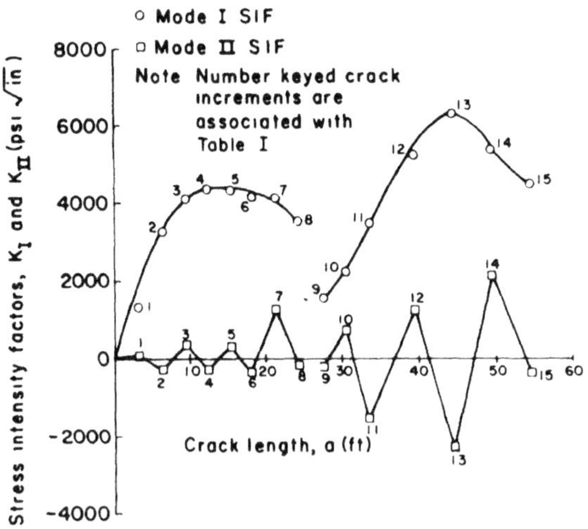

Fig. 4.20. Computed stress-intensity factors for each crack tip location [38].

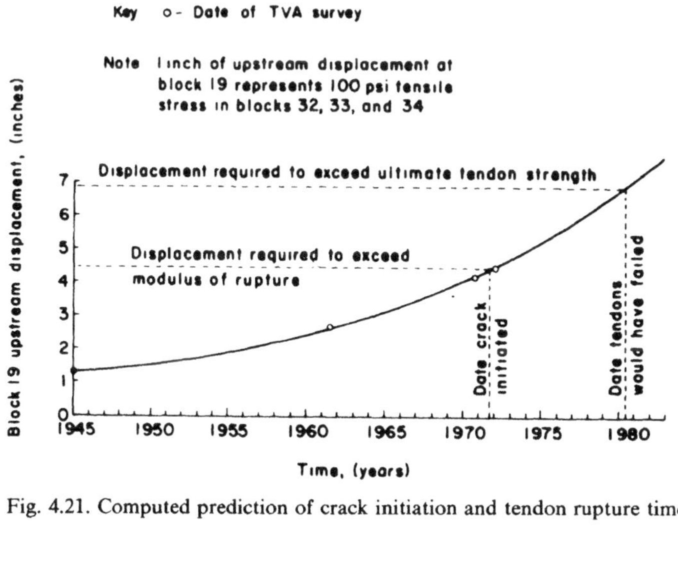

Fig. 4.21. Computed prediction of crack initiation and tendon rupture times [38].

dam at its midspan in the finite element analysis (Block 19) resulted in a tensile stress of about 100 psi (690 kPa) in the crack region (Blocks 32, 33, 34). This information was combined with TVA survey results to produce Figure 4.21 which shows a predicted crack initiation late in the Fall of 1971. This date is between the time of discovery, Summer of 1972, and the previous inspection date in 1967.

The TVA used two interim measures to stabilize the cracking monoliths, pressure grouting of the crack, and installation of post-tensioned tendons bridging the crack. The two-dimensional analysis with the mesh shown in Figure 4.18 predicted negligible change in stress-intensity factors from those two measures; that is, neither measure hindered or aided crack growth. In fact, had the TVA not used another measure to isolate the cracking monoliths from the crack driving forces, it was predicted that the tendons would have ruptured in about 1981, Figure 4.21.

Finally, efforts to predict the current extent of the crack were only partially successful for two reasons. First, a small error was discovered late in the case study in the thermal loading case effects on the dam's foundation. This resulted in probably incorrect stress intensity factors and trajectory for the last three or four crack steps. It is for this reason that the crack propagation study was not continued up to and into the foundation. Second, the trajectory prediction was for only one cross-section through a crack that had a front length of as much as 150 feet (45.8 m) and was propagating through a structure with changing geometry along this length. Clearly, a fully three-dimensional crack propagation analysis was warranted in this case. However, the capability for such an analysis using three-dimensional analogues of the crack propagation and remeshing algorithms previously described is only now becoming a possibility.

The Fontana Dam case study was performed before the completion of FEFAP. This program completely integrated the linear fracture model and automatic remeshing algorithms in a user-friendly, interactive-adaptive graphics package. Before proceeding to an example use of FEFAP, certain of its characteristics will be described.

The first of these is FEFAP's element library. Concrete is modeled by quadrilateral or triangular quadratic isoparametric elements. Main steel reinforcement is modeled by quadrilateral elements which provide both axial and bending stiffness. Three-noded linear elements are used for shear reinforcement. Bond between the concrete and steel and aggregate interlock along cracks are modeled by an isoparametric version of the interface element developed by Goodman et al. [40].

Although originally written for materially linear analyses, the code has been modified to account for concrete nonlinearity. A simple model which would encompass both the biaxial and triaxial state of stress (for plane strain and axisymmetric analyses) for the failure criterion and the prediction of the secant Young's modulus and Poisson's ratio was selected and implemented. The secant values are needed because load is always applied in a single increment. The selected failure criterion, proposed by Ottosen [41,42], is based on a four-parameter equation containing explicitly all the three stress invariants. The steel model is elastic-perfectly plastic, while steel-concrete bond is governed by the equation developed by Nilson [43]. Finally, the aggregate interlock model requires special comment. It has been recognized that the shear stiffness along a crack is a portion, α, of the initial uncracked shear modulus. Theoretically, α should be inversely proportional to the crack opening, but such a relative displacement is not directly derived in the general smeared cracked model. The present model takes advantage of the discrete nature of the crack to directly compute the crack opening and accurately evaluate the shear stiffness along the crack using the empirical equation derived by Fenwick and Paulay [44].

To model discrete crack propagation in plane and axisymmetric structures, the following capabilities were implemented:

(1) An automatic element and node generation capability for regular meshes.
(2) An automatic nodal adjustment for the singular elements, and direct extraction of the stress intensity factors.
(3) A graphical display capability on a TEKTRONIX 4010 series terminal.
(4) Forms of loading: a) nodal, b) edge, c) initial nodal displacement, d) gravity, and e) thermal.
(5) A mesh optimizer for bandwidth minimization and direct nodal renumbering [13].
(6) A higher efficient In/Out of core skyline-banded equation solver [45].
(7) A completely interactive means of operation within the program.
(8) Automatic, discrete crack nucleation at arbitrary points and angles on an edge or in the interior of a domain, as specified by the analysis.
(9) Automatic, discrete crack propagation capability with mesh adjustment along the propagating crack.
(10) User interaction with the code to perform interactively final minor adjustments to each regenerated mesh.

The distinguishing characteristic of FEFAP is a set of routines which will control the mesh modification caused by an arbitrary extension of a crack. The general strategy followed for the crack extension through a finite element mesh is as follows:

(1) From the initial direction of the crack axis and the predicted angle of crack extension with respect to the crack direction determine the angle of crack propagation in global coordinates.
(2) Replace the quarter-point nodes to their initial midside position, to remove the local singularity.
(3) Define a new crack tip node whose coordinates are determined from the length and angle of crack extension.
(4) Define a new node adjacent to the old crack tip node.

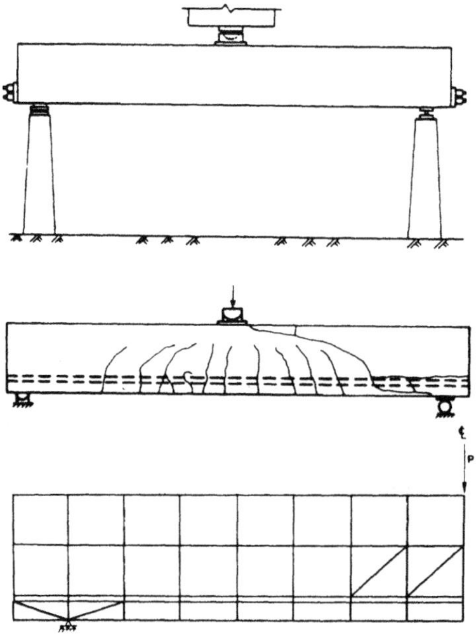

Fig. 4.22. Example analysis of reinforced concrete beam. (top) Experimental arrangement, from [46]; (center) Crack pattern observed on one face of beam at failure; (bottom) Initial idealization [47].

(5) Search the previous singular elements, to determine which one is going to be crossed by the crack.
(6) If the new crack tip node falls inside this element, extend the crack to it and go to 8, otherwise simply extend the crack through the entire length.
(7) Locate the next element to be crossed by the crack and go to 6.
(8) Define the new nodes from which the stress intensity factors will be evaluated.
(9) Adjust the midside nodes to the quarter-point position, where needed.
(10) Display the modified mesh, to allow the user to interactively perform final adjustments.
(11) Compute the stiffness matrices of those elements perturbed or newly created by the mesh modification.

The analysis of the beam shown in Figure 4.22a will be used as an example problem. Although it has been shown that the response of such a structure is largely independent of cracking model and its smeared or discrete implementation, this example will demonstrate the characteristic capabilities of FEFAP. The example beam is Specimen OA-1 of the test series of Bresler and Bertero [46]. A detailed description of the beam and

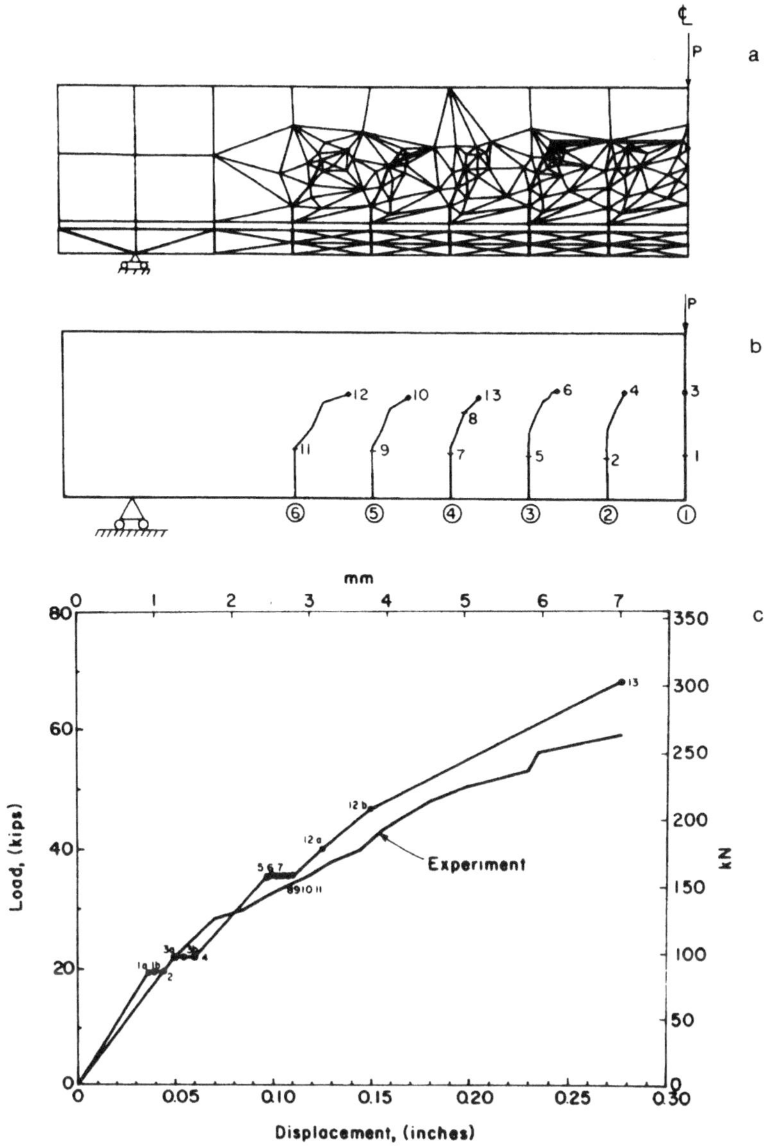

Fig. 4.23. Analysis results. a) Final mesh; b) Predicted crack trajectories; c) Comparison with experimental load-deflection curve [47].

of a range of analyses performed in a parameter study using FEFAP is given in [12,47].

The particular analysis described here included the nonlinear concrete constitutive model, shear transfer along the crack, and a supposed concrete K_{Ic} of 600 psi in. (660 $MNm^{-3/2}$). The initial mesh is shown in

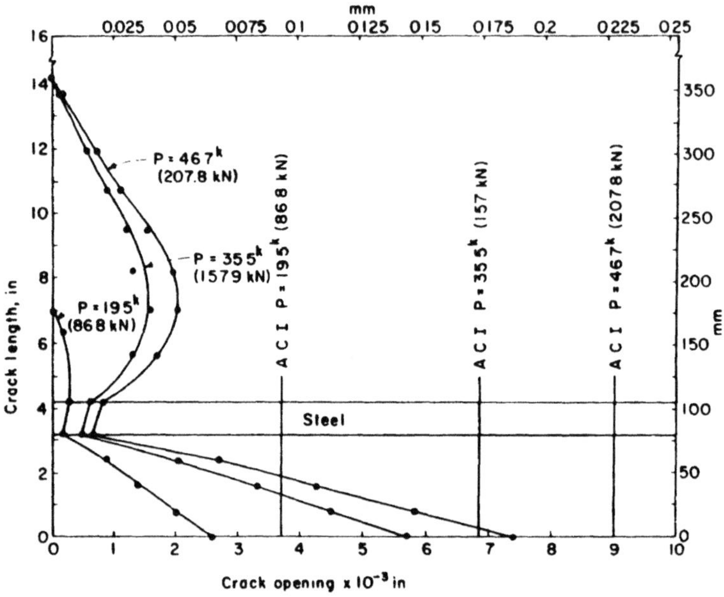

Fig. 4.24. Computed COD profile of centerline crack [47].

Figure 4.22c, and it includes interface elements along both sides of the steel layer to model bond. Figure 4.23 presents results in the form of a final mesh, predicted trajectories for six cracks (compare to observed pattern shown in Figure 4.22b), and a comparison between predicted and observed load-deflection curves. An additional result which could not be produced directly by a smeared crack analysis is shown in Figure 4.24. Here the COD profile of the center-line crack at various load levels is shown. The arresting effect of the steel and bond slip at the crack plane are clearly shown. The COD at the bottom face predicted by the ACI 318-77 Code is also shown for comparison.

4.4 The nonlinear model

The apparent reason for the inapplicability in many cases of linear elastic fracture mechanics to cracking in concrete lies in the nature of the process zone. Theoretically, a K_{Ic} approach is valid if the size of the process zone is small compared to the size of the area in which the singular term in the theoretical stress distribution is the dominant contribution to the stress field. This situation is investigated using the example problem shown in Figure 4.25a. Figures 4.25a, b, and c show the theoretical singular term contribution to the stress component normal to the crack plane at incipient instability for three materials in the config-

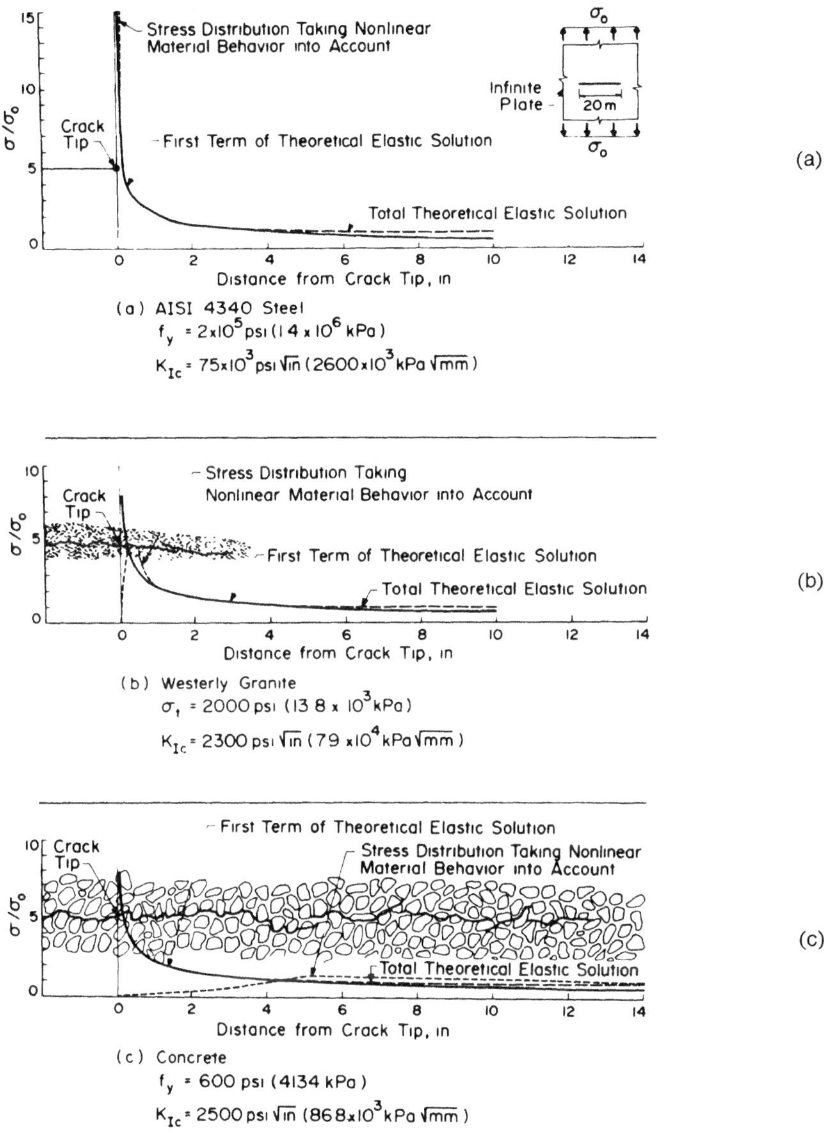

Fig. 4.25. Theoretical crack tip stress distribution with effect of process zone formation. (top) Steel; (center) Rock; (bottom) Concrete.

uration shown in Figure 4.25a. Also shown are the approximate process zone sizes and their first order effect on stress redistribution.

Clearly, for the metal example shown in Figure 4.25a, a K_{Ic} approach to fracture initiation would be valid. A K_{Ic} approach also has been shown to be applicable to many rocks. Figure 4.25b shows that while the process zone is much larger in this case, the singular term still dominates.

A number of recent investigations [16,17,18,21] have conclusively shown that these example situations are quite different from that which occurs in concrete and even mortar. Figure 4.25c qualitatively represents process zone length in concrete as implied by the testing of Catalano [16] and Petersson [18]. In their tests on unreinforced concrete beams stresses were still being carried over what appeared to be a visible crack. This confirmed previous experimental observations by Evans and Marathe [48], in that what appeared as a crack on a particular viewing surface of a specimen was in fact a "fictitious" [21] crack in the fracture mechanics sense. Even though a discrete, visible crack was seen to exist on the specimen surface, this crack was not fully formed through the specimen thickness. The surface dimpling that is a manifestation of flow in the process zone of a metal is replaced by microcracking due to relatively high volumetric stresses in the process zone in concrete. If the crack tip is viewed as the point behind which no stress is transmitted, the process zone in Figure 4.25c is seen to extend a relatively large distance in front of the crack tip. Further, at the scale of the example problem it is obvious that the singular stress term has little to do with the stress at the tip of the crack.

In a metal the stress state in the process zone depends on the constitutive relationship past yield. Similarly, the stress state in the process zone in concrete is presumed to depend on a post-peak stress-strain relationship in tension. Such relationships have been measured [48,18] and have been incorporated into the process zone model qualitatively shown in Figure 4.25c and implemented into FEFAP [19,20].

Stress v.s. COD models. It has been observed in direct tension tests [18,21] that surface microcracks, though they have not yet coalesced into a continuous, through-thickness crack, can give rise to an effective crack opening displacement. Apparently, a single, continuous crack occurs when the total effective crack opening displacement reaches a characteristic value. For typical concretes this value appears to be about 0.01 inch (0.025 mm) [18,16]. Even after this amount of opening, however, some stress continues to be transmitted across the crack, and in direct tensile tests of specimens characteristic stress versus COD relationships have been measured [18]. Such relationships form the basis of the discrete nonlinear cracking model developed by Gerstle and the first author [19,20].

Some typical relationships are shown in Figure 4.26. They are implemented into FEFAP in the following manner. During the automatic remeshing associated with each crack increment, the program inserts interface elements into the crack. These elements [49,6] prohibit overlapping of the crack sides while allowing their relative opening and sliding.

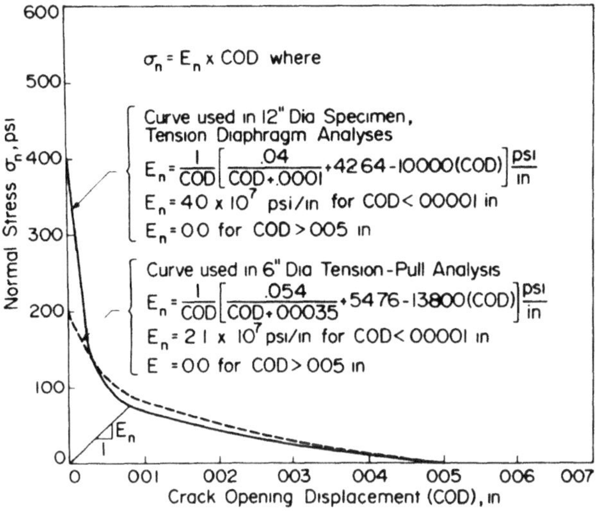

Fig. 4.26. Typical normal stress versus COD relationships for discrete process zone in concrete [19].

The element stiffness relationship is of the form:

$$\begin{pmatrix} \sigma_N \\ \sigma_S \end{pmatrix} = \begin{bmatrix} E_N & 0 \\ 0 & E_S \end{bmatrix} \begin{pmatrix} \text{COD} \\ \text{CSD} \end{pmatrix} \tag{4.26}$$

where σ_N is the stress normal to the crack, σ_S the shear stress along the crack, E_N the element normal stiffness, and E_S the element shear stiffness. The elements are isoparametric and of quadratic order and can be made to possess variable normal and shear stiffness along their length. In FEFAP the secant normal stiffness, E_N, at any point along the crack is assigned by a relationship such as shown in Figure 4.26. The COD computed in a previous analysis step determines the secant E_N and, consequently, the process zone normal stress, σ_N. The shear stress transferred across the crack is modeled in a similar manner using the relationship developed by Fenwick and Paulay [44]. In that relationship the transferred shear stress varies nonlinearly with COD and linearly with CSD as shown qualitatively in Figure 4.27.

A nonlinear crack propagation algorithm. Nonlinear, mixed-mode (both shear and normal stress transfer in the process zone) crack propagation modeling using FEFAP is accomplished through the use of interface

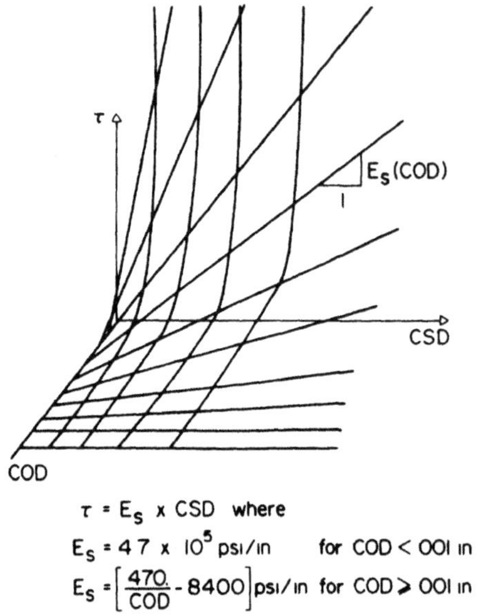

Fig. 4.27. Typical shear stress versus COD and CSD relationship for process zone in concrete.

elements which reproduce one of the characteristic stress-versus-COD models. However, the singular elements used in linear analysis are also used. These elements are still automatically placed at the tip of the process zone, that is, at the tip of the fictitious crack. There are two reasons for their continued use. One wants to be able, in a single analysis, to make the transition from the nonlinear model involving development of a process zone to a linear analysis in which the process zone length becomes a negligible fraction of the total crack length. This transition is facilitated if use of the singular elements is maintained in the automatic remeshing algorithms. The second reason, described below, involves a vestigial use of the stress intensity factors computed from the singular element nodal displacements.

Theoretically, the stress intensity factors computed at a fictitious crack tip should always be zero. A nonzero stress intensity factor would indicate that the crack should actually have propagated beyond the existing fictitious crack tip. The second requirement in the nonlinear crack propagation algorithm is that the stress-versus-COD constitutive models, Figures 4.26 and 4.27, be satisfied. These requirements are satisfied in the present nonlinear analysis option of FEFAP in an iterative manner.

In a crack increment and given the typical local mesh configuration shown in Figure 4.28a as a starting point the following steps are performed iteratively:

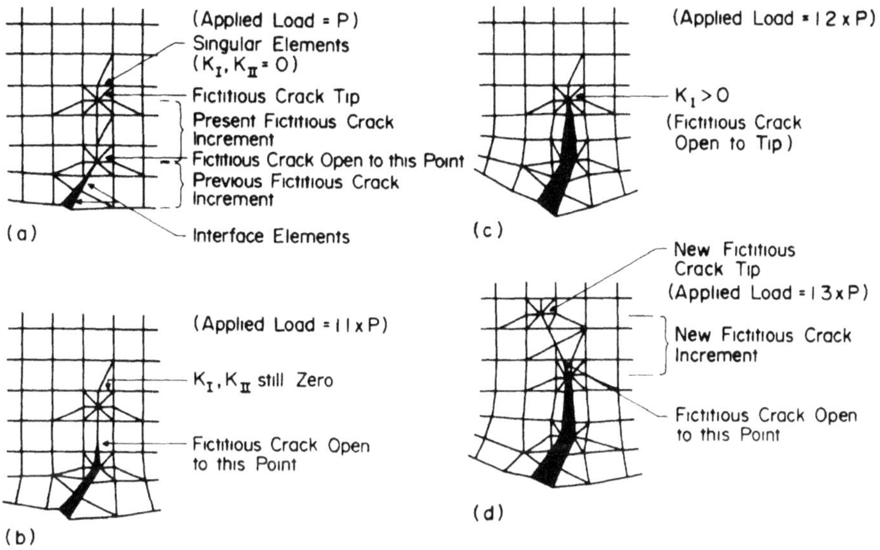

Fig. 4.28. Typical crack propagation step for nonlinear model.

(1) At the current load level iterations are performed until the interface model is satisfied to a specified tolerance along the entire process zone length.
(2) If K_{I} and K_{II} calculated at the assumed fictitious crack tip are nominally zero, then the fictitious crack has not yet opened along its entire length to its assumed tip. The load may be increased by the user, and interactions again performed, until convergence is achieved (see Figure 4.28b).
(3) Step 2 is repeated until, at a given load level, K_{I} and/or K_{II} become nonzero after convergence (see Figure 4.28c).
(4) Using one of the mixed-mode interaction equations (4.16) or (4.20), FEFAP automatically calculates the direction of propagation of the next crack increment based upon the nonzero (but small) value of K_{I} and K_{II}. FEFAP then automatically remeshes (see Figure 4.28d) to allow the fictitious crack to propagate further. The length of each fictitious crack propagation increment is prespecified by the user.
(5) The process is repeated for the next crack increment by repeating step 1 through 4.

The logic of the nonlinear analysis option of FEFAP is shown in the flow chart of Figure 4.29. The nonlinear option is highly interactive; after each iteration certain quantities are displayed on the terminal screen. These include K_{I} and K_{II} for each crack, the load factor, and values of preselected nodal displacements and reactions. In this way, the user can observe convergence trends. At any point he can decide to change the

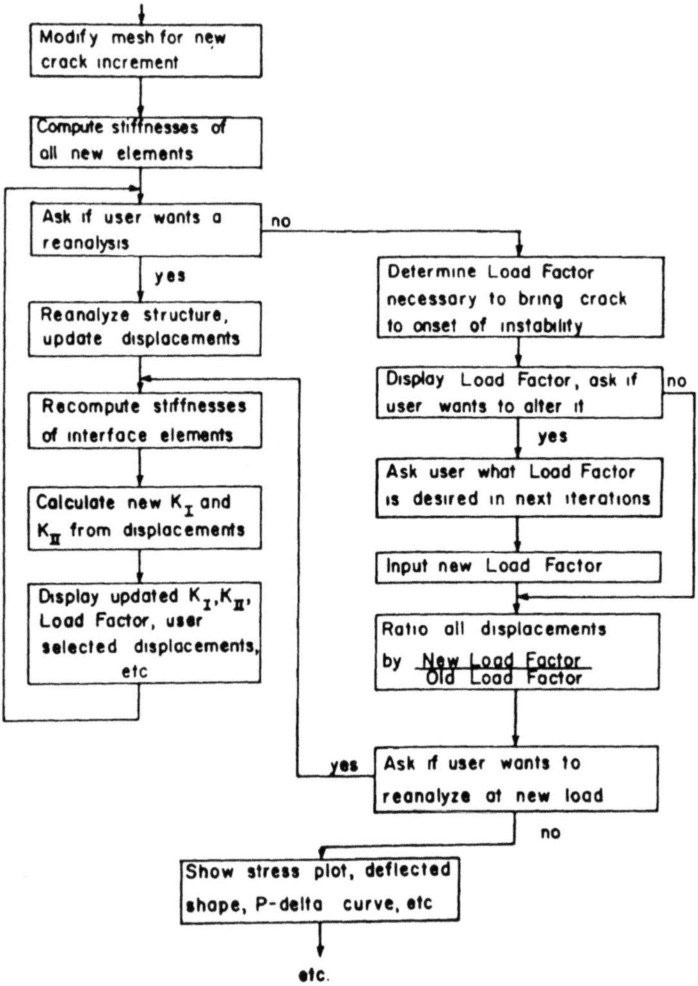

Fig. 4.29. Flow chart of algorithm for nonlinear crack propagation in FEFAP [19].

load with the present configuration, to reanalyze at the current load, or to continue on to the next crack increment. This adaptive flexibility, permitted by the use of interactive graphics, has proven invaluable in the types of problems to be exemplified next.

Example problems. Two problems involving discrete, nonlinear crack propagation modeling will be described. Although the problems differ by nearly two orders of magnitude in crack length scale, they will be seen, surprisingly, to be intimately related by the same cracking model.

The first example is an application to the micromechanics of bond slip

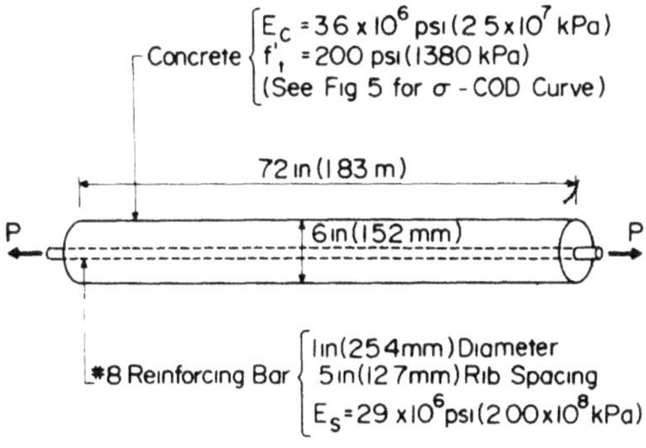

Fig. 4.30. Tension-pull specimen used by Broms and Raab [49].

by way of the structure shown in Figure 4.30. Many experiments have been performed on such "tension-pull specimen" configurations which consist of a single reinforcing bar surrounded by a concentric cylinder of concrete. The particular specimen shown in Figure 4.30 was tested by Broms and Raab [49].

The tension-pull specimens is a simplified model of the situation which occurs on the tension side of a reinforced concrete beam. Also, the tension-pull specimen can be thought of as a portion of a steel reinforced diaphragm under tension, or, under certain circumstances, as an element in a shear wall.

The present analysis was carried out with the use of the axisymmetric option in FEFAP. Because an axisymmetric analysis was performed it was impossible to model discrete longitudinal splitting cracks which sometimes occur in this specimen type. Only radial cracks were modeled. A portion of the specimen shown in Figure 4.30 was modeled using the mesh of Figure 4.31. Details of the idealization are described in References [19,20].

The cracking of the tension-pull specimen progressed as shown in Figure 4.31. The cracks shown are called "secondary cracks." Secondary cracks always nucleate at the steel and propagate radially outward from the steel. As load is applied to the specimen, the first secondary crack forms at a very low load at the location where the steel exits the concrete, as seen in Figure 4.31a. Each secondary crack was seen to form at the location where a rib bore on the concrete. This type of behavior has been observed by Goto [50], as seen in Figure 4.32. As the load was increased, secondary cracks propagated, and at the same time additional secondary cracks nucleated at ribs further from the end of the specimen, as seen in Figures 4.31b through 4.31d. At the load corresponding to Figure 4.31d, the tensile stress in the concrete at the midlength of the specimen reached the tensile strength of the concrete. At this point, a new primary crack

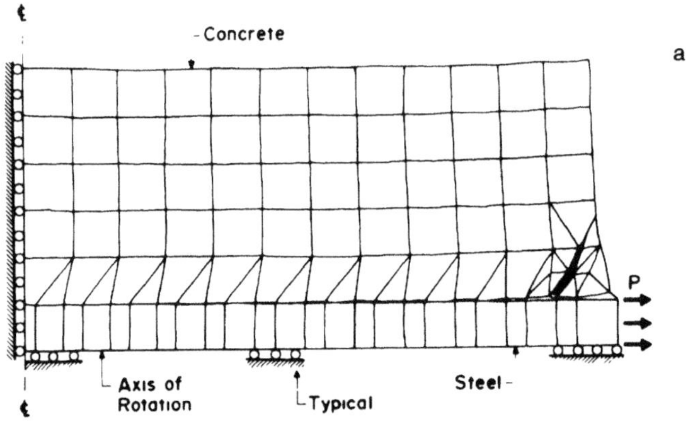

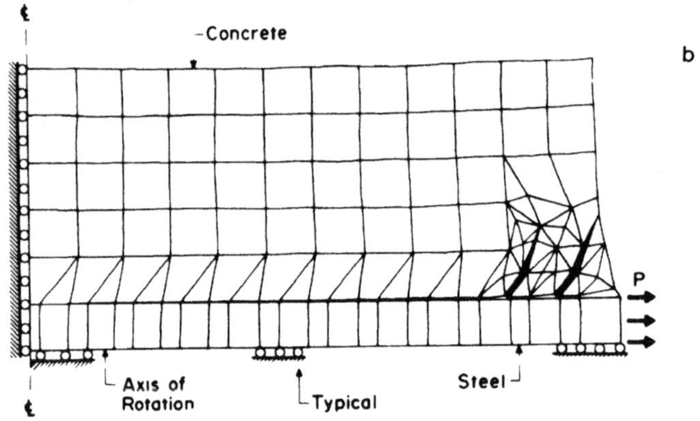

Fig. 4.31. FEFAP modeling of secondary crack formation in portion of specimen shown in Figure 4.30 [19].

was formed through the entire concrete cross section. When this primary crack opens sufficiently, secondary cracks would form in response to the free surface created by the primary crack, in the same way that they formed at the end of the tension-pull specimen initially. The secondary cracks always slant toward the closest primary crack, as seen in Figure 4.32. The behavior shown in this characteristic length of a tension pull specimen would be repeated sequentially, beginning with both ends, until the pattern was developed along its entire length.

In [19,20] it is shown that, under certain conditions, the crack propagation pattern of Figure 4.31 is independent of cover and ultimate primary crack spacing. This implies that the additional "slip" of the bar as defined in Figure 4.33 is characteristic only of bar and concrete properties. The nonlinear curve of Figure 4.33 was developed from analyses like those shown in Figure 4.31. The nonlinear slip component of the curve which is a consequence of secondary cracking can be viewed

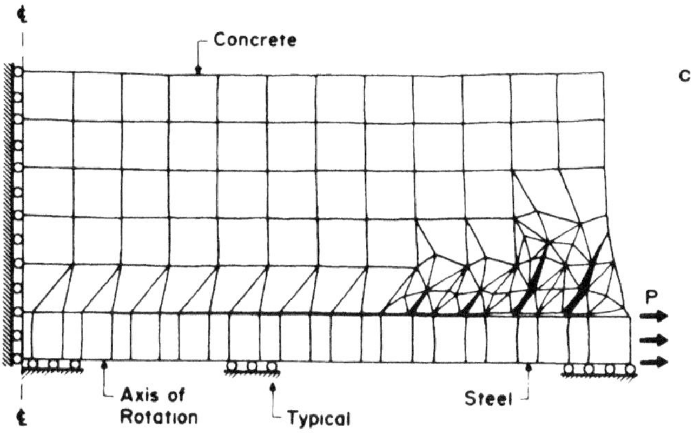

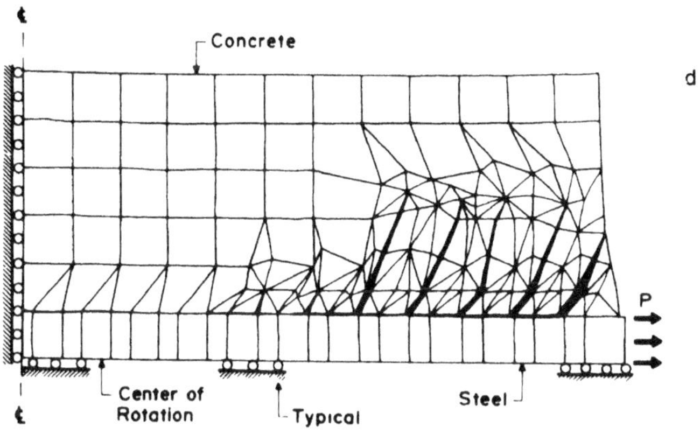

Fig. 4.31. For caption see opposite page.

as a "tension softening" of the concrete-steel composite structure. For the purposes of a discrete crack analysis this softening behavior can be lumped into an interface element which is inserted into the reinforcing bar at each location where the bar crosses a primary crack. The interface element has zero thickness and a normal stiffness, E_N, which depends on the stress level in the bar. The particular relationship derived for the test specimen of Figure 4.31 is shown in Figure 4.33.

This tension-softening, or bond-slip, element was used in the analysis of the reinforced concrete diaphragm in the next example problem.

The structure shown in Figure 4.34 can be viewed as the Griffith problem for reinforced concrete. It is also similar to the diaphragm which has been repeatedly analyzed by Bazant and Cedolin [15,51]. This problem will be used to compare, in the presence of bond-slip, the linear and nonlinear fracture models previously described.

As a reference case, the diaphragm was first analyzed used the linear

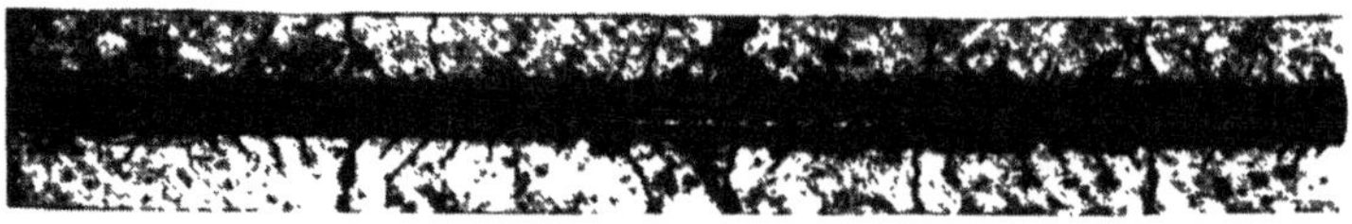

Fig. 4.32. Observed primary and secondary cracking in specimen similar to that shown in Figure 4.31 [50].

fracture model assuming no bond-slip. This produced a typical deformed mesh of the type seen in Figure 4.35. As the figure shows, the bars crossing the crack pinched the crack sides completely together at each location where they crossed the crack, but the crack was free to open between reinforcing bars. For each analysis, the crack tip was assumed to be midway between two bars. The singular crack tip elements were used to determine K_I, and the load necessary to cause the crack to propagate was determined by comparing K_I to K_{Ic}.

A plot of applied load versus crack length is seen in Figure 4.36. The crack is seen to be almost metastable. This result is expected when consideration is taken of the fact that the stress field surrounding the crack tip is unaffected by a change in crack length, because as the crack crosses each bar, a local stress redistribution takes place in which the force previously carried by the concrete is transferred only to the reinforcing bars in the immediate vicinity of the crack tip. The crack tip stress field is therefore essentially independent of crack length.

Figure 4.37 shows a typical deformed mesh for the diaphragm in which bond-slip was allowed and LEFM was still assumed to prevail. The total crack opening displacement pattern is seen to be the superposition of a small amplitude, oscillating component, due to the pinching effect of the reinforcing bars, superimposed upon a larger amplitude, approximately elliptical component, allowed by bond-slip.

The tension-softening elements have opened according to the bond-slip relationship described earlier. The analysis was nonlinear because of the necessity to satisfy this nonlinear relationship. Iterations were performed until the load level was found which would cause K_I to equal K_{Ic}, while at the same time satisfying the bond-slip law of each tension-softening element.

The resulting load versus crack length curve is seen in Figure 4.36. The load necessary to propagate the crack is less than for the case in which bond-slip was neglected, and the crack is initially unstable, until a crack length of about 50 inches (1.27 m) is reached, after which the crack appears to become essentially metastable.

Now comes the surprise. Even for the longest crack length of 57 inches

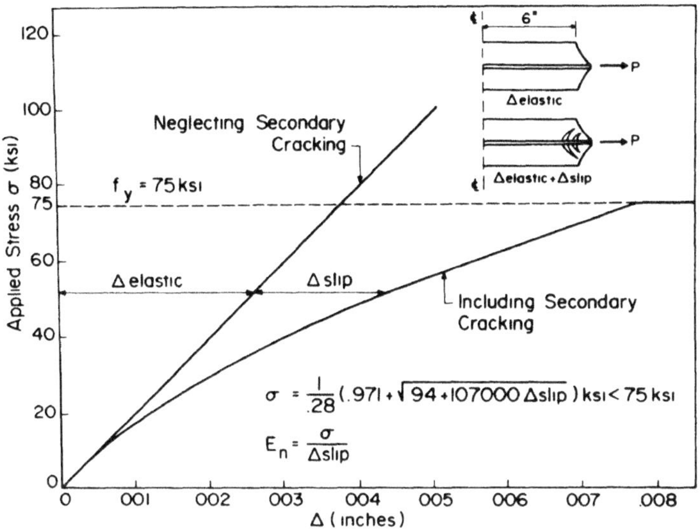

Fig. 4.33. Computed applied steel stress versus elongation for specimens shown in Figure 4.31 [19].

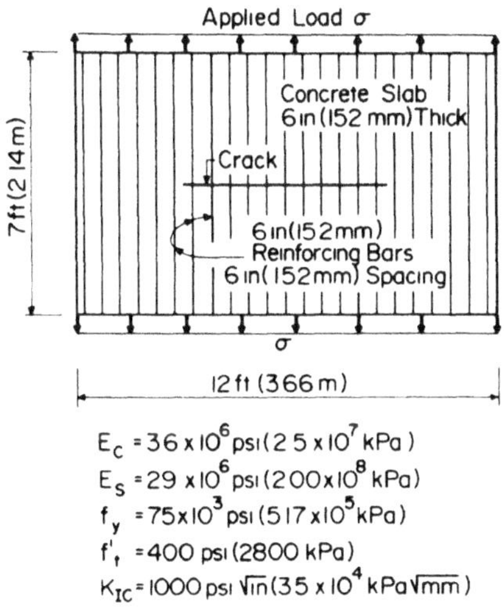

Fig. 4.34. Center-cracked reinforced concrete panel.

(1.45 m), the maximum crack opening is only about .002 inch (.05 mm) (as opposed to a maximum opening of about .0006 inch (.015 mm) in the no bond-slip case). This indicates that the use of LEFM is inappropriate in this problem, even though the crack length is several feet in length. This is because the process zone is as long as the crack itself. Therefore, a final analysis was performed in which the crack process zone was discretely modeled using the nonlinear fracture model described earlier

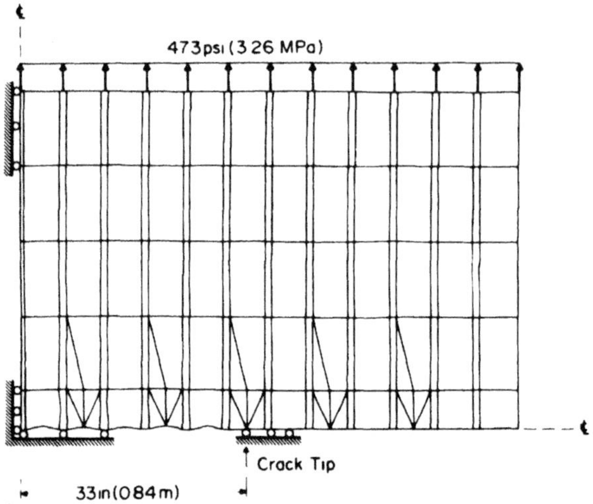

Fig. 4.35. Typical deformation pattern assuming LEFM but no bond slip.

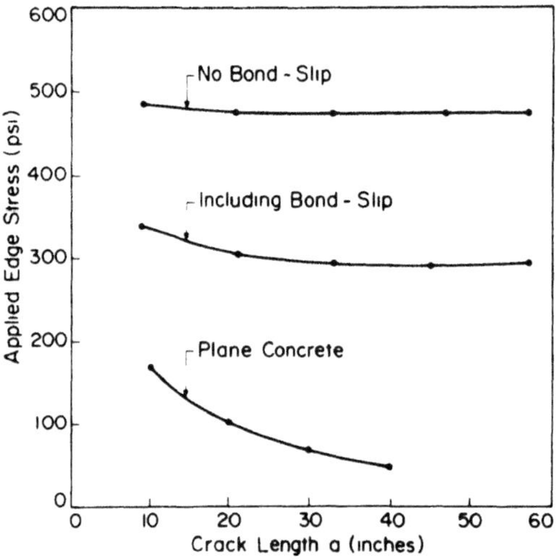

Fig. 4.36. Applied stress versus crack length for panel shown in Figure 4.34. For nonlinear process zone model arbitrary criterion shown in Figure 4.40 is used to determine "crack" length.

and used for the much shorter secondary cracks associated with the bond-slip example.

The bond-slip model of Figure 4.33 and the stress-COD curve of Figure 4.33 and the stress-COD curve of Figure 4.26 were used in the nonlinear crack analysis performed next. In this case a load was applied, then interactions were performed until the corresponding process zone

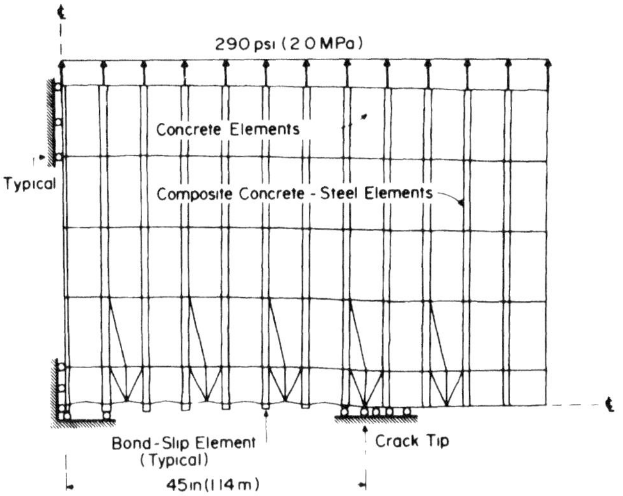

Fig. 4.37. Typical deformation pattern assuming LEFM and allowing bond slip according to the model shown in Figure 4.33.

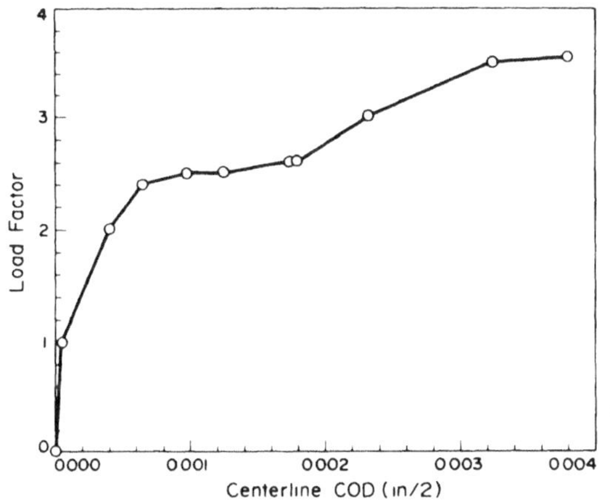

Fig. 4.38. Load factor versus centerline COD from analysis using the process zone model of Figure 4.26 and bond slip model of Figure 4.33.

was fully developed. Key events in this analysis can be discerned with the aid of Figure 4.38 which is a plot of load-factor versus one-half the COD at the center of the specimen. For a load factor of unity (167 psi; 1.15 MPa), the process zone has begun to propagate from the initial crack tip. The zone continues to propagate under increasing load until a load factor of about 2.5 at which level the specimen "unzips" to an average COD of about .004 inch (.10 mm). That is, the process zone propagates com-

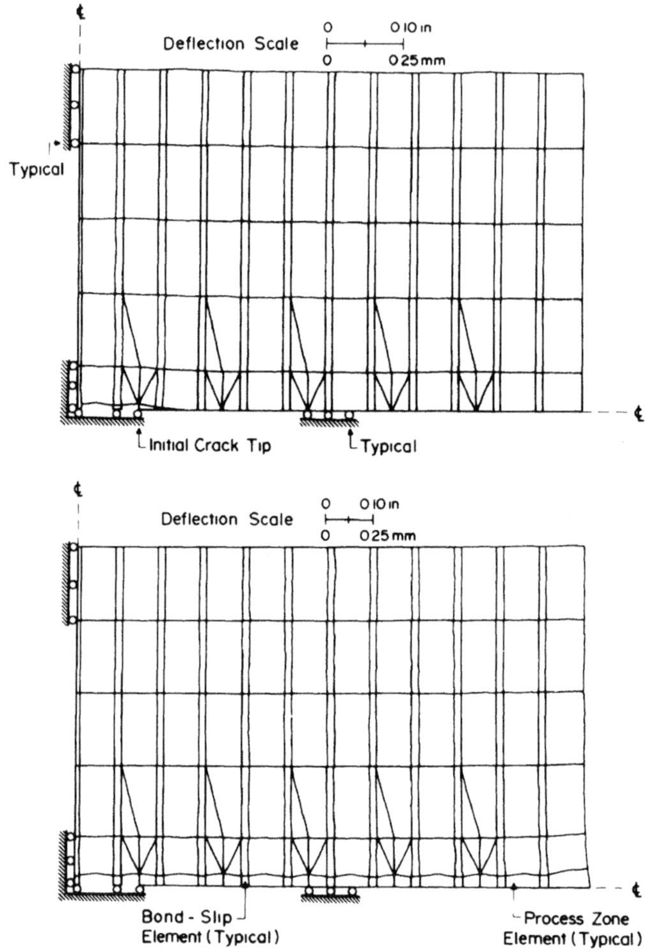

Fig. 4.39. Process zone formation. a) Stable configuration at load factor of one; b) "Unzipped" configuration at load factor of three.

pletely across the diaphragm from its last stable position as indicated in Figure 4.39a and relieves virtually all the stress carrying of the concrete along the "crack" plane. Figure 4.39b shows the process zone formation at a load factor of 3. At this load, no stress is being transmitted across the process zone since the COD over the zone is greater than 0.005 inch (.13 mm).

In the range between load factors of 2.5 to 3.5, the curve in Figure 4.38 reflects the bond-slip relationship of Figure 4.34. Yielding begins between load factors of 3.5 and 3.55 resulting in a pronounced flattening of the load-displacement curve.

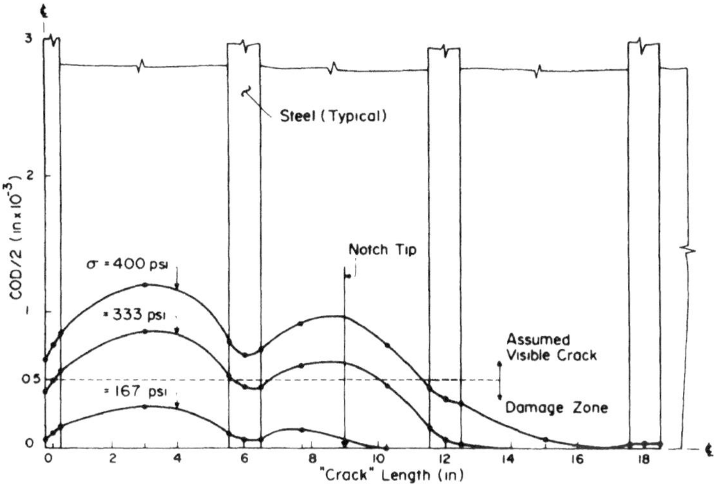

Fig. 4.40. Detail of process zone formation at convergence for load factors of one and two, and during iteration at a load factor of three.

Direct comparison of the results of this analysis with the linear analysis results by way of Figure 4.36 is difficult, for two reasons. First, what is "crack length" in the nonlinear analysis? Except for stable process zone growth over a short distance, the concrete virtually transmutes from being notched to being fully cracked in a very small load interval. One can invoke an arbitrary criterion such as that of the "visible" crack, a process zone with, say, a COD of 0.001 inch (.025 mm). With this arbitrary definition one can enter Figure 4.40 and see that the stable process zone formation would yield visible crack growth of about 3 inches (76 mm).

The second difficulty is that the curves in Figure 4.36 resulting from LEFM approaches cannot be quantitatively compared to the nonlinear result curve. Recall that the LEFM curves are directly related to the assumed K_{Ic}, here equal to 1 ksi $\sqrt{\mathrm{in}}$ (1.1 $\mathrm{MNm}^{-3/2}$). The plateau in the nonlinear result is directly related to the assumed tensile strength in the process zone model. In the authors' opinion, no rigorous relationship between K_{Ic} and f_t' has yet been established for concrete and verified by testing.

4.5 Crack propagation modeling: The future

The authors have described models which incorporate the flavor of the modern finite element techniques, the rigor of linear and nonlinear fracture mechanics, and the excitement and power of interactive/adaptive computer graphics. Traditional models will, no doubt, continue to be used effectively where they are safe. The authors feel strongly, however, that the discrete crack/fracture mechanics approach should become a useful and, in many cases, necessary alternative.

The techniques described in this chapter are certainly not the only ones available for modeling of crack propagation in concrete. Alternative approaches can be based on other numerical methods, theories, and algorithms. However, it is the authors' strongly held opinion that, regardless which model is pursued, interactive computer graphics will play a decisive role in determining the viability of any program in the marketplace of "real world" problems. The continuing, rapid revolution in graphics hardware capability and software development and the ever-increasing cost-effectiveness of large, virtual-memory mini-computers are the driving forces in the evolution of sophisticated crack propagation programs.

Nowhere will this be more evident than in the area of fully three-dimensional modeling. The present high cost of performing analysis of three-dimensional structures is due to the human effort required to define and check geometrical data, element topology, boundary conditions, and material properties. In fact, the complexity of error detection, or even slight modification, with three-dimensional meshes can substantially reduce the cost effectiveness of a program. The user falls back onto a two-dimensional or axisymmetric model that is "good enough," sacrificing the realism of the three-dimensional problem in the face of the reality of tremendous labor cost.

However, interactive/adaptive preprocessing can eliminate a large percentage of such cost while simultaneously placing the engineer back in control of computer analysis. During the course of an analysis to predict rigorously the size, shape, and orientation of a propagating crack in a complex concrete structure it will be necessary to generate, display, and modify a mesh many times. This can only be done in a high level interactive graphics environment. Given the changes in the speed and cost of computers and the increasing use of interactive graphics, the authors see the use of truly three-dimensional crack propagation codes, with the versatility of existing two-dimensional programs, as a certainty in only a few years.

References

1. Clough, R.W., The Stress Distribution of Norfork Dam, Structures and Materials Research, Department of Civil Engineering, Series 100, Issue 19, University of California, Berkeley (1962).
2. Ngo, D. and Scordelis, A.C., Finite Element Analysis of Reinforced Concrete Beams, Journal of the American Concrete Institute, Vol. 64, No. 14, pp. 152–163 (1967).

3. Nilson, A.H., Nonlinear Analysis of Reinforced Concrete by the Finite Element Method, Journal of the American Concrete Institute, Vol. 65, No. 9, pp. 757–776 (1968).
4. Mufti, A.A., Mirza, M.S., McCutcheon, J.O., and Houde, J., A Study of the Behavior of Reinforced Concrete Elements, Structural Concrete Series No. 70-5, McGill University (1970).
5. Rashid, Y.R., Analysis of Prestressed Concrete Pressure Vessels, Nuclear Engineering and Design, Vol. 7, No. 4, pp. 334–344 (1968).
6. Ngo, D., A Network-Topological Approach for the Finite Element Analysis of Progressive Crack Growth in Concrete Members, Ph.D. Dissertation, Division of Structural Engineering and Structural Mechanics, University of California, Berkeley, UC-SESM 75-6 (1975).
7. Dodds, R.H., Darwin, D., Smith, J.L., and Leibengood, L.D., Grid Size Effects with Smeared Cracking in Finite Element Analysis of Reinforced Concrete, Structural Engineering and Engineering Materials SM Report No. 6, The University of Kansas, Lawrence (1982).
8. Mindess, S. and Diamond, S., The Cracking and Fracture of Mortar, Fracture in Concrete (W.F. Chen and E.C. Ting, eds.), Proceedings of an ASCE Session, Hollywood, Florida, American Society of Civil Engineers, NY (1980).
9. Haber, R.B., Shephard, M.S., Abel, J.F., Gallagher, R.H., and Greenberg, D.P., A General Two-Dimensional Graphical Finite Element Processor Utilizing Discrete Transfinite Mappings, International Journal for Numerical Methods in Engineering, Vol. 17, No. 7, pp. 1015–1044 (1981).
10. Perucchio, R., Ingraffea, A.R., and Abel, J.F., Interactive Computer Graphic Preprocessing for Three-Dimensional Finite Element Analysis, International Journal for Numerical Methods in Engineering, Vol. 18, No. 6, pp. 909–926 (1982).
11. Perucchio, R. and Ingraffea, A.R., Interactive Computer Graphic Preprocessing for Three-Dimensional Boundary-Integral Element Analysis, Computers and Structures, Vol. 16, No. 1–4, pp. 153–166 (1983).
12. Saouma, V.E., Finite Element Analysis of Reinforced Concrete: A Fracture Mechanics Approach, Ph.D. Dissertation, School of Civil and Environmental Engineering, Cornell University (1981).
13. Gibbs, N.E., Poole, W.J., and Stockmeyer, P.K., An Algorithm for Reducing the Bandwidth and Profile of a Sparse Matrix, SIAM Journal of Numerical Analysis, V. 13, pp. 236–250 (1976).
14. Mindess, S., The Cracking and Fracture of Concrete: An Annotated Bibliography, 1928–1980, Materials Research Series Report No. 2, I.S.S.N. 0228–4251, Department of Civil Engineering, The University of British Columbia, Vancouver, BC, 93 pp. (1981).
15. Bazant, Z.P. and Cedolin, L., Blunt Crack Band Propagation in Finite Element Analysis, Journal of the Engineering Mechanics Division, ASCE, Vol. 105, No. EM2, pp. 297–315 (1979).
16. Catalano, D. and Ingraffea, A.R., Concrete Fracture: A Linear Elastic Fracture Mechanics Approach, Department of Structural Engineering Report 82-1, School of Civil and Environmental Engineering, Cornell University, Ithaca, NY (1982).
17. Arrea, M. and Ingraffea, A.R., Mixed-Mode Crack Propagation in Mortar and Concrete, Department of Structural Engineering Report 81-13, School of Civil and Environmental Engineering, Cornell University, Ithaca, NY (1981).
18. Petersson, P-E., Crack Growth and Development of Fracture Zones in Plain Concrete and Similar Materials, Report TVBM-1006, Division of Building Materials, Lund Institute of Technology, Lund, Sweden (1981).
19. Gerstle, W., Ingraffea, A.R., and Gergely, P., The Fracture Mechanics of Bond in Reinforced Concrete, Department of Structural Engineering Report 82-7, School of

Civil and Environmental Engineering, Cornell University, Ithaca, NY, 144 pp. (1982).
20. Gerstle, W., Ingraffea, A.R., and Gergely, P., Tension Stiffening: A Fracture Mechanics and Interface Element Approach, Proceedings International Conference on Bond in Concrete, Paisley, Scotland, pp. 97–106 (1982).
21. Hillerborg, A., Modeér, M., and Petersson, P-E., Analysis of Crack Formation and Crack Growth in Concrete by Means of Fracture Mechanics and Finite Elements, Cement and Concrete Research, Vol. 6, pp. 773–782 (1976).
22. Barsoum, R.S., On the Use of Isoparametric Finite Elements in Linear Fracture Mechanics, International Journal of Numerical Methods in Engineering, Vol. 10, No. 1, pp. 25–37 (1976).
23. Freese, C.E. and Tracey, D.M., The Natural Isoparametric Triangle Versus Collapsed Quadrilateral for Elastic Crack Analysis, International Journal of Fracture, Vol. 12, p. 767 (1976).
24. Shih, C.F. de Lorenzi, H.G., and German, M.D., Crack Extension Modeling with Singular Quadratic Isoparametric Elements, International Journal of Fracture, Vol. 12, pp. 647–651 (1976).
25. Ingraffea, A.R. and Manu, C., Stress-Intensity Factor Computation in Three Dimensions with Quarter-Point Crack Tip Elements, International Journal of Numerical Methods in Engineering, Vol. 12, No. 6, pp. 235–248 (1978).
26. Saouma, V., Ingraffea, A.R., Gergely, P., and White, R.N., Interactive Finite Element Analysis of Reinforced Concrete: A Fracture Mechanics Approach, Department of Structural Engineering Report 81-5, School of Civil and Environmental Engineering, Cornell University, Ithaca, NY (1981).
27. Ingraffea, A.R. and Heuze, F.E., Finite Element Models for Rock Fracture Mechanics, International Journal for Numerical and Analytical Methods in Geomechanics, Vol. 4, pp. 25–43 (1980).
28. Ingraffea, A.R., Discrete Fracture Propagation in Rock: Laboratory Tests and Finite Element Analysis, Ph.D. Dissertation, University of Colorado (1977).
29. Erdogan, F. and Sih, G.C., On the Crack Extension in Plates Under Plane Loading and Transverse Shear, ASME Journal of Basic Engineering, Vol. 85, pp. 519–527 (1963).
30. Sih, G.C., Some Basic Problems in Fracture Mechanics and New Concepts, Engineering Fracture Mechanics, Vol. 5, p. 365 (1973).
31. Sih, G.C., Strain-Energy-Density Factor Applied to Mixed Mode Crack Problems, International Journal of Fracture, Vol. 10, pp. 305–321 (1974).
32. Sih, G.C. and Macdonald, B., Fracture Mechanics Applied to Engineering Problems – Strain Energy Density Fracture Criterion, Engineering Fracture Mechanics, Vol. 6, pp. 361–386 (1974).
33. Sih, G.C., Surface Layer Energy and Strain Energy Density for a Blunted Notch or Crack, Proceedings International Conference Prospects of Fracture Mechanics, Delft, pp. 85–102 (1974).
34. Saouma, V., Ingraffea, A.R., and Catalano, D., Fracture Toughness of Concrete – K_{Ic} Revisited, Journal of the Engineering Mechanics Division, ASCE, Vol. 108, No. EM6, pp. 1152–166 (1982).
35. Beech, J. and Ingraffea, A.R., Three-Dimensional Finite Element Stress Intensity Factor Calibration of the Short Rod Specimen, International Journal of Fracture, Vol. 18, No. 3, pp. 217–229 (1982).
36. Blandford, G.E., Ingraffea, A.R., and Liggett, J.A., Automatic Two-Dimensional Quasi-Static and Fatigue Crack Propagation Using the Boundary Element Method, Department of Structural Engineering Report 81-3, School of Civil and Environmental Engineering, Cornell University, Ithaca, NY (1981).
37. Sloan, R.C. and Abraham, T.J., TVS Cuts Deep Slot in Dam, Ends Cracking Problem, Civil Engineering, Vol. 48, No. 1, pp. 66–70 (1978).

38. Chappell, J.F. and Ingraffea, A.R., A Fracture Mechanics Investigation of the Cracking of Fontana Dam, Department of Structural Engineering Report 81-7, School of Civil and Environmental Engineering, Cornell University, Ithaca, NY (1981).
39. Hussain, M.A., Pu, S.L., and Underwood, J.H., Strain Energy Release Rate for a Crack Under Combined Mode I and Mode II, Fracture Analysis, ASTM, STP560, pp. 2–28 (1974).
40. Goodman, R.E., Taylor, R.L., and Brekke, T.L., A Model for the Mechanics of Jointed Rock, Journal of the Soil Mechanics Division, ASCE, SM3, pp. 637–659 (1968).
41. Ottosen, N.S., A Failure Criterion for Concrete, Journal of the Engineering Mechanics Division, ASCE, Vol. 103, No. EM4, (1977), pp. 527–535.
42. Ottosen, N.S., Constitutive Model for Short-Time Loading of Concrete, Journal of the Engineering Mechanics Division, ASCE, Vol. 105, No. EM1, pp. 127–141 (1979).
43. Nilson, A.H., Bond Stress-Slip Relations in Reinforced Concrete, Department of Structural Engineering Report No. 345, Cornell University, Ithaca, NY (1971).
44. Fenwick, R.C. and Paulay, T., Mechanics of Shear Resistance of Concrete Beams, Journal of the Structural Division, ASCE, Vol. 94, No. ST10, pp. 2325–2350 (1968).
45. Chang, S.C., An Integrated Finite Element Nonlinear Shell Analysis System with Interactive Computer Graphics, Ph.D. Dissertation, School of Civil and Environmental Engineering, Cornell University (1981).
46. Bresler, B. and Scordelis, A.C., Shear Strength of Reinforced Concrete Beams, Structures and Materials Research, Department of Civil Engineering, Series 100, Issue 13 (SESM 61–13), Institute of Engineering Research, University of California, Berkeley (1961).
47. Saouma, V. and Ingraffea, A.R., Fracture Mechanics Analysis of Discrete Cracking, Proceedings of IABSE Colloquium on Advanced Mechanics of Reinforced Concrete, Delft, the Netherlands, pp. 413–436 (1981).
48. Evans, R.H. and Marathe, M.S., Microcracking and Stress-Strain Curves for Concrete in Tension, Matériaux et Constructions, Vol. 1, No. 1, pp. 61–64 (1968).
49. Broms, B. and Raab, A., The Fundamental Concepts of the Cracking Phenomenon in Reinforced Concrete Beams, Department of Structural Engineering Report No. 310, Cornell University, September (1961).
50. Goto, Y. and Otsuka, K., Experimental Studies on Cracks Formed in Concrete Around Deformed Tension Bars, reprinted from: The Technology Reports of the Tohoku University, Vol. 44, No. 1, pp. 49–83 (1979).
51. Bazant, A. and Cedolin, L., Fracture Mechanics of Reinforced Concrete, Journal of the Engineering Mechanics Division, ASCE, Vol. 106, No. EM6, pp. 1287–1306 (1980).

M. ELICES 5

Fracture of steels for reinforcing and prestressing concrete

5.1 Introduction

This chapter presents an ordered account of the available information on the fracture of steels used for reinforcing and prestressing concrete. Such information is essential to an understanding of the fracture behaviour of reinforced concrete. Reasons of limit space prevent a state of the art survey but, from the literature and references cited, it should be possible to obtain a global view of the current status of the field.

Any study of the fracture of reinforced and prestressed concrete requires not only a knowledge of the characteristic properties of the constituent components, steel and concrete, but also of the behaviour of their interface since it is the bond between the two which governs their interaction. In fact failure of a structural member may be due not to fracture of the steel or concrete but rather to a break down in the interfacial bond. Once again limited space prevents coverage of bond behaviour, even though it is fundamental to a complete understanding of the fracture of concrete. The interested reader might, however, consult the proceedings of the recent conference on Bond in Concrete [11] which cover a wide spectrum of the available knowledge on this subject.

First of all a brief description of the steels used in reinforcing and prestressing concrete is given. Different types of fracture in reinforcing steels are considered next and the importance of defects, principally surface, are discussed. There is very little available information relevant to the case of surface defects since these generally require a three dimensional treatment. Recent solutions to such problems have employed sophisticated numerical techniques, and since reinforcing steels generally fail after appreciable plastic deformation an elastoplastic analysis would be required. This and the three dimensionality of cracks (which tend to originate in the interior) make a quantitative treatment extremely difficult. For this reason the first part will be essentially descriptive. The chapter ends with two sections devoted to the initiation and growth of subcritical cracks; one considers fatigue phenomena, which are of increasing importance in design, and the other environment sensitive cracking which is of increasing interest as construction takes place in more aggressive conditions and with higher design stresses.

The theme of this chapter has now begun to have sufficient coherence to form a branch within the study of the fracture mechanics of civil engineering materials. There are, however, many gaps which are still to be bridged and these are identified within the text. Previously unpublished information is also presented.

Steels for concrete reinforcement. On its own, concrete has a high compressive strength but relatively little resistance to tension and bending. These deficiencies in the tensile and bending strengths may be overcome by the addition of steel reinforcement. The reinforcing steel must have not only adequate tensile properties but also develop a strong bond with the concrete since load is transmitted to the steel through shearing stresses at the interface. This bond which is purely mechanical arises from surface roughness and friction between the steel and concrete.

Steels used in the reinforcement of concrete may be classified according to the method of production as: hot rolled, cold worked or special steels (eg. hardened and tempered steels). In the hot rolling process, the steel is passed through a series of rolls which appreciably reduce the cross section of the steel. In doing so, the comparatively coarse grains of the steel are deformed. The resulting refinement and homogeneity of the metallurgical structures are proportional to the amount of rolling.

Hot rolled mild steel bars may have either a plain or deformed surface while high tensile bars tend to have a deformed surface produced during hot-rolling and consisting of a transverse rib pattern with or without a longitudinal rib.

High strength may be achieved through subsequent cold working of hot rolled bars. The cold working process is usually performed by twisting round ribbed or square bar and controlling the pitch to give optimum strength and bond characteristics.

Steels for prestressing concrete. Prestressing of concrete structures is used to put the greater part of the cross-section of concrete into permanent compression and allow more effective use of the material under externally applied loads, particularly in bending. There are two general types of prestressing: Pretensioning is a method of prestressing in which the tendons are stressed before the surrounding concrete is cast, whereas in post-tensioning the tendons, usually in pre-formed ducts, are stressed after the surrounding concrete has set and hardened.

Because the tendon forces necessary to put massive cross-sections of concrete into the required degree of compression are very large, very high strength steels are often used for tendons and also for some parts of the anchorage systems. High-tensile steel for prestressing usually takes one of three forms: wires, strands or bars.

Cold drawn wires and strands have a carbon content of about 0.8 per cent (eutectoid) which is mainly responsible for their high tensile strength. The usual manufacturing process involves heating to a temperature in the austenite region, followed by an appropriate cooling procedure to give a metallurgical structure suitable for subsequent cold working. Strength is further improved by cold working carried out by drawing the wire through a series of dies. A final stress relieving heat treatment is given to improve some of the mechanical properties. The stress relieving treatment may be performed under prescribed conditions of deformation to improve stress relaxation.

Smooth and deformed high tensile alloy steel bars and wires are available for prestressing. They are made from high carbon alloyed steel or from medium carbon silicon chromium steel. Bars are hot rolled and generally cold stretched to raise the yield point and to render the bars more linear elastic at stress levels below the yield point. After cold stretching they are frequently stress relieved to improve the ductility and stress-strain characteristics. Steel wires and bars of the silicon chromium type are hot rolled to their final surface condition and then quenched and tempered. Figure 5.1 shows possible shapes of hot rolled deformed bars.

In both pre-tensioning and post-tensioning, the most common method for stressing the tendons is jacking. In post-tensioning, jacks are used to pull the steel with the reaction taken against the hardened concrete; in pretensioning, jacks pull the steel with reaction against end bulkheads or moulds. It is not possible to compile all the systems of prestressing, new systems are being developed and existing ones modified from time to time. Nevertheless, as most steel fractures occur during tensioning, it is expedient to review some of these systems:

The anchorage for the BBRV system is arranged by passing each wire through a hole in a special block and then forming a button-head at the end of the wire by upset cold forging. The wires are cut to the required tendon length before the buttonheads are formed. The buttonheads are located in shaped depressions on the back of the anchorage block.

The PCS Freyssinet system generally uses multi-wire cables or strands. The anchorage system involves passing the strands through a tapered female cone, into which the male cone is forced at the time of jacking and stressing. The conical surface of the male cone carries fluted grooves, and each strand is gripped in a groove as it is stressed and the male cone is embedded in the female anchorage plate.

The CCL anchorage system involves passing the wires through holes in the anchorage block and the swaging of barrel-and-cone wedge grips onto individual strands behind the blocks at the time of stressing.

The anchorages at the end of each bar are provided, generally, by

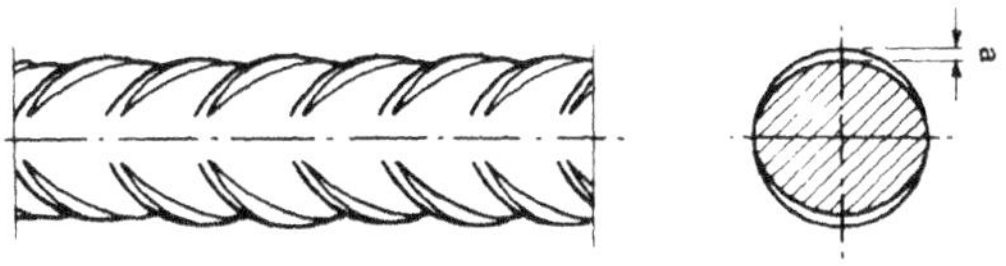

Round deformed rod, hot rolled.

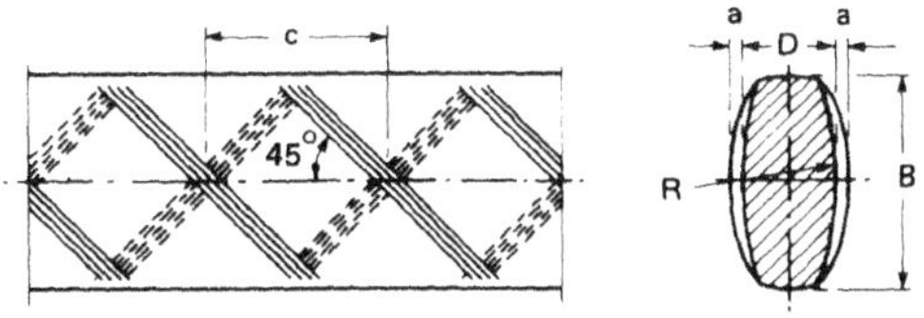

Oval deformed rod, hot rolled.

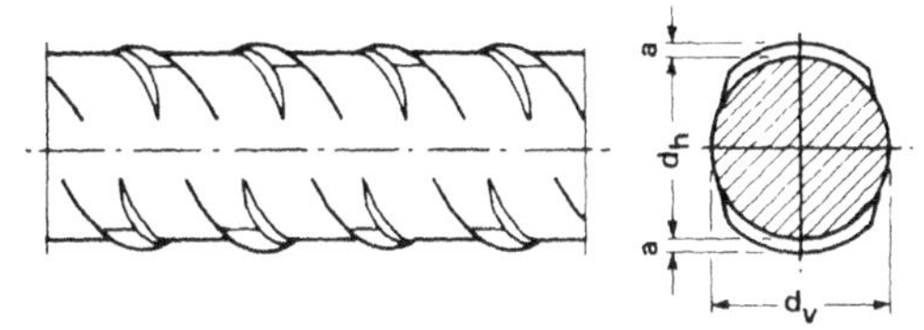

Round deformed bar, hot rolled.

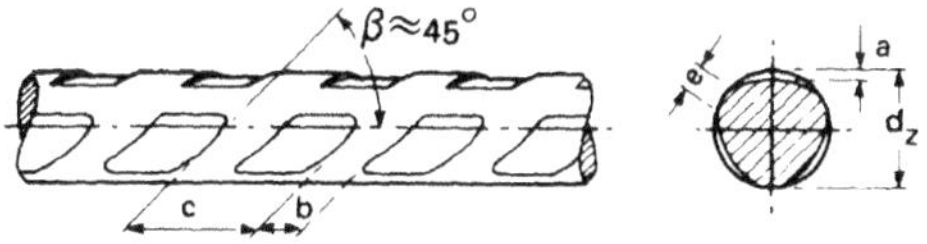

Indented wire, cold drawn.

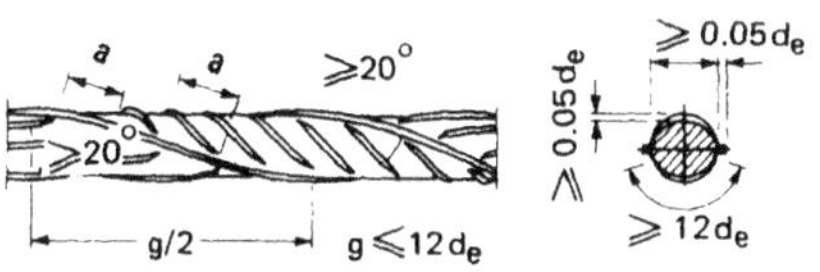

Round deformed bar in twisted condition.

Fig. 5.1. Possible shapes of hot rolled deformed bars. (Courtesy FIP [37]).

threaded assemblies with a nut bearing onto an end plate. Stressing is carried out by hydraulic jacks gripping the threaded end behind the nut and loading onto stools over the end plate. An essential point is the proper threading of the ends to take a special nut capable of developing as nearly as possible the full strength of the bar. By using tapered threads, about 98% of the bar strength is developed.

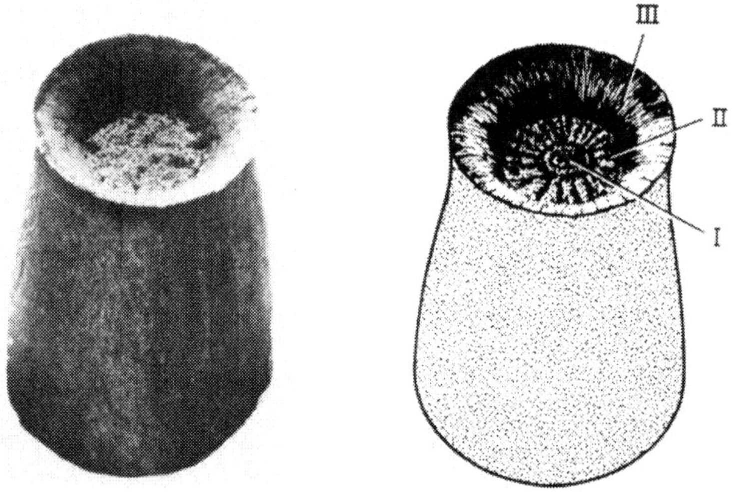

Fig. 5.2. Schematic illustration of cup-cone fracture.

5.2 Fracture

Tensile loading for unnotched specimens. In both reinforced and prestressed concrete the loading made in the reinforcement is one of tension. However, in practice, tendons can be subjected to transverse forces arising from contact or, in the case of prestressing reinforcement, forces from the jack during tendon stressing. This complex stress state can induce rupture and this is discussed at the end of the section. Consider first, however, the simplest case of a wire or rod under tension.

The basic features of ductile fracture that should be modeled in an analysis of tensile bars have been discussed by G.C. Sih [74–76]. Evidence from the fracture surfaces of broken specimens shows that there are three distinct stages of fracture development, each of which possesses a different surface appearance and instability at the macroscopic level. These stages are designated as I, II and III in Figure 5.2. Fracture originates inside region I which represents slow or stable crack growth. It then changes to rapid or unstable crack propagation indicated by region II. Final separation of the material begins as the crack turns away from the normal to the free surface. This slanted fracture surface is usually referred to as the shear lip zone III. The boundary separating stages I and II represents a transition from slow to rapid fracture, while the time elapsed between stages II and III is so short that the load and deformation do not change appreciably during this period. Details and comments on crack propagation beyond elastic limit and shear lip formation are given in [75], where the three-dimensionality of these phenomena is emphasized.

The microscopic aspects of these fractures, at ambient temperatures and under moderate rates of loading, are well documented [64]. The most frequently observed fracture in the smooth bar tensile test, occurs as a result of void nucleation, growth and linkage, following the formation of the central crack. Acting as an internal notch, this crack tends to concentrate the deformation at its tip in narrow bands of high shear strain. These shear bands are at angles of 50° to 60° to the transverse direction. Under the combined action of the tensile stress and the resulting shear strain, "sheets" of voids are nucleated in these bands, growing and elongating until coalescence occurs, producing local fracture of the "void sheet". The crack zigzags back and forth across the plane of the minimum section by void sheet formation as it extends outward radially, reducing the unfractured section of the specimen. This crack grows in a direction perpendicular to the axis of the specimen until it approaches the surface of the specimen. When instability occurs, it then propagates along localized shear planes at roughly 45° to the axis to form the "cone" part of the fracture. Some typical cup-cone fractures of steels for reinforcement and prestressing concrete are shown in Figure 5.3.

The mechanism of the final cup-cone fracture of ductile tensile bars is not well understood. Recently, an analysis of this instability based upon the mechanics of progressive rupture and the tearing modulus concept was suggested by Paris and coworkers [56,48]. As fracture instability is approached, the specimen is fully plastic in the necked region around the crack and the flow stress on the remaining ligament is nearly constant. Crack propagation decreases the area of remaining uncracked ligament and lowers the plastic limit load. Assuming fixed grips, this results in a certain amount of elastic shortening of the unnecked portions of the specimen. During the instant of unstable fracture, the movement of the testing machine grips is nearly zero, and thus the total length of the specimen remains nearly constant. Therefore, for an increment of crack growth, if the amount of elastic shortening is greater than the amount of plastic elongation associated with crack growth then instability results. When the amount of plastic elongation and elastic shortening are equated, then all intrinsic material properties, as distinguished from properties of geometric configuration of the specimen, can be separated to different sides of an equation. The side of material properties is termed T_{material} while the side of geometric configuration is termed T_{applied} [56]. Kong and Paris found, for a round tensile bar of length L diameter D, with a circular center crack of diameter $d = 2a$.

$$T_{\text{material}} = \frac{\mathrm{d}J}{\mathrm{d}a}\frac{E}{\sigma_0^2} \leqslant \frac{4Ld}{D^2} = T_{\text{applied}} \tag{5.1}$$

where σ_0 is the most appropriate flow stress. Also, the tearing instability

Fig. 5.3. Typical cup-cone fractures. Reinforcing steel; a – smooth, b – deformed steel for prestressing; c – cold drawn; d – quenched and tempered.

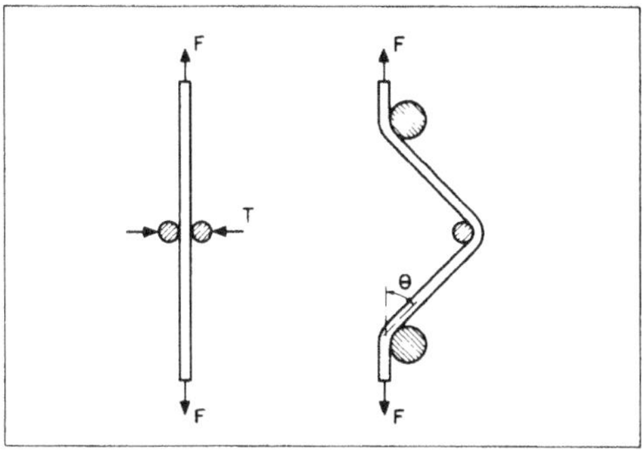

Fig. 5.4. Tensile tests with lateral constriction.

of a single microflaw in uniform plastic stress field can be obtained [48]:

$$T_{\text{material}} = \frac{\mathrm{d}J}{\mathrm{d}a} \frac{E}{\sigma_0^2} \leqslant 2.6 = T_{\text{applied}} \tag{5.2}$$

In consequence, after a certain amount of deformation, a microflaw, in the center of a tensile bar might begin to grow (in particular the beginning of growth might be associated with J_{IC} or a similar parameter). If growth begins and T_{material} is less than 2.6, Equation (5.2) implies instability from the beginning of growth and perhaps a sudden completely flat fracture occurs, a so-called star fracture. However, if T_{material} is greater than 2.6 Equation (5.2) implies that microflaw growth will at first be stable. But as the crack grows, its diameter d increases and Equation (5.1) implies that at some point T_{applied} may respondingly increase to exceed T_{material} and instability should ensue. Since the slip field associated with Equation (5.1) is 45° slip from the crack tip to the outside of the bar, this instability mechanism accounts for the change from flat stable growth to oblique-shear, forming the cup and cone.

As mentioned at the beginning of this chapter, the usual fracture of reinforcing and prestressing steel wires is of the cup and cone type. However, inside the anchorages and in places associated with a lateral compression a shear type fracture may occur, with an important lowering of the fracture load [37,42,51]. This type of rupture has been reproduced in the laboratory by Maupetit et al. [51], as well as by the author and coworkers, during tensile tests with a transverse load or by deviating the wire, as shown in Figure 5.4. The wires tested were made of bainitic, oil quenched and tempered and cold drawn carbon steels. Only the last type of steel exhibited shear fractures. A marked decrease of the fracture load as the lateral compression increases can be seen in Figure 5.5. Francois et al. [51] conclude that this test may be used to characterize the resistance

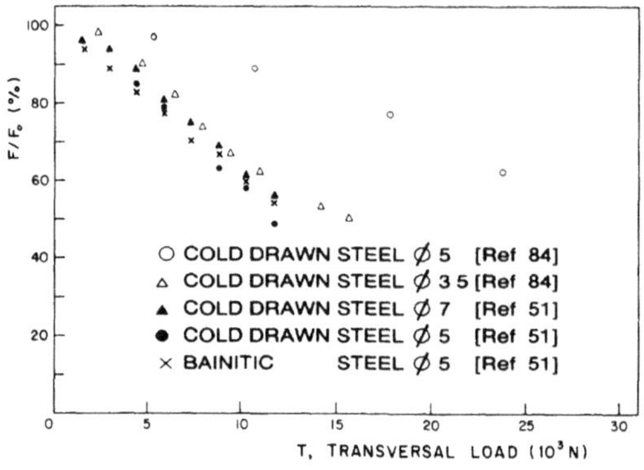

Fig. 5.5. Tensile strength under lateral load, [51], [84].

of wires to shear fracture and also for selection, or quality control, of steel wires for some prestressed concrete applications.

Bending [42] for unnotched specimens. The bend test, which is used to determine the ductility of wire, is designed to subject the test specimen to a large amount of deformation. Such tests are standardized. Under test conditions, the tensile strain on the outer surface of the wire is very high, 50% when the diameter of the mandrel equals the diameter of the wire. Steel wire that has an elongation of only 2% as measured in the conventional tensile test, may withstand such bending without fracturing. The usual bend test is made by bending the test specimen back and forth over mandrels considerably larger than the wire diameter. Such tests have shown that the number of reverse bends has a well defined relationship to the true strain at the point of fracture in the tensile test [41]. The bend test on steel wire also correlates with the reduction of area as measured in the tensile test [40].

The fact that steel wire will elongate as much as 50% in the bend test without necking-down is due to the stress distribution in the bend test. Only one side of the specimen is under tension while the opposite side is subject to compressive stress. While the tensile side is trying to neck, the compression side is trying to increase in cross-section. The end test is considered to be a good measure of quality because the maximum tensile strain takes place on the surface of the wire where defects are most likely to be present.

Torsion [42] for unnotched specimens. The torsion test has been used in the wire industry for evaluating the uniformity of wire. In contrast to the tensile test, the shearing stress in the torsion test is equal to the tensile stress, for this reason the fracture generally takes place on a plane of maximum shearing stress.

During some tests, after only about one twist, a sudden decrease in torque may be registered due to the formation of a longitudinal split in the wire parallel to the wire axis. If the test had been interrupted at that point, and a transverse section made through the split, examination would have shown that the split extended from the surface down to a point approaching the axis of the specimen and a very irregular torque-twist curve is developed until the final fracture takes place.

Another type of fracture is the helical fracture, generally found in specimens that fracture initially with a transverse shear fracture. The helical fracture is believed to be the result of the sudden release of torque after the initial fracture and the subsequent impact load on the remaining part of the test specimen.

The quality of steel wire is normally evaluated by the number of torsions to failure. The stress distribution in the torsion test is significantly different from that of the tension test and, because of this stress condition, the quality of the wire surface can considerably influence the test results.

Theoretical approach for notched specimens. Brittle fractures are sometimes observed in steels used to reinforce or prestress concrete. These are characterized by little or no deformation before fracture and are usually initiated at some surface defect; perhaps a notch or crack caused by fatigue or stress corrosion. In these cases the failure load can be well below the elastic limit so that, as mentioned above, the deformation may be very small. Consequently this type of fracture is very difficult to anticipate. In this section, the theoretical approach to the simplest cases is summarised and, then, results from laboratory tests are discussed.

The case of a surface crack in a linear elastic regime is considered first. The treatment is then extended over cover elasto-plastic behaviour. A summary of most of the information summarised below may be found in three civil engineering doctoral theses (Astiz [4], Athanassiadis [7] and Valiente [82]).

The investigation of numerous fractures which have occurred in service, as well as in laboratory tests, suggests the defects which initiated failure probably had a thumb-nail shape and that they may be of two types. The first pertains to cracks with a more or less elliptical boundary and the second to shallow cracks which extends around the circumference of the cylinder.

A number of crack shapes are illustrated in Figure 5.6. The treatment of such cracks can then be simplified by comparing the shapes to elliptical areas which are symmetrical with respect to the meridional plane and lie in a plane perpendicular to the edge of the tendon. With the

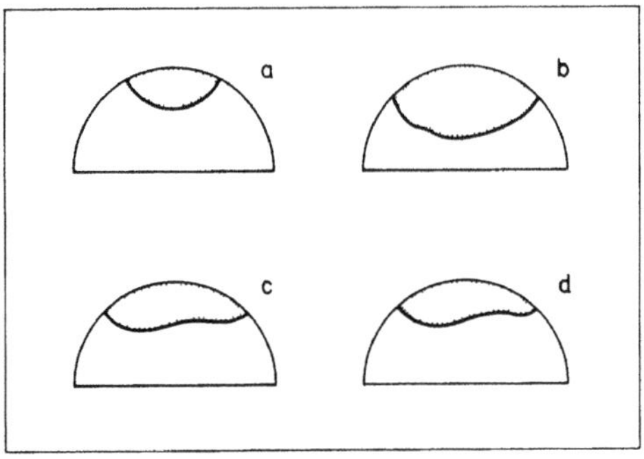

Fig. 5.6. Crack shapes; a – fatigue, b – stress corrosion, c and d from [7].

notation of the figure the most frequently used relationships between the different parameters are

$$a/D \leqslant 0.5 \qquad 0 < a/b < 1$$

Athanassiadis proposes a more complex expression for the crack boundary

$$\frac{R^2\alpha^2}{l^2} + \frac{(R-r)^2}{a^2} = 1 \tag{5.3}$$

where r and α are polar coordinates in the plane of the crack, l the length of the crack on the exterior boundary, a the depth and R the radius of the tendon. The difference between this curve and the ellipse is not significant for the usual cracks since both curves match satisfactorily the shapes of the cracks. Results obtained on both shapes are then comparable.

Once the crack has been modeled the problem becomes one of calculating the stress intensity factor and other fracture parameters of linear elastic fracture mechanics. The problem has been solved by various methods; the three dimensional finite element techniques [4,6] and boundary integral evaluation [7].

The finite element method, due to the high complexity of any three dimensional problem, has to be optimized in order to minimise the computational costs. A new prismatic non compatible singular element has been used in conjunction with the virtual crack extension method to obtain accurate results with coarse meshes.

The boundary integral method consists of transforming partial differential equations, which are valid within the volume of a solid, into integral equations over the surface. The great advantage of the technique is the necessity to discretize only the surface of the body so that a three dimensional problem is reduced to one in two dimensions.

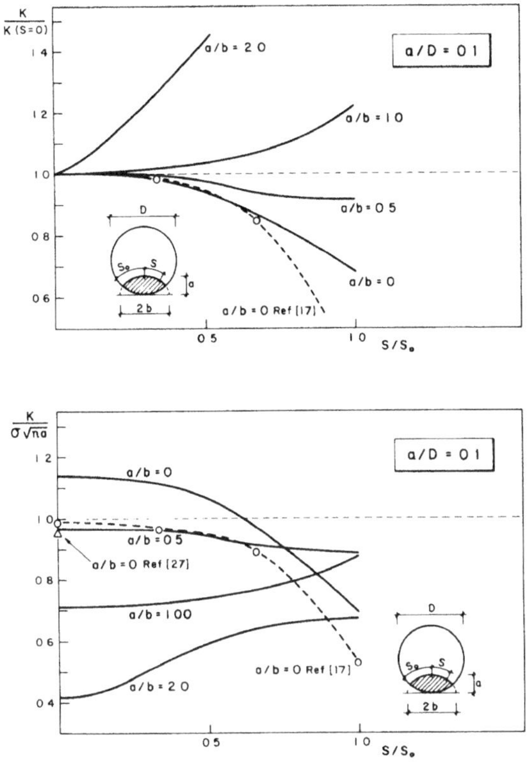

Fig. 5.7a. Variation of the stress intensity factor along the crack border.

Both techniques yield numerical results which have to be presented in a tabular form for application. The most commonly used fracture mechanics parameters are the stress intensity factors, $K_I(\theta)$, along the crack front (allowing a prediction of the crack growth shape) and the strain energy release rate, G, which is related to the square of $K_I(\theta)$ and from which the compliance can be calculated.

The stress intensity factor is taken to be locally

$$K_I(\theta) = \lim_{r \to 0} \sqrt{2\pi/r}\, E' u_3 = M(\theta)\sigma\sqrt{\pi a} \tag{5.4}$$

in a direction perpendicular to the crack front, with $E' = E/4$ in plane stress, i.e. at the surface of the wire, and $E'' = E/4\ (1 - \nu^2)$ in plane strain, i.e. inside the wire.

Values of $K_I(\theta)$ for different values of a/b and a/D are shown in Figure 5.7a and 5.7b. Interpolation will provide values of $K_I(\theta)$ for most practical cases. The following conclusions may be drawn from these results [6]:

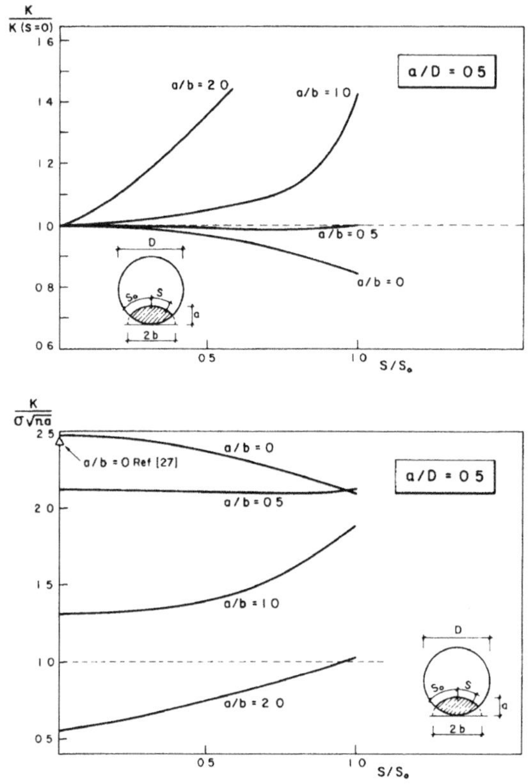

Fig. 5.7b. Variation of the stress intensity factor along the crack border.

(1) The variation of $K_{\mathrm{I}}(\theta)$ with a/b. The maximum value of K_{I} depends upon a/b. For flat fronted cracks (say $a/b < 0.7$) the maximum occurs in the center, $\theta = 0$, and crack growth would tend to make the front bow outwards. Acute cracks, on the other hand ($a/b > 0.8$), have the maximum values of $K_{\mathrm{I}}(\theta)$ at the edges and subsequent crack growth would be such as to flatten the crack front. It would appear that some values of a/b could give rise to K_{I} values which are constant along the crack so that growth takes place by the production of similar, iso $- K$, crack fronts.
(2) The variation of $K_{\mathrm{I}}(\theta)$ with a/D. The effect of crack depth on the value of K_{I} is most pronounced in flat cracks (small values of a/b) than in pointed cracks, for which it is hardly apparent.

Two particular crack geometries have been studied by a number of authors; these are straight-fronted and circular cracks. Some of the published results for these shapes are shown in Figures 5.7a and 5.7b and compared with those obtained from the numerical techniques mentioned above.

Stress intensity factors have been calculated by Blackburn [17] by finite element methods for straight edged crack and found that K_I, due to tensile or bending loads, was highest at the centre and lower than that due to a semi-elliptical notch in a slab of the same thickness under the same load. An average value for the stress intensity factor has been calculated by Daoud [27] from strain energy release rate determinations. Values of G were obtained using a finite-element representation of the bar and by measuring the compliance of the bar experimentally. For crack depths of less that one-half diameter, average K_I are found to be lower than existing results of rectangular bars having the same relative crack length. Also, average stress intensity factors were determined for single-edge-crack solid and hollow round bars loaded in tension by Bush [22]. These factors were calculated from experimental compliance measurements made over a range of dimensionless crack depths from 0.05 to 0.65. A comparison was made with data in the literature for rectangular bars. Results for a round bar falls below that for a rectangular bar, as already noted.

Results for surface cracks of circular shape in cylinders loaded in tension have been obtained by Fan [36] using the alternate method. It is basically an iterative method that couples two solutions; the exact solution for a penny-shaped crack in an infinite elastic body and a series solution for a cylinder with finite size under an arbitrary lateral surface load. From the alternate action of these basic solutions, an analytical expression for $K_I(\theta)$ can be obtained.

Both the finite element and boundary integral equation methods yield numerical results, which, for application, may be presented in tabular or graphical form. The programs give the potential energy P of the cracked wire from the calculation of the displacements and stresses at each node for different crack configurations. The strain energy rate G is deduced from the potential energy as

$$G = \sup\left(-\frac{\delta P}{\delta S}\right) \tag{5.5}$$

δS being a virtual increase of the crack surface. An average value of G deduced from the local values of K_I was also computed as

$$G^* = \frac{1}{2s}\int_{-s}^{+s}\frac{K_I^2}{E'}\,\mathrm{d}S \tag{5.6}$$

S being, as always, the curvilinear coordinate of a point on the crack front.

Figure 5.8 shows the variation of G with a/b and a/D. It can be seen

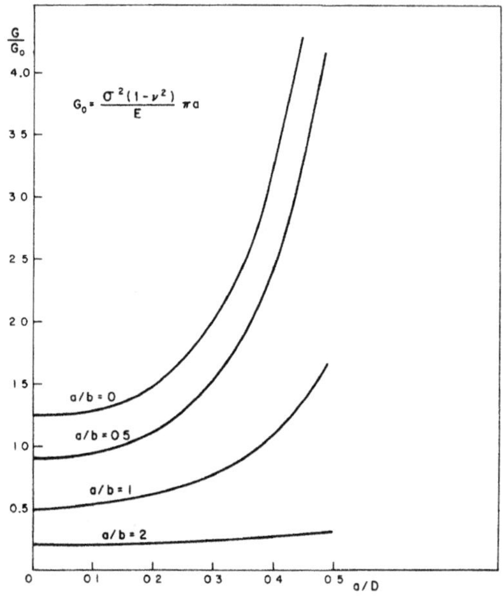

Fig. 5.8. Energy release rate for different crack configurations.

that, for a given crack depth (a/D), the strain energy release rate, G, increases rapidly with the crack shape parameters; being greater for flatter, less curved crack fronts (small values of a/b).

$K_I(\theta)$ and G values for the same crack geometries subjected to bending have also been obtained [6,9,17,22,27,36]. A study by Athanassiadis [7] and Maupetit et al. [51] is of interest since the case of a cracked specimen subjected to lateral forces is considered. This could be particularly useful in account for the stress state within the anchorage zone or for complex loading in general [51,62].

This section, covering theoretical attempts to solve the problem of the fracture of cracked tendons, is concluded with a summary of efforts made at extending the treatment to more realistic situations. In doing so, brief comment on the roles of plasticity and notches will be made.

At ambient temperature steels used for reinforcing and prestressing concrete exhibit some plastic deformation, even in the presence of cracks. While experiments shows that deep cracked cold drawn prestressing steel does behave in a linear elastic manner, both low yield reinforcing steels and quenched and tempered prestressing steels do not obey linear elastic fracture mechanics. For the latter cases, Valiente [84] has developed two limiting solutions; one based on a generalisation of the work of Hahn [46] for pipes and the other on limit analysis.

The crack tip opening displacement may be determined using the

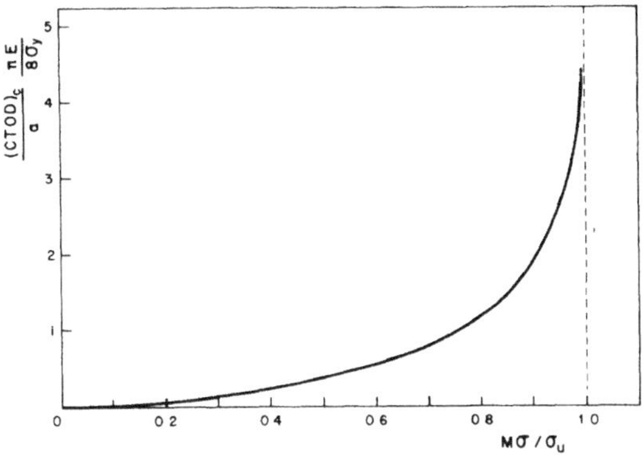

Fig. 5.9. Variation of CTOD as a function of $M\sigma/\sigma_Y$

Dugdale-Bilby-Cottrell-Swinden (DBCS) model [48]:

$$\text{CTOD} = \frac{8\sigma_y}{\pi E} a. \ln. \sec. \frac{\pi M\sigma}{2\sigma_u} \tag{5.7}$$

in which M is the geometrical factor for an elliptical crack, σ_u is a flow stress whose value is usually taken somewhere between σ_y (the yield stress, if it is well defined) and σ_R the tensile strength and the other notation has its usual meaning. If it can be assumed that crack growth will occur when CTOD reaches a critical value $(\text{CTOD})_c$, and that this is a material property, then the criterion for crack growth can be represented on Figure 5.9. It is then apparent that if the term, $(\text{CTOD})_c \pi E/8a\sigma_y$, is high (either because the crack depth is small or the resistance high), the point on the curve representing the condition for fracture in a given material with a given crack,corresponds to a value of $\sigma M/\sigma_u$ of about one. In these cases the fracture criterion can be taken as

$$M\sigma = \sigma_u \tag{5.8}$$

This result although overextended to cases of large scale plasticity, fits quite accurately with some experimental results.

Valiente's limit analysis solution is based on perfectly plastic behaviour. Equilibrium of forces across the ligament, Figure 5.10, yield the following failure criterion

$$F\sigma = \sigma_y \tag{5.9}$$

where σ is the applied stress, σ_y the yield stress and F is given by

$$F^{-1}\left(\frac{a}{D}, \frac{a}{c}\right) = \frac{2}{\pi} \text{ arc sin}\left(1 - \frac{2z}{D}\right) + \frac{4}{\pi}\left(1 - \frac{2z}{D}\right)\left(\frac{z}{D} - \frac{z^2}{D^2}\right)^{1/2} \tag{5.10}$$

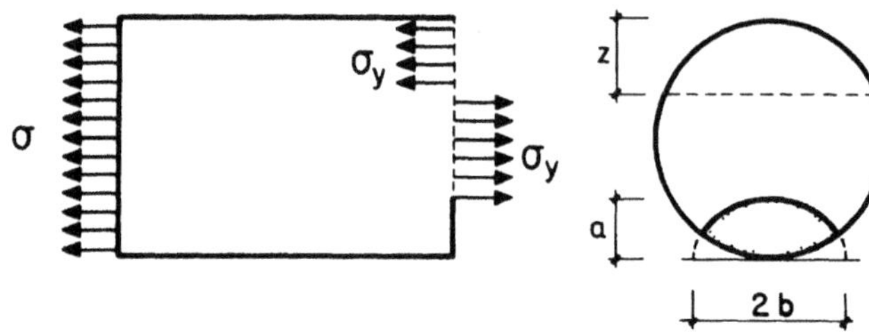

Fig. 5.10. Limit analysis for a cracked bar.

which is obtained from moment equilibrium and z is shown in Figure 5.10.

Another complication which arises when attempts are made to establish failure criteria for more realistic cases is the consideration of the notch root radius. The formulae mentioned above were valid for mathematically sharp cracks. However, unstable fracture is, sometimes, initiated at a notch of finite tip radius. In such cases corrections suggested by Creager and Paris [25] may be used to determine the elastic stress field in the vicinity of the notch. The problem of deriving a failure criterion, however, is not solved since fracture tends to involve significant plastic deformation. It is then necessary to adopt a two parameter approach.

For the sake of completeness and as a reference for future work attention is drawn to fracture toughness measurements made on rectangular notched bars subjected to tensile loading [8,29].

Finally, cold worked steels show a marked degree of anisotropy which results from the fabrication process. It is much easier to propagate cracks in the longitudinal rather than the transverse direction. From the toughness point of view, this is a useful property since cracks tend to deviate from the plane perpendicular to the edge and seek a plane of propagation parallel to the edge. There is a consequent reduction in the stress intensity factor and the crack may even be arrested. Anisotropy adds further complexity to the analysis and, at the time of writing, the author knows of no solution to the problem of a surface crack in a transversely anisotropic cylinder.

Experimental results for notched specimens. In this section various aspects of the stress intensity factor are considered; firstly, the experimental determination and then the critical value at which crack propagation occurs. The only direct experimental determinations of the stress intensity factors for a cracked cylinder are those obtained from the photoelastic stress freezing technique [5,83]. The results from such tests were in broad agreement with those from finite element and boundary integral analyses. Indirect methods involving compliance measurement have also

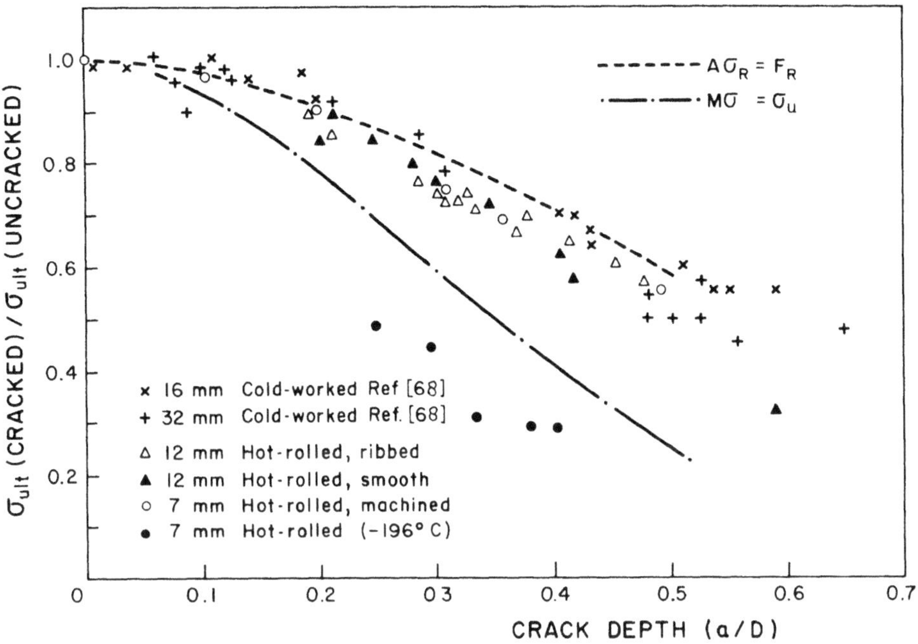

Fig. 5.11. Effect of crack depth on tensile strength.

yielded K_I values which agree well with theory [4,22,27]. Critical values of K_I have been determined from fatigue tests on cracked specimens and the theoretical expressions just described. Fracture toughness values K_c are, thus, available for reinforcing and prestressing steels. This section continues with a discussion of the tensile behaviour of cracked cylindrical specimens taken from such steels.

The simplest and most direct method of evaluating the effect of cracks on the rupture load and ductility consists of performing tensile tests on specimens with different crack depths. Results of such tests on various steels are shown in Figure 5.11 [31,68,82]. The materials tested were, cold worked bars of 16 and 32 mm. diameter and hot rolled smooth and ribbed bars of 12 mm. diameter.

The great toughness of these steels is immediately apparent, that is to say, the presence of small cracks causes little reduction in the failure load. This effect should be borne in mind when the growth of subcritical cracks is considered: say under the action of fatigue loading, stress corrosion or corrosion fatigue. The results of Figure 5.11 can be explained with the aid of models based on the plastic behaviour of the materials. Two extreme solutions have been superimposed upon the experimental results; one, crude approach, assumes that failure occurs when the rupture stress is developed across the whole section area A so

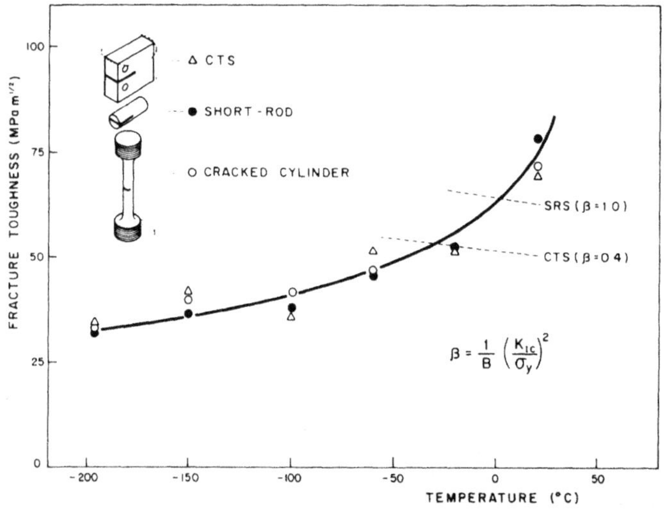

Fig. 5.12. Fracture toughness of hot rolled steel for concrete reinforcement.

that the stress distribution is then uniform. The rupture load F_R is then given by:

$$F_R = \sigma_R A = F_{R0} A / A_0 \tag{5.11}$$

where F_{R0} is the failure load of the uncracked specimen. This provides an upper bound which is not far from experimental values. A lower solution is obtained by extrapolating the fracture mechanics results into the elastoplastic regime using the DBCS model. This gives $M\sigma = \sigma_y$. An intermediate criterion based on limit analysis, for example, would give results which lie between those mentioned above and which are closer to the experimental observations.

If the experiments, described above, are carried out at low temperatures (−196°C) linear elastic behaviour is observed on a force-displacement record. Such behaviour suggests that valid K_{IC} results may then be obtained. Compact tension, short-rod and cracked cylindrical specimens machined from 57 mm. diameter hot rolled steel bars were used to determine the apparent K_{IC} values shown in Figure 5.12 [31,54]. It is clear that, above −30°C, compact tensile specimens do not yield valid K_{IC} results. Short rod specimens can be used to determine K_{IC} up to temperatures of about 0°C.

If it can be assumed that stable crack growth does not occur at low temperatures, so that once initiated cracks propagate rapidly, the linear elastic fracture mechanics expressions for $K_I(\theta = 0)$, using the appropriate geometrical parameters, can be used to estimate the variation

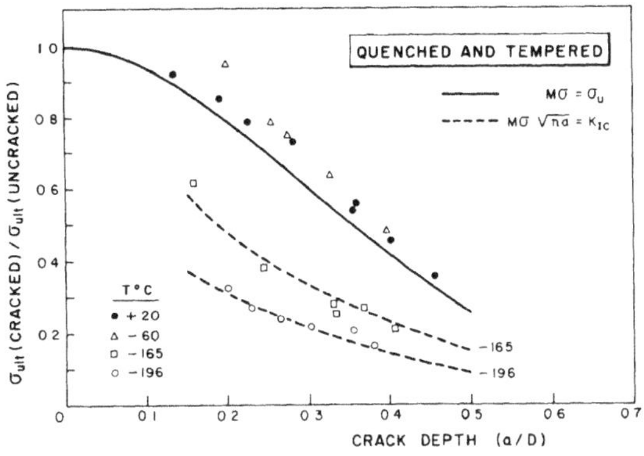

Fig. 5.13. Effect of crack depth on tensile strength. Quenched and tempered steel. $\sigma_y = 1380$ N/mm^2.

of failure load with crack depth. At liquid nitrogen temperature (−196°C) a value of $K_{IC} = 32$ MNm$^{-3/2}$ provides a good estimate of failure loads over the experimental range of (a/D) values.

The same steps, outlined above, are followed in this discussion of the behaviour in prestressing steels. Quenched and tempered steels are considered first and then cold drawn steels.

Figure 5.13 illustrates the variation of normalised values of rupture load with crack depth at temperatures ranging from ambient down to −196°C. These results were obtained from tests on quenched and tempered tendons with a 9 mm. diameter and a 0.2% proof stress of 1380 N/mm^2 [31,54,82].

For temperatures down to −60°C it will be observed that results fit an elastoplastic failure criterion $M\sigma = \sigma_u$ even though the steels have a high yield stress and exhibit little plastic deformation. At very low temperatures, the usual fracture mechanics expressions predict the failure loads, at least, within the experimental range of (a/D).

With a view to determining a valid K_{IC} value, liquid nitrogen temperature experiments were performed on short-rod specimens taken from the longitudinal direction in 9 mm. tendons and also on conventional cylindrical specimens. Values of $K_{IC}(L) = 29$ MNm$^{-3/2}$ and $K_{IC}(T) = 39$ MNm$^{-3/2}$ were obtained; where L refers to the longitudinal and T to the transverse direction. The difference in the values suggests a weak anisotropy. The variation of fracture toughness in the longitudinal and transverse directions with temperature is shown in Figure 5.14 [31,54]. Plane strain fracture toughness values for longitudinal specimens are considered to be valid up to a temperature of −150°C. The $K_{IC}(T)$ values were obtained using the techniques described in the previous section while $K_{IC}(L)$ values were determined following Barker's recommended proce-

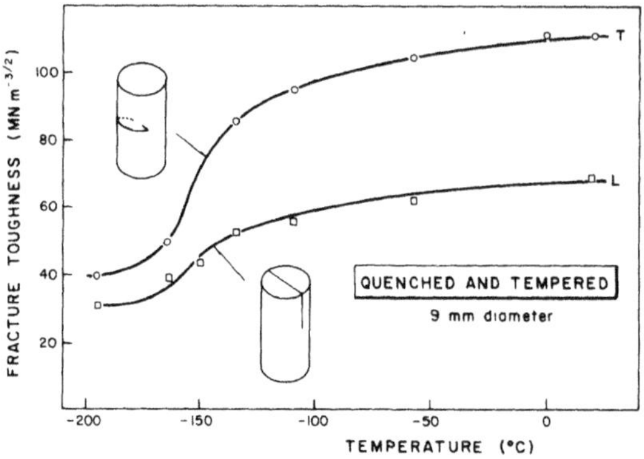

Fig. 5.14. Fracture toughness of quenched and tempered steel for prestressing concrete.

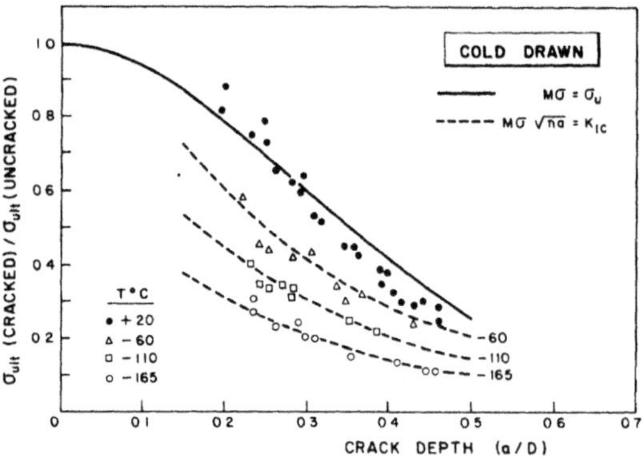

Fig. 5.15. Effect of crack depth on tensile strength. Cold drawn steel. $\sigma_y = 1590\ \mathrm{N/mm^2}$.

dure [12]. The range of validity of transverse specimens is less than for longitudinal ones.

Analogous results for cold drawn eutectoid steels of 7 mm. diameter and 0.2% proof stress $\sigma_{0.2} = 1590\ \mathrm{MNm}^{-3/2}$ are shown in Figures 5.15 and 5.16 [31,54,82]. This material exhibits a greater degree of anisotropy which arises from the fabrication procedure. This frequently results in a mode of rupture in which the crack tends to leave the plane perpendicular to the edge and propagate in a longitudinal direction. Tensile fractures for quenched and tempered (little anisotropy) and cold drawn wires are compared in Figure 5.17. The strong anisotropy makes interpretation of experimental results more difficult for cold drawn wires but Figure 5.15 does suggest that the $M\sigma = \sigma_u$ criterion can be used at ambient temperature and that, at low temperatures, linear elastic fracture mechanics may be applied, at least, for $a/D > 0.15$. Fracture toughness values for longitudinal and transverse cracked specimens are shown in Figure

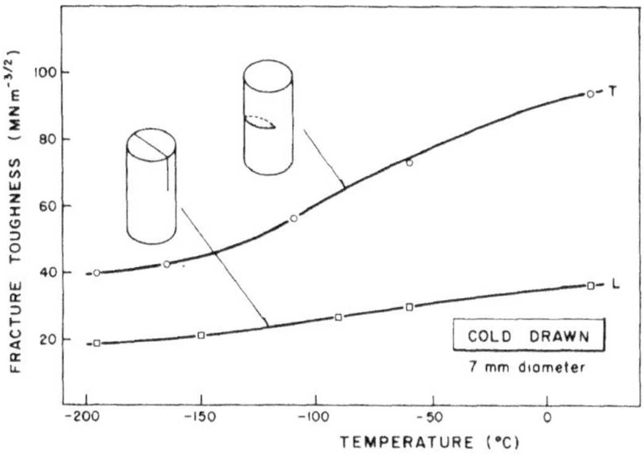

Fig. 5.16. Fracture toughness of cold drawn steel for prestressing concrete.

Fig. 5.17. Tensile fractures of cracked wires for prestressing concrete. (left) quenched and tempered. (right) cold drawn.

5.16. Longitudinal values may be valid upto −60°C. There does not seem to be the marked transition which was obtained with quenched and tempered steel. It should be remembered, however, that the results are based on expressions for isotropic materials and that crack propagation tended not to be confined to the plane perpendicular to the edge. Other attempts to measure the fracture toughness of cold drawn wires have been made by McGuinn [52,53].

5.3 Fracture under extreme conditions

Parameters such as strain rate and temperature are known to affect the fracture behaviour of reinforcing and prestressing steels. In this section, the effects of extreme values of these parameters are discussed. Such extreme values may occur when a structure is subjected to impact loading, in the case of strain rate. Fires or storage of liquid natural gas in prestressed concrete are examples of extreme temperatures conditions.

Effect of strain rate. Very few, well documented, tests on the influence of strain rate on the fracture behaviour of reinforced concrete are available. Results of tests carried out on reinforced concrete structures under static and dynamic loads show that the behaviour is mainly influenced by the properties of the steel under test. Most of the case studies available have been based on reinforced concrete, this is why the behaviour of reinforcing steel is better understood. The investigations were performed on service steels and there is no data on the cracked or notched dynamic response. It is known that if steel specimens are surface machined for the application of strain gauges it is difficult to obtain realistic values of strain distribution. Tests performed by Limberger et al. [50] at different strain rates showed significant differences in permanent strain distribution after fracture along lightly machined specimens and others which were ribbed.

The few available results [2,47,77] allow the following general comments to be made upon the ductility. Cold worked reinforcing steels exhibit an unexpected behaviour. Increasing the loading rate increases the elongation. This phenomenon is much less marked in hot rolled steels. The effect of strain rate upon the stress strain curves of a number of steels is illustrated in Figure 5.18. Prestressing steels suffer a loss of ductility of between 15% and 20% [2]. The reduction in area decreases for all types of steel as the strain rate increases but the reduction is small being of the order of 2% to 5%.

As far as the relationship between strength and strain rate is concerned, a slight increase has been observed with increasing strain rate in the range $\dot{\varepsilon} = 5.10^{-5}$. Reinforcing steels show a 10% increase in the limit of proportionality and in the elastic limit but the increase in the rupture load is somewhat less at about 5%. Increments for prestressing steels are still less, between 2% and 4%.

Experiment has shown that the ductility parameters (uniform and permanent elongation after rupture) and strength (limit of proportionality, yield and stress and failure load) are related to the strain rate through a logarithmic law over the range $\dot{\varepsilon} = 10^{-5}$ to 10 s^{-1},

$$a = A + B \log \dot{\varepsilon} \qquad (5.12)$$

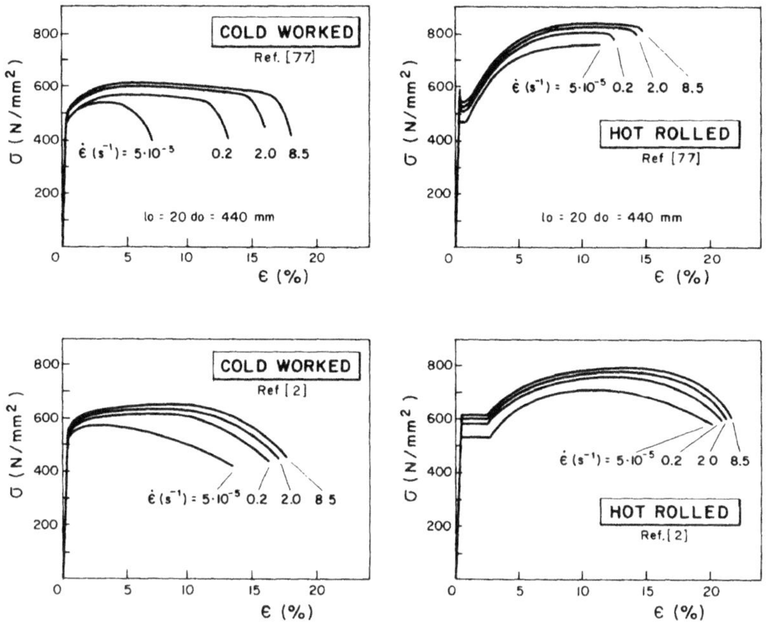

Fig. 5.18. Stress-strain curves at different strain rates. Influence of the quality of the steel.

where a represents one of the parameters mentioned above and A and B are material properties of the steel. It is also possible to use an expression which has only one dependent variable C[2] of the form

$$\frac{f_{\text{dyn}}}{f_{\text{static}}} = 1 + C \log \frac{\dot{\varepsilon}}{\dot{\varepsilon}_{\text{static}}} \tag{5.13}$$

values of A, B and C are given in the references.

Several tests on reinforced concrete beams have been performed to demonstrate the influence of the steel properties on the behaviour of the tested structure under high strain rates [50]. Beams with a high percentage of reinforcement, both tensile and compressive, failed by fracture of the reinforcement in the tensile zone. The concentration of cracks in the plastic hinge implies a high strain concentration in the reinforcement. The increase of the inelastic strains in dynamic beam tests is as significant as that in the tensile tests with increasing strain rate. The ductility, as measured in a bending moment-rotation diagram, of beams reinforced with cold drawn steel increases significantly with the rate of deflection, whereas the beam reinforced with hot rolled steel are rather insensitive to the rate of straining.

The properties of the reinforcement matrix interface over a range of strain rates is also required for prediction of the fracture behaviour of reinforced concrete structures [63,77]. To study the influence of rate of straining on the bond strength between the reinforcement and concrete pull-out tests have been performed. No significant influence of straining rate on bond strength was observed for plain bars and prestressing strands. However deformed bars showed a significant increase in bond strength with loading rate. The rate effect was also found to be greater for low strength concrete specially at small displacements between reinforcing steel and concrete. This result can be interpreted such that the effective bond length of a deformed bar decreases with increasing loading rate. This rate sensitivity of deformed bars is believed to occur also because of the bearing of the ribs on concrete which produces crushing and splitting of the concrete which is expected to be rate dependent.

Effect of high temperature [24]. Fracture properties for all types of steel are affected by temperature. Composition and processes of manufacture affect the behaviour of reinforcing steels. With temperatures of about 350°C recrystallization of the micro-structure commences which leads to a loss of the work-hardening effects. This explains the differences observed between cold-drawn or cold-twisted reinforcement and high tensile alloy steel.

Two types of tests are carried out which attempt to reproduce the mechanical conditions of the reinforcement in concrete elements under fire attack; stress-strain tests at the desired temperature, by this method the temperature dependent rupture strength can be measured, and creep tests to measure the critical temperature, at which $\dot{\varepsilon} \simeq 10^{-4}\ s^{-1}$. Rapid elongation of the steel at the critical temperature normally leads to undue rate of deflection and subsequent collapse of the structural member.

For design purposes the maximum strength, may be taken from Figure 5.19 as recommended by CEB [24].

Similarly, the benefits of the thermomechanical treatment of prestressing steels disappear when the treatment temperature is again reached. This explains the differences in the performances at high temperatures of cold-drawn stabilized steel and quenched and tempered steel of the same original strength.

Figure 5.19 shows the decrease in ultimate tensile strength of prestressing steels when the temperature increases. Two curves, which may be considered to be boundary values for the tensile strength have also been included for design purposes.

To predict fracture of reinforced concrete at elevated temperatures, the behaviour of concrete, reinforcing steel and prestressing steel has been investigated. In comparison with the knowledge accumulated for these materials little is known about bond at high temperature. An extensive experimental study has been initiated by Rostasy and co-

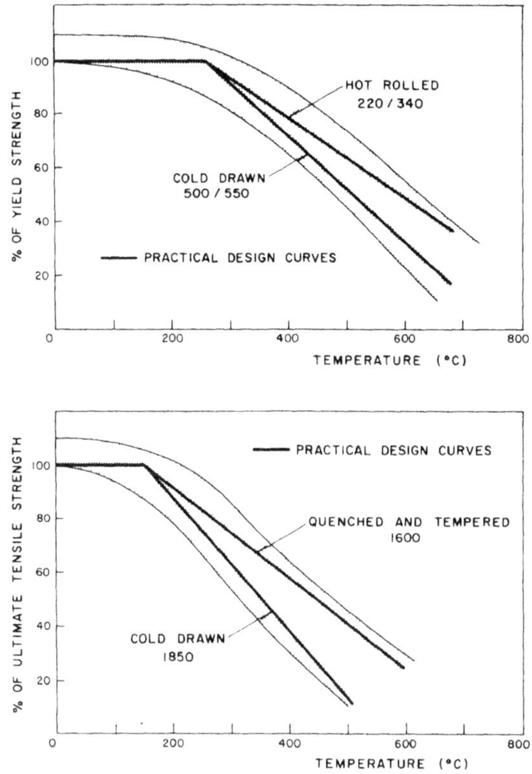

Fig. 5.19. Influence of temperature in tensile strength. Ref. (24).

workers [66,67]. The test results can be applied to estimate the safety of a bond anchorage against slip failure. It is possible to relate the load-displacement relationship τ, Δ from a bond test, to the compressive strength of concrete, β_w

$$\frac{\tau(T)}{\beta_w(T)} = A + B\Delta^C \tag{5.14}$$

where the bond stress has been subdivided into an adhesive resistance and a shear resistance. The adhesive resistance A disappears as the temperature exceeds 300°C. The coefficients B and C, describing the shear resistance, are independent of the compressive strength. Thus the bond stress is proportional to the strength of concrete, as the temperature rises. From this expression the distribution of bond stresses along the anchorage length of a bar and the safety against slip failure in case of fire can be determined.

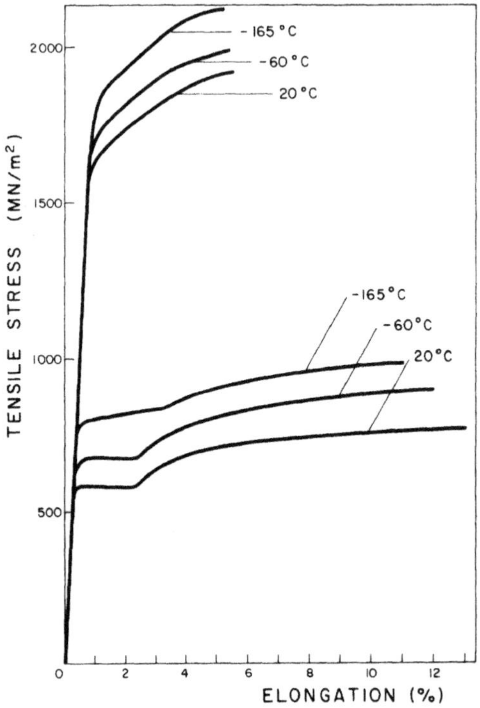

Fig. 5.20. Stress-strain diagrams of steels for reinforcing and prestressing's concrete, at different temperatures.

Effect of low temperatures. Prestressed concrete is a material normally used at ambient temperature conditions but it can also be used structurally at cryogenic temperatures. The satisfactory performance of prestressed concrete in cryogenic environments, coupled with its inherently sound structural characteristic, makes this material ideally suited to a variety of applications in the storage of cryogenic liquids. A state of the art report on the behaviour of prestressing steel and systems, reinforcing steel and concrete at low temperatures has been recently published, Elices, Faas, and Rostasy [33] in which some information relative to steel fracture has been included.

Tensile tests on unnotched samples have shown that steels for prestressing and reinforcing steel exhibit an increase in ultimate tensile strength as the testing temperature is reduced. This behaviour is depicted in Figure 5.20. It can be seen that, the elongation is not appreciably reduced.

A cause of concern, in the use of these steels at low temperatures, is the possible sensitivity to the presence of cracks and notches. This behaviour has been illustrated in Figure 5.11, for reinforcing steel, at

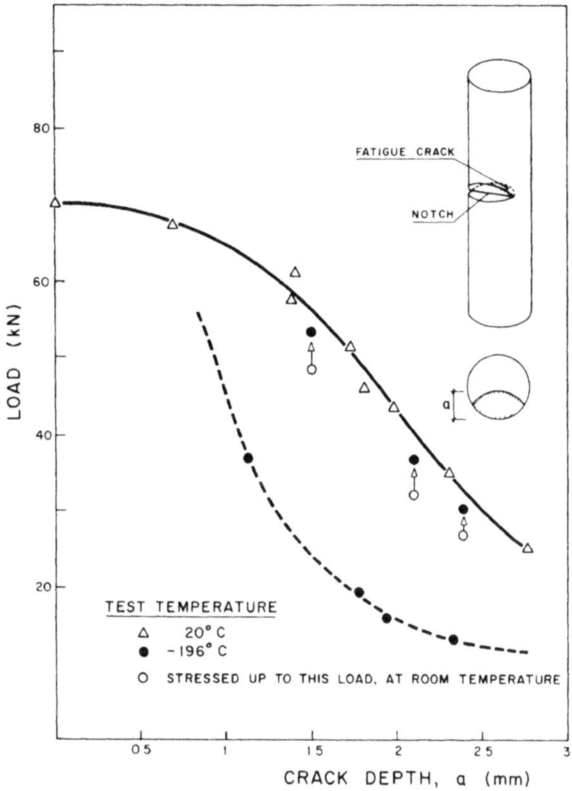

Fig. 5.21. Effect of prestretching before cooling.

−196°C, and Figures 5.13 and 5.15 for different types of prestressing steel. Fracture toughness has also been measured and results are shown in Figures 5.12, 5.14 and 5.16. In spite of its decrease at low temperatures steel tendons are tough enough, since the tendons are stretched at ambient rather than at very low temperatures. The behaviour which should be simulated in laboratory tests is that of a cracked tendon stretched at ambient temperature and subsequently loaded. The results of such an experiment are shown in the Figure 5.21 and it can be seen that a reasonable toughness is still achieved.

As already mentioned, for the design of bond anchorages and also for crack control, the bond strength, as well as the bond stress-displacement relation, must be known. Although many tests have been done at room temperature, only few tests have been performed at low temperatures [43,86].

Tendon anchorages are sensitive zones where fracture may be produced and this risk may be increased at low temperatures. For cryogenic applications it is important to note that, whether or not the cold strength

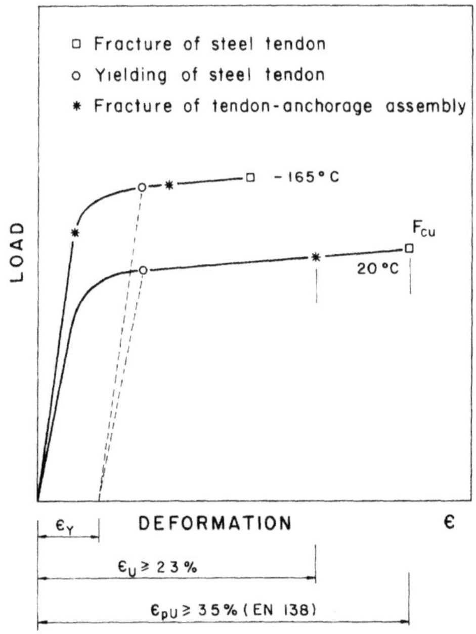

Fig. 5.22. Requirement for tendon-anchorage assembly (Ref. 33).

of the prestressing tendon is utilized in structural analysis, the tendon anchorage assembly does not fail in a brittle fashion. This means that the failure load of the tendon-anchorage assembly, for any value of the temperature, should exceed the cold yield strength of the prestressing steel. This basic requirement is shown in Figure 5.22. A failure load below the yield strength is accompanied by a brittle type failure.

5.4 Fatigue

Although fatigue of reinforced and prestressed concrete has not proved to be a problem in the past, the fatigue lives of many future structures will require assessment. The reasons for this are: (1) Refinements in the method of structural analysis and load evaluation, which imply reduction of safety factors and allowable live loadings (2) improved tensile strength of reinforcing steels not being accompanied by a similar improved fatigue strength and (3) the growing use of prestressed and reinforced concrete structures which in service are predominantly subjected to loads of repeated nature or aggressive environments that may reduce fatigue strength.

If design is to be based on the fracture of steel tendons, fracture mechanics techniques can be used to predict stress range-life data. These techniques complement the traditional approach based on SN curves, allowing a distinction betwen initiation and propagation times. Another advantage of the method is that it allows a proper investigation of the effect of any of the different factors which influence the fatigue behaviour of reinforcement; type of bar, mean stress, bar diameter, etc., which has often confused experimental studies. Further, this method has been used to predict the fatigue life of reinforced concrete beams and provide an acceptable lower bound of the fatigue life.

A summary of methods of analysis is beyond the scope of this chapter. Rather a resume of the applications of fracture mechanics to the fatigue failure of tendons is intended. It is also hoped to provide data which will be of use to the designer. Classical aspects of this section are based on two excellent reviews by Tilley [79,80]. The fracture mechanics approach uses data from Lovegrove and Salah [68–70] for reinforcing steels and from the author's laboratory for prestressing steels [30,71,72].

Fatigue crack initiation. It has been shown for many materials that there exists a critical value of the stress intensity factor range below which fatigue cracks either remain dormant or grow extremely slow at experimentally undetectable rates. This threshold cyclic stress intensity range, ΔK_{th}, is often defined as the maximum ΔK value at which no detectable growth occurs in 10^7 cycles. Since crack length monitoring equipment is usually accurate to better that 10^{-4} m, a test endurance of 10^7 cycles gives a maximum growth rate of 10^{-11} m./cycle.

Several methods have been used to establish these threshold values experimentally. The traditional approach to design under cyclic loading has involved S–N (stress range versus number of cycles to final failure) data. This approach does not separate out crack initiation and propagation stages and cannot be applied to the determination of structural life-expectancy with a crack-like defect of a known size. Nevertheless, it is accepted that for low amplitudes, some 90% of the total fatigue life is expended in developing the initial intrusion and these results can be used to estimate fatigue crack initiation as well as to establish lower limits to fatigue performance. The next section considers some results from Tilley's reviews [79,80].

Of the various types of fatigue assessment, uniaxial tests conducted on uncladded bars in repeated tension give the most conservative fatigue data. However, bending tests on reinforced concrete beams more closely reproduce service conditions and give a realistic assessment of the effect of factors that are influenced by the concrete. Each type of testing has merit and the selection of the one or the other is a question of convenience in relation to the aspect of fatigue being studied rather than any more fundamental issue. The slightly longer lives for bending tests can be justified because the highest stresses are restricted to the parts of the bar

Table 5.1
Values of K ($\times 10^{26}$) [80].

Type of loading	16 mm diameter	32 and 40 mm diameter
Axial	7.5	1.1
Bending	30.9	3.1

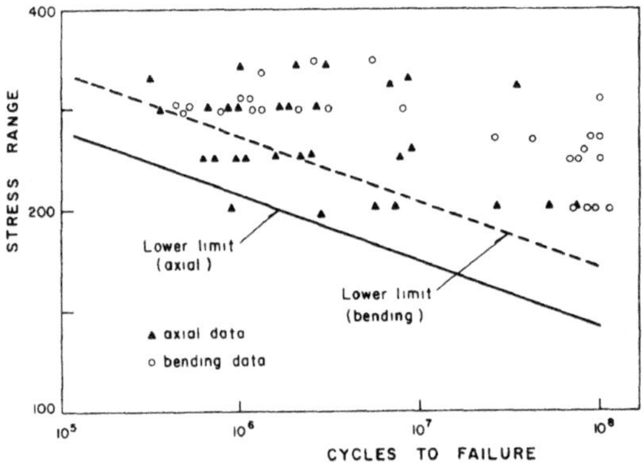

Fig. 5.23. Fatigue performance of 16 mm bars, according to G.P. Tilly and D.S. Moss [80].

furthest from the neutral axis and in the vicinities of cracks in the concrete. The likelihood of these locations coinciding with the worst defects in the bars is lower than in axial tests in which the bar is subjected to a uniform nominal stress so that fracture may be initiated from the worst defect.

The long endurance fatigue behaviour of six types of high strength reinforcement bars (cold-worked and hot rolled) has been studied and quantitative relationships involving the main variables have been evaluated. The performances have been expressed by a power law [80].

$$S^m N = K \tag{5.15}$$

When S is expressed in N/mm^2, the value of m has been shown to be close to 9. This holds for both axial tests in air and bending in concrete. The values of K are shown in Table 5.1, depending on type of test and bar characteristics.

Results of axial as well as bending test [80] are shown in Figure 5.23. A regression analysis of $\log N$ on $\log S$ was performed and a lower 95% confidence limit is shown in the figure. There is no evidence of the fatigue limit which is usually considered to develop at about 2×10^6 cycles.

Fatigue strength is dependent on the type of reinforcement bar and this is thought to be due to differences in initiation times. Plain bars with smooth surfaces exhibit the highest strengths. The presence of ribs introduces high local stresses and reduces the crack initiation phase. An increase in diameter of deformed bars produces a very pronounced reduction in fatigue strength which is an order of magnitude more than the size effect associated with plain cylindrical specimens; the strength at 10^7 cycles for 40 mm, diameter is about 25% less than for 16 mm. The explanation usually given for size effects is that bigger sections have a statistically greater likelihood of containing large flaws. For plain specimens most of the life is spent in initiation and the differences in propagation times for different diameter sections are minimal. For ribbed bars the initiation phase is reduced due to the presence of local stress concentrations at the ribs. Furthermore, Gurney [44,45] has shown that in the vicinity of a stress concentration the minimum size of flaw required to permit crack growth reduces with increased thickness of specimen. In consequence it can be argued that the initiation life of larger diameter bars is reduced.

Experimental S–N curves have been obtained for prestressing steels and Goodman or Smith diagrams may be obtained from these. As already mentioned, interpretation is made difficult when initiation and propagation times are difficult to separate. They do, however, provide much useful information for the designer.

A theoretical approach based on the concepts of fracture mechanics has been used by Lovegrove and Salah [68,69,70] and Verpoest [85] to predict crack initiation. Experimental work on formation of fatigue cracks in Torbar [68] has shown that cracks normally initiate at the root of the transverse ribs on the surface of the bar. At the peak of the first load cycle a plastic zone may be formed at the root of the rib. According to the abovementioned authors [69] the maximum depth of this zone in the bar, a_p, depends on the peak nominal stress σ_{max}, the stress concentration factor K_t the radius of the rib root ρ and the yield stress of the steel σ_y and can be estimated by the following expression:

$$a_p = \frac{\rho}{4}\left(\frac{k_t\sigma_{max}}{\sigma_y}\right)^2 - 1 \tag{5.16}$$

The critical fatigue crack is assumed to form at the deepest irregularity at the root of the rib, a_r (values of $a_r = 0.07$ mm, have been quoted [70] for Torbar steel). The initial size of the crack, a_0, may be defined as

$$a_0 = a_p + a_r$$

Therefore, the fatigue crack initiation life is defined here as the number of cycles required to sharpen the trough of the deepest surface irregularity in the vicinity of the rib and/or damage the plastic zone. The fatigue crack initiation life, N_i, is assumed to be related to the range of the stress intensity factor ΔK_0 at the tip of the initial crack by the following equation

$$N_i = B_1(\Delta K_0)^{m_1} \tag{5.17}$$

in which B_1 and m_1 are material constants which have been evaluated [70] for Torbar steel and found to be: $B_1 = 1.448 \times 10^8$ and $m_1 = 2.701$, when MN and m units are used. No similar values are yet available for other reinforcing steels.

Fractographic examination indicate that fatigue under axial loading tends to initiate at surface defects rather than in the vicinity of the ribs. This contrast with the behaviour under bending fatigue which exhibited fractures having initiation associated with the ribs [80].

An investigation, currently in progress in the author's laboratory, suggests that the ΔK_0 range for prestressing steels is a function of the stress ratio $R = \sigma_{min}/\sigma_{max}$. For example, values of $\Delta K \simeq 20$ MN m$^{-3/2}$ for $R = 0.28$, $\Delta K \simeq 11$ MN m$^{-3/2}$, for $R = 0.67$ and $\Delta K_0 \simeq 8$ MN m$^{-3/2}$ for $R = 0.9$ have been obtained, for 7 mm diameter wires with $\sigma_{0.2} = 1400$ N/mm^2 and ultimate tensile strength of 1620 N/mm^2.

Verpoest et al. [85] have developed a model which allows a qualitative and quantitative explanation of the influence of surface characteristics and bulk properties on the fatigue limit (i.e. the maximum stress amplitude at which cracks, that are present before or created during the fatigue test, do not grow up to final fracture) and fatigue threshold were measured on 2 mm diameter wires ranging from $\sigma_n = 1320$ N/mm^2 to $\sigma_n = 2218$ N/mm^2. In this range, fatigue thresholds decreases almost linearly from 7 MN m$^{-3/2}$ to 3.5 MN m$^{-3/2}$.

Fatigue crack propagation. For many materials the rate of fatigue-crack growth can be expressed by the Paris-Erdogan equation

$$\frac{da}{dN} = C(\Delta K)^n \tag{5.18}$$

where, N is the number of cycles, C is a material constant and n is an exponent, usually about 3. Recent research has demonstrated that in a typical log-log plot of fatigue crack growth rate da/dN versus stress intensity factor range ΔK, the abovementioned law is only valid for the intermediate range of growth rates, typically $10^{-8} - 10^{-6}$ m/cycle. The variation of growth rate da/dN with ΔK is actually sigmoidal in form, defined in a range of ΔK values bounded at its extremes by K_{IC} and threshold parameter ΔK_0.

The results of fatigue tests under constant amplitude load on cracked specimens taken from Torbar have shown that the rate of fatigue crack propagation da/dN may be described by a bi-linear relationship, to take into account the behaviour of very small cracks:

$$\frac{da}{dN} = C_1(\Delta K)^{n_1} \quad \text{for} \quad \Delta K < 9\ \text{MN.m}^{-3/2} \tag{5.19a}$$

$$\frac{da}{dN} = C_2(\Delta K)^{n_2} \quad \text{for} \quad \Delta K > 9\ \text{MN.m}^{-3/2} \tag{5.19b}$$

where, material constants are given Table 5.2.

The above theory, crack initiation plus crack propagation, was used to generate theoretical S–N data for concrete beams reinforced with Torbar. It was found [69] that the theory represents a reasonable lower bound for all the experimental data. In order to demonstrate the use of the theory, to investigate the effect of individual factors on fatigue performance of reinforced concrete beams, tests have been performed in which each of the influencing factors was varied individually. It was found that the stress concentration at the root of the rib is the most effective factor in explaining the variation in performance, 1% increase in the stress concentration factor at the rib results in 1.4% decrease in the fatigue limit at 10^7 cycles. The second most effective factor is the yield strength of the bar, indicating that provided the initiation and propagation properties of the steel are constant, improvement in the yield strength can improve the fatigue performance of the bar. Surface roughness, as measured by the maximum depth of surface irregularities, and the minimum stress have approximately identical effects, 1% increase results in about 0.15% decrease in the fatigue limit at 10^7 cycles. Minimal effects due to the change in bar size were detected. It is emphasized that the same is not true for initiation.

Experimental results of fatigue crack growth in eutectoid cold drawn steels for prestressing concrete [71] fit Paris law over a wide range of ΔK, as shown in Figure 5.24. Values of material constants are given in Table 5.2.

Figure 5.25 contains results obtained at different frequencies, with different waveforms and with different stress ratios. It seems that fatigue crack growth for prestressing steel wires is independent of waveform and frequency, at least inside the tested intervals. Crack growth also appears to be independent of stress ratio R, for positive values of R.

Integration of Paris law has been used for life prediction of prestressing wires without any defect except those produced during steel

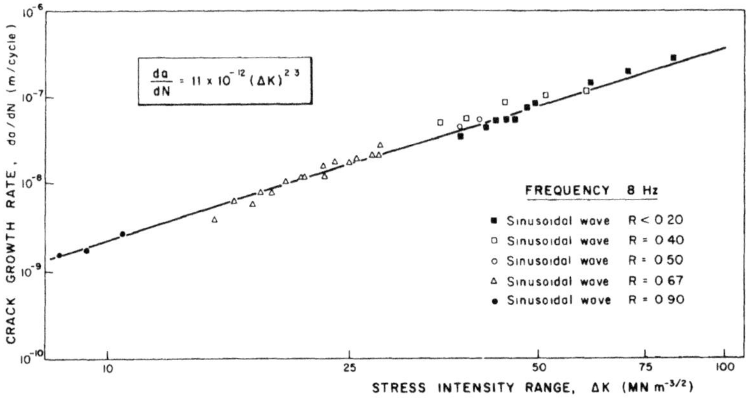

Fig. 5.24. Crack growth rate for different R values.

TABLE 5.2
Values of C and n (Units: m, MN).

Steel	$\sigma_{0.2}$ (N/mm²)	C	n	References
Torbar	435	3.83×10^{-29}	20.9	[70] *
Torbar	435	3.16×10^{-12}	3.1	[70]
Cold-drawn	1400	11.08×10^{-n}	2.3	[71]
Cold-drawn	1520	13.96×10^{-12}	2.3	[71]
Cold-drawn	1560	16.05×10^{-12}	2.3	[71]

* For $\Delta K < 9$ MN.m$^{-3/2}$

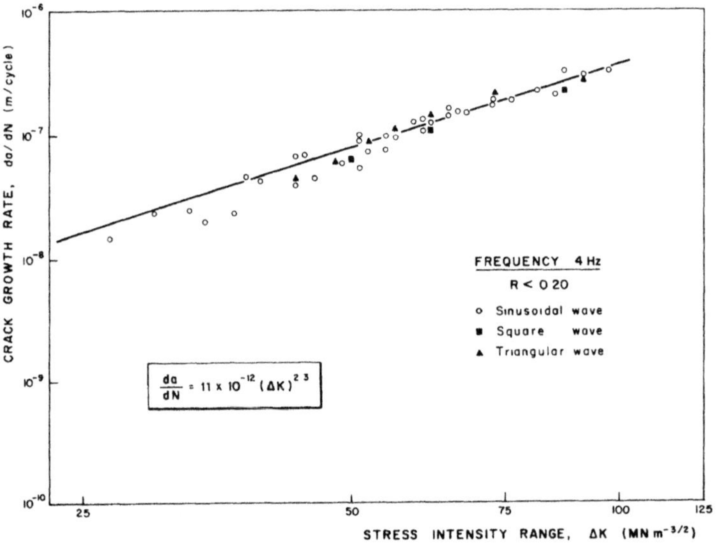

Fig. 5.25. Crack growth rate for different wave forms and frequencies.

processing. Fatigue tests results, carried out on smooth samples [30] with stress ranges high enough to ensure that initiation times are negligible as compared with propagation times, have been predicted successfully [71]. The agreement found with theoretical predictions seems to validate this approach, based on LEFM, although some plasticity has been developed, this aspect deserves further research. Another aspect, worthy of study, is the effect of material anisotropy, which has already been mentioned in the section on the fracture of cracked specimens.

Literature on fatigue failure of cables is scarce. Birkenmaier [15,16] presented a method of determining a safe value for the permissible stress range in large parallel wire tendons based on a series of systematic fatigue tests on the wire alone. Basu and Chi [13] investigated, using the methodologies of LEFM, the fatigue behaviour of bridge cables resulting from the stress reversals caused by vortex shedding. An attempt was made to compare the results of fatigue life obtained from the formulation with the available experimental data.

There is, likewise, little published information on the role of anchorages in the fatigue problem, even though these are known to be weak points [15,30]. At present, such problems are tackled empirically, there being no available theoretical treatment of fatigue in multiaxial stress systems.

There has been comparatively little work on random fatigue of reinforcement bars. The only reports the writer is aware of are the reviews of Tilly [79,80] and references therein. The section continues with a brief summary of the ideas contained in them. In this work, random loading data are expressed in terms of r.m.s. stress ranges so that they can be compared with constant amplitude data. This is based on assumptions that there are no interactive effects between different stress levels, life is wholly in propagation, and $n = 3$.

Axial tests have been performed on reinforcement bars subjected to narrow-band random loading having an approximated Rayleigh spectrum. The spectra was characterized, for convenience, by the RMS values of stress and it was realized that estimation of behaviour under variable-amplitude loading using the Palmgren-Miner concept was strongly dependent, for welded joints, on the method used to represent constant-amplitude data at long endurances. Taking an endurance limit at 2×10^6 cycles gives an optimistic or upper bound representation. Assuming the same law, $S^m N = K$, for all endurances, gives a pessimistic or lower-bound. It seems appropriate to use $S^m N = K$ to 10^7 cycles with the stress exponent raised to $m + 2$ at $N > 10^7$, because it has been shown that, for welded joints as well as for continuous bars, this gives a reasonable allowance for the contribution of low stresses which become damaging after some crack propagation has occurred.

5.5 Environment sensitive cracking

When steel is completely surrounded by concrete it may be expected that it is unlikely to suffer corrosion. Nevertheless, the ingress of aggressive ions at defects in the concrete may result in a breakdown of the passivity promoted by the relatively high pH environment that would otherwise exist. The high strength of steel for prestressing concrete, coupled to the high tensile stress to which it is subjected, raise questions as to the possibility of stress corrosion cracking since a wide range of high strength steels are well known to be susceptible to stress corrosion, particularly in the presence of chloride ions, which are found in many applications of concrete structures. Indeed, fractures thought to be due to stress corrosion have been reported in a wide range of prestressed components. Most of this information is to be found in the proceedings of three symposia organised by F.I.P. [14,18,32], a monograph by CUR [26] and reviews by Phillips [60] and Nurnberger [55].

Despite this evidence, the situation is by no means alarming and in the last F.I.P. symposium it was concluded that: "In well designed and detailed structures, which are constructed with sufficient case, the necessary conditions for stress corrosion to occur should not exist". Nevertheless, laboratory investigations have shown that prestressing steel may fracture by stress corrosion in environments that may occur when steel is unprotected by concrete. Under these conditions all steels are susceptible to stress corrosion cracking.

Thus, since a potential problem exists some comment should be made on the fracture of the steels. The section will then continue with reviews of the two most characteristic aspects of environment sensitive cracking; stress corrosion and corrosion fatigue.

Stress corrosion cracking. Just as had been the case in fatigue, the study of stress corrosion cracking in reinforcing and prestressing steels passed, first, through an empirical stage in which experimental data were accumulated and, now, it is approaching the stage where fracture mechanics can be used to obtain a more quantitative description of the phenomenon.

Experiments performed in the first phase consisted of subjecting the tendons to a predetermined load, immersing in an aggressive medium and recording the time to failure. In this way stress-time to failure curves were obtained for different media. These curves, which are analogous to S–N fatigue curves, provide much useful information on material behaviour but, just as in fatigue, there are difficulties in separating times for crack initiation and propagation. Consequently test results tend to show a lot of scatter, Figure 5.26, and the test method proposed by the F.I.P., for the detection of hydrogen embrittlement, then requires a minimum of

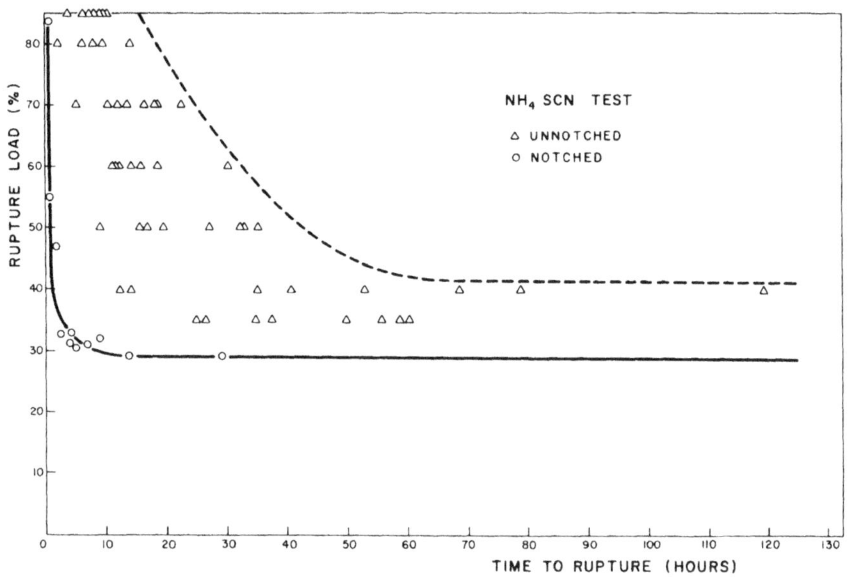

Fig. 5.26. Stress-time to fracture results for smooth specimen.

twelve specimens [39]. The tests are, however, simple, easy to perform, relatively cheap and provide much useful information [19]. For example the F.I.P. test in NH_4SCN is useful for characterizing the sensitivity of prestressing steels to hydrogen embrittlement and can be recommended for quality control of a given steel grade and to compare various prestressing steels of the same grade. Parkin's constant strain rate test [58] has also provided much valuable data on stress corrosion in prestressing steels.

The application of fracture mechanics techniques, although promising, is not without its difficulties. As well as the usual requirement of linear elasticity, there is also the need to incorporate new electrochemical parameters such as the pH of the medium and the potential between the tendon and a reference electrode. The mission of one of these has produced confusion by attempting to compare results under different conditions. The stress field in a cracked specimen is well defined by the stress intensity factor K. Bearing in mind that, at the tip of the crack, local variations in pH and potential may occur, it might be more meaningful to examine phenomena in a (K, pH_e, E_e) space, where the subscript e refers to local values, rather than in a global (K, pH, E) space.

As well as the difficulties in determining times to initiate and propagate cracks there is the additional problem of defining an initial crack size a_0. Results on cold drawn eutectoid steels [58] show that further

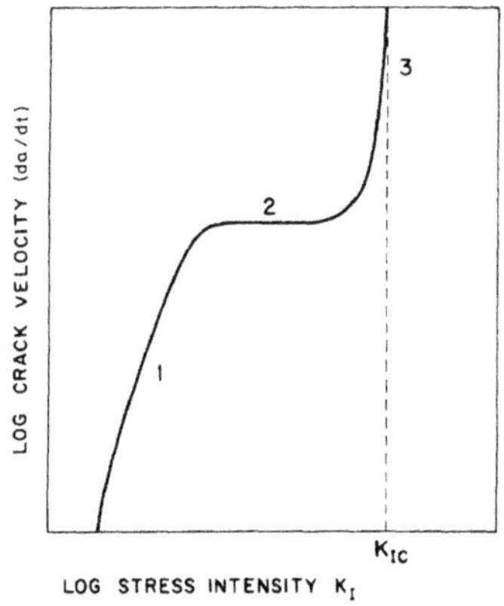

Fig. 5.27. Crack velocity versus stress intensity.

complexity is added by the presence of variables pH and E. Even under identical conditions of stress and pH, the initiation times vary with potential E; enhanced cracking results at potentials below about -900 mV (SCE), with an intermediate region in which cracking is absent or less severe than at other potentials. Cracking at higher potentials occur when the pitting potential is exceeded and appears to be dissolution-related. The surface condition, clearly, must play an important part in determining the initiation time and it has been shown that residual fabrication stresses can exert an enormous influence upon the crack initiation time, t_i [35].

The propagation phase is easier to investigate. It is known that the da/dt versus K curve can be characterised by three regions, just as in fatigue, Figure 5.27. In the first region, for K values close to K_{ISCC}, there is a rapid variation of the crack propagation rate with K. Then, in the second phase, this is much reduced and may even be insignificant. This would seem to be the case in quenched and tempered steels and, in general, in those steels which show weak anisotropy in fracture. Finally, in the third zone, in which K approaches the critical value, the growth rate increases once again until fracture. In heavily cold worked steels the crack propagation is hindered by the anisotropy of the tendon. The crack, then, tends to leave the plane perpendicular to the edge and propagate in a plane parallel to the edge. From the point of view of strength, this may be regarded as advantageous; the increase in strength being caused by a reduction in the stress intensity factor.

A number of papers have been published on this topic [20,34,52,53] and all have included experiments on tendons with transverse cracks which have been grown from notches by fatigue. The stress intensity factor, used under linear elastic conditions, has been described in Section 5.2. The K_c values were obtained by fitting a curve to data from tests in

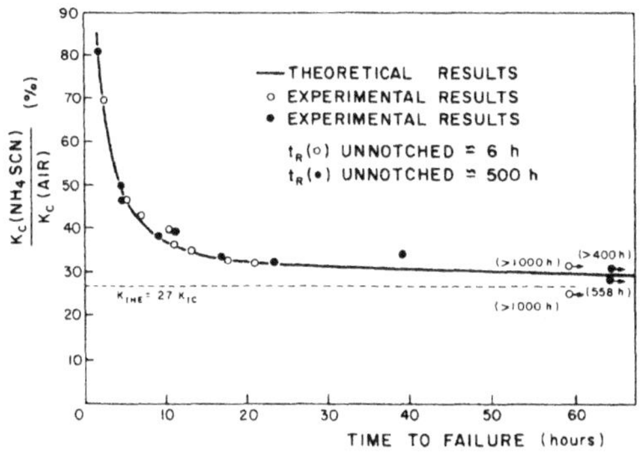

Fig. 5.28. Stress-time to fracture results for cracked specimens.

air giving failure load as a function of a/D also described earlier. One conclusion that may be drawn is that cracked prestressing steels behave in a very similar manner under test. Thus, two steels which have times to failure as different as 500 and 5 hours, as measured in the F.I.P. test, behave in an identical manner when tested with a crack, Figure 5.28. Furthermore, it has been shown that the limiting value in NH_4SCN is practically constant for all commercial steels and is of the order 30% of K_c. This limit is a function of the hydrogen fugacity which varies from one medium to another [53]. A knowledge of these limiting values in practical media would allow a determination of the design life of a prestressed member subjected to such aggressive media. While detailed work has not yet been done to establish limiting environmental conditions for cracking, it seems that these might be expected to be broadly similar to those previously found for axially precracked specimens [52] despite the pronounced anisotropy of this material.

Corrosion fatigue. Although fatigue has not proved to be a problem to date, loading cycles and corrosive conditions are becoming increasingly severe so that the margin of reserve strength is progressively being reduced. Nowadays, it is widely accepted that fatigue failure of off-shore concrete structures would most likely occur in the reinforcement tendons. Browne and Domone [21] state that, in composite reinforced or prestressed concrete – sections, the levels of the stress in the steel are a greater percentage of the ultimate stress than for concrete and it is therefore generally sufficient to consider fatigue properties of the steel as controlling the fatigue performances of the structural element.

Just happened with fatigue and stress corrosion, this problem has been tackled in a classical manner and it is only recently that it has been viewed in the light of fracture mechanics. The traditional investigation involved fatiguing beam in aggressive media, almost always seawater, and

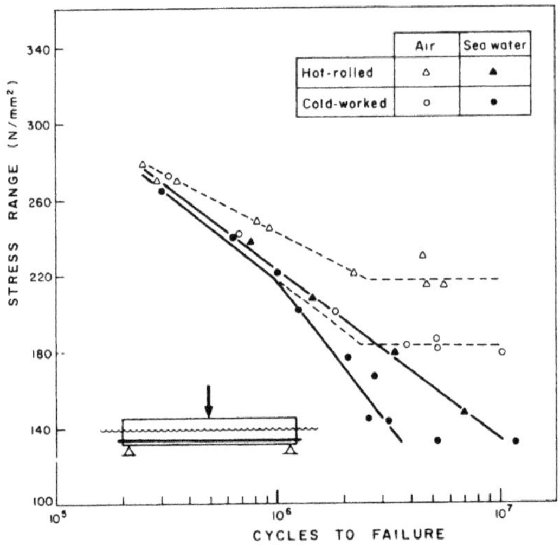

Fig. 5.29. Corrosion fatigue of steel for reinforcement. S-N curves.

recording the number of cycles to failure. A number of articles which follow this approach have been published recently [3,10,59,78]. Bending tests in a liquid environment can present experimental difficulties because it has been found that the overall stiffness of a reinforced beam increases during cycling so that stresses in the reinforcement cannot be calculated accurately [59]. This is believed to be due to cracks being wedged open by debris, a mechanism which does not occur during tests in air. A more recent paper, by Roper [65], studies the effects of differences in materials and exposure conditions on the fatigue endurance of reinforcing steels by testing reinforced concrete beams, simply supported and centrally loaded (Figure 5.29). All the beams, reinforced with hot-rolled steel, tested in sea water as well as NaCl solution (3%), showed no endurance limit before 10^7 cycles, as is suspected for corrosion-fatigue phenomena. The behaviour for beams manufactured with cold-worked steel reinforcement is not so clear and it was reported that a finite life region, with two different slopes, is followed by a long life region. Galvanizing improved the fatigue properties of concrete beams in all cases. For beams tested in air, the existence in the case of galvanized reinforcement of long life regions at higher stress ranges is noteworthy but most significant are the improvements in fatigue properties when beams are tested in the presence of sea water.

Once again, application of fracture mechanics concepts to the problem of corrosion fatigue, in this case, leads to difficulties in separating the periods of initiation from those of propagation; especially the lack of precision in the meaning of initiation. The electrolyte plays a decisive

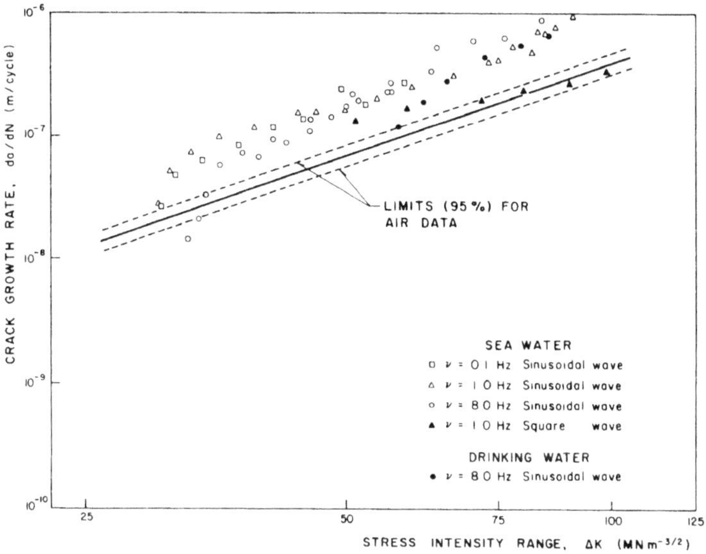

Fig. 5.30. Corrosion fatigue of prestressing steel. Crack growth rate.

part in crack initiation; so much so that it is believed that initiation will always occur, i.e. $\Delta K_0 = 0$, and that it is just a matter of waiting long enough. It should be remembered that, during its working life, a structure could suffer 10^8 cycles. Crack initiation time is reduced as corrosion pits readily become initiation sites for fatigue cracks. Pit development is time dependent rather than frequency dependent; the longer the corrosion fatigue life, the greater the reduction in fatigue endurance.

The pattern of crack growth rate in corrosion-fatigue is not so simple as in fatigue or stress corrosion alone. The effect of the corrosive medium is to increase the fatigue crack growth rate and this can depend on frequency, wave form and stress ratio, R, among other parameters. Recent, as yet unpublished results [72] for prestressing steels are shown in Figure 5.30. These are based on 7 mm diameter, cold drawn eutectoid steel with a $\sigma_{0.2} = 1400$ N/mm^2. The experiments were performed on precracked tendons with crack depths ranging from 1.5 to 3.5 mm. As may be deduced from the figure, the rate of crack propagation increases in a liquid medium such as in seawater or drinking water. Effects of frequency are rare or insignificant in the range 0.1 to 8 Hz. There does, however, seem to be a detectable effect of wave form and values obtained with a square wave are similar to those in air.

The advantages of the treatment of corrosion fatigue with fracture mechanics techniques are still to be realised in current design codes. So that both API [1] and F.I.P. [38] state that the resistance of a structure is adequate, if the stress range in the straight rebars is less than 140

N/mm^2. FIP also imposes a minimum stress level of 0,4 σ_y. Veritas [28] specifies both a Wohler curve and an endurance limit. With these criteria, allowing more cycles in the useful design life of the structure would require the specification of lower endurance limits.

References

1. ACI Report 357R-78, Guide for the design and construction of fixed offshore concrete structures, American Concrete Institute (1978).
2. Ammann, W., Muhlematter, M., Bachmann, H., Stress-strain behaviour of non-prestressed and prestressed reinforcing steel at high strain rates. Symp. Concrete Structures under impact and impulsive loading. RILEM.CEB.IABSE.IASS. Bundesanstalt fur Materialprufung (BAM) (1982).
3. Arthur, P.D., Earl, J.C. Hodgkiss T., Fatigue of reinforced concrete in sea water, Concrete. Vol. 13, n° 5 pp. 26–30 (1979).
4. Astiz, M.A., Estudio de una fisura superficial en un alambre de acero de alta resistencia, Universidad Politécnica de Madrid. Escuela de Ingenieros de Caminos (1976).
5. Astiz, M.A., Elices, M., Morton J., Valiente, A., A photoelastic determination of stress intensity factors for an edge-crack rod in tension, Proceedings Soc. Exp. Stress Analysis. Michigan Meeting (1981).
6. Astiz, M.A., Elices, M., An elastic singularity element for two and three dimensional crack problems (to be published).
7. Athanassiadis, A., Stabilité, tenacite, propagation des fissures dans les fils et barres en acier, Report de recherche LPC 89 (1979).
8. Ayling, E., Bevan, L., Load distribution in a single-edge-notch tensile specimen, Int. J. Fracture Mech. 8 pp. 341–342 (1972).
9. Bannister, J.L., Steel reinforcement and tendons for structural concrete, Concrete, July, August (1968).
10. Bannister, J.L., Fatigue and corrosion fatigue of Tor bar reinforcement, The Structural Engineer. Vol. 56 A, pp. (1978).
11. Bartos B. (Ed.), Bond in concrete, Applied Science Publishers pp. 466 (1982).
12. Barker, L.M., Short rod and shor bar fracture toughness specimen geometries and test methods for metallic materials, ASTM STP 743 pp. 456–475 (1981).
13. Basu, S., Chi, M., Analysis study for fatigue of bridge cables, IABSE Colloquium. Fatigue of steel and concrete structures Lausanne 1982.
14. Bijl, C.L., Lamers, L.R., Wind, G. (Eds), 1st FIP stress corrosion symposium, Royal Dutch Blast Furnaces and Steelworks Ltd. (1971).
15. Birkenmaier, M., Narayab R., Fatigue resistance of large high tensile steel stay tendons, IABSE Colloquium. Fatigue of steel and concrete structures Lausanne 1982.
16. Birkenmaier, M., Fatigue resistance tendon for cable-stayed construction, IABSE Proceed. P.30/80, May 1980.
17. Blackburn, W.S., Calculation of stress intensity factors for straight cracks in grooved and ungrooved shafts, Eng. Fracture Mechanics, Vol. 8 731–736 (1976).
18. Blekkenhorst, F., Lamers, L.R., Wind, G. (Eds), 2nd. FIP stress corrosion symposium, Hoogovens IJmuiden, B.V. (1974).
19. Brachet, M., Bilan de quelques années d'observation des phenomenes de corrosion sous tension des fils en acier a hautes caracteristiques, Annales de l'ITBTP vol. 267. pp. 81–100 (1970).

20. Brachet, M., Raharinaivo, A., Stress corrosion cracking of reinforcing bars, Int. Conf. SCC and HE of iron base alloys. NACE pp. 766–773 (1973).
21. Browne, R.D., Domone, P.L., Permeability and fatigue properties of structural marine concrete at continental sheef depths, Int. Conf. Underwater Construction Technology. Cardiff (1875).
22. Bush, A.J., Stress intensity factors for single-edge-crack solid and hollow round bars loaded in tension, J. Testing and Evaluation, JTEVA, Vol. 9, n° 4, 216–223 (1981).
23. CEB Bond action and bond behaviour of reinforcement, Bulletin d'information N 151 (1982).
24. CEB Design of concrete structures for fire resistance, Bulletin d'information N. 145 (19) (1982).
25. Creager, M., Paris, P. Elastic field equations for blunt cracks with references to stress corrosion cracking, Int. J. Fracture Mechanics, Vol. 3 n°4. pp. 247–252 (1967).
26. CUR, Cases of damage due to corrosion of prestressing steel, Netherlands Committee for Concrete Research pp. 96 (1971).
27. Daoud, O.E.K., Cartwright, D.J., Carney, M., Strain energy release rate for a single-edge-cracked circular bar in tension, J. Strain Analysis. Vol. 13, n°2, 83–89 (1978).
28. Det Norske Veritas, Rules for the design, construction and inspection of offshore structures, Appendix D. Concrete Structures (1977).
29. Dixon, J.R., Strannigan, J.S., McGregor, J., Stress distribution in a tension specimen notched on one edge, J. Strain Analysis, Vol. 4, n° 1, pp. 27–31 (1969).
30. Elices, M., Sànchez-Gàlvez, V., Fatiga de alambres de pretensado, Hormigón y Acero, V. 125, pp. 85–99, 1977.
31. Elices, M., Mestre, A., Planas J., Valiente, A., Low temperature fracture properties of steels for reinforcing and prestressing concrete (to be published).
32. Elices, M., Sànchez-Gàlvez, V., 3rd, FIP stress corrosion symposium, FIP and Berkeley Univ. (1982).
33. Elices, M., Faas, W., Rostasy, F., Wiedemann, G. Cryogenic behaviour of materials for prestressed concrete, FIP/5/11 84 pp. (1982).
34. Elices, M., Sànchez-Gàlvez, V., Entrena, A., SCC testing of cold drawn steel wires in NH_4SCN solution. K_{ISCC} measurements, 3rd FIP Symposium on stress corrosion (1982).
35. Elices, M., Maeder, G. and Sànchez-Gàlvez, V. Effect of surface stress on hydrogen embrittlement of prestressing steels, Br. Corros. J., Vol. 18, No. 3 (1983).
36. Fan Yuan-Xun, Tian-you Fan and Da-Jun Fan, The approximate analytical solution for both surface and embedded cracks in cylinder with finite size, Eng. Fracture Mech. Vol. 16 n° 1, pp. 55–67 (1982).
37. FIP, Report on Prestressing steel: 1, types and properties (1976).
38. FIP, Recommendations for the design and construction of offshore sea structures, FIP Recommendations (1977).
39. FIP, Report on Prestressing Steel: 5. Stress Corrosion Cracking resistance test for prestressing tendons (1980).
40. Godfrey, H.J., The fatigue and bending properties of cold-drawn steel wire, Trans. ASM, pp. 133–168 (1941).
41. Godfrey, H.J., Compression and tension test of structural alloys (Discussion), Proceed. ASTM Vol. 41, pp. 573–574 (1941).
42. Godfrey, H.J., The mechanical properties of steel wire: A seminar, The Wire Association International (1979).
43. Goto, Y., Miura, T., Experimental studies on properties of concrete cooled to about −160°C, Tohoku University Report. Vol. 44, n° 2 pp. 357–385 (1979).
44. Gurney, T.R., The influence of thickness on fatigue strength of welded joints, Proceed. 2nd Int. BOSS Conf. pp. 523–534 (1979).

45. Gurney, T.R., Johnston, G.O., Revised analysis of the influence of defects on the fatigue strength of transverse non load carrying fillet welds, Welding Res. Inst. 3 pp. (1979).
46. Hahn, G.T., Sarrate, M., Rosenfield, A.R., Criteria for crack extension in cylindrical pressure vessels, Int. J. Fracture Mechanics, Vol. 5 pp. 187–210 (1969).
47. Hobbs, R.E., Ghavami, K., The fatigue of structural wire strands, Int. J. Fatigue pp. 69–72, April 1982.
48. Knott, J.F., Fundamentals of Fracture Mechanics, Butterworths, pp. 153, 1973.
49. Kong, B. Paris, P.C., On the cup and cone fracture of tensile bars, ASTM STP 677 pp. 770–780 (1979).
50. Limberger, E., Brandes, K., Herter, J., Influence of mechanical properties of reinforcing steel on the ductility of reinforced concrete beams with respect to high strain rates, Symp. Concrete Structures under impact and impulsive loading, RILEM.CEB. IABSE. IASS, Bundesanstalt fur Materialprufung (BAM) (1982).
51. Maupetit, P., Olivie, F., Raharinaivo, A., Francois, D., Shear fracture of prestressing plain carbon steel wires under complex loading, Int. J. Fracture, pp. 725–727 (1977).
52. McGuinn K.F., Griffiths J.R., Rational test for stress corrosion crack resistance of cold drawn prestressing tendon, Br. Corros. J., Vol. 12 No. 3, pp. 152–157 (1977).
53. McGuinn, K.F., Elices, M., Stress corrosion resistance of transverse precracked prestressing tendon in tension, Br. Corrs. J. Vol. 16, pp. 187–195 (1981).
54. Mestre A., Comportamiento mecânico de aceros de armar y pretensar en condiciones criogénicas, Tesis Doctoral (1982).
55. Nurnberger, U., Analyse und Auswertung von Schadensfallen and Spannstahlen Strasenbau und strasenverkehrstechnik. Heft 308 (1980).
56. Paris, P.C., Tada, H., Zahoor, A., Ernst, H., The theory of instability of the tearing mode of elasto-plastic crack growth, ASTM STP 668, pp. 5–36 (1979).
57. Parkins, R.N., Stress Corrosion cracking. The slow strain rate technique, ASTM. STP 665 (1979).
58. Parkins, R.N., Elices, M., Sanchez-Gâlvez, V., Caballero, L., Environment sensitive cracking of prestressing steels, Corrosion Science. Vol. 22, pp. 379–405 (1982).
59. Peyronnel, J.P., Trinh, J., Experimental study on the behaviour of concrete structural element in sea water, Annales de l'ITBTP. Vol. 360, pp. 42–57 (1978).
60. Phillips, E., Survey of corrosion of prestressing steel in concrete water retaining structures, Australian water Resources Council. Tech. Paper n°9 (1975).
61. Raharinaivo, A., Utilisation des concepts de la mecanique de la rupture pour l'exploitation des essais classiques de corrosion sous tension. Corrosion, Traitements, Protections, Finition, Vol. 20, pp. 276–284 (1972).
62. Rehm, G., Nurnberger, U., Patzak, M., Keil und Klemmverankerungen fur dynamish beanspruchte Zugglieder and Hochfesten Drahten, Bauingenieur 51 pp. 287–298 (1977).
63. Reinhardt, H.W., Concrete under impact loading. Tensile strength and bond, Heron. Vol. 27, n°3, pp. 2–48 (1982).
64. Rogers, H.C., The effect of material variables on ductility, Ductility ASM pp. 31–62 (1968).
65. Roper, H., Reinforcement for concrete structures subject to fatigue, IABSE Colloquium. Fatigue of steel and concrete structures. Lausanne 1982.
66. Rostasy, F.S., Kepp, B., Time dependence of bond, Institut fur Baustoffe, Massivbau und Brandschutz, Braunschweig (1982).
67. Sager, H., Rostasy, F.S., The effect of elevated temperature on the bond behaviour of embedded reinforcing bars, Technische Universität. Braunschweig (1982).
68. Salah el Din, A.S., Lovegrove, J.M., Formation and growth of fatigue cracks in reinforcing steel for concrete, Fatigue of Engineering Materials and Structures, V. 3, pp. 315–323 (1980).

69. Salah el Din, A.S., Lovegrove, J.M., Fracture mechanics predictive technique applied to fatigue IABSE Colloquium. Fatigue of steel and Concrete Structures. Lausanne 1982.
70. Salah el Din, A.S., Lovegrove, J.M., Fatigue of cold worked ribbed reinforcing bar. A fracture mechanics approach, Int. J. Fatigue. pp. 15–26, January 1982.
71. Sanchez-Gàlvez, V., Elices, M., Valiente, A., Fatigue crack propagation in steel prestressing wires, IABSE Colloquium, Fatigue of Steel and Concrete Structures. Lausanne 1982.
72. Sànchez-Gàlvez, V., Valiente, A., Elices, M., Corrosion fatigue of prestressing steels, III Congreso Nacional y I Iberoamericano de Corrosión y Protección, Madrid (1983).
73. Shupack, M., A survey of the durability performance of Post-Tensioning Tendons, J.ACI. Vol. 75, pp. 501–510 (1978).
74. Sih, G.C., A three-dimensional strain energy density factor theory of crack propagation, Mechanics of Fracture 2. Edited by G.C. Sih, Noordhoff International Publishing, Leyden, pp. 15–53 (1975).
75. Sih, G.C., Fracture Toughness Concept, ASTM STP 605, pp. 3–15 (1976).
76. Sih, G.C., Experimental fracture mechanics: Strain energy density criterion Mechanics of Fracture. 7, Edited by G.C. Sih, Martinus Nijhoff Publishers, pp. 17–56 (1981).
77. Suaris, W., Shah, S.P., Mechanical properties of materials, Symp. Concrete Structures under impact and impulsive loading, RILEM. CEB. IABSE. IASS., Bundesanstalt fur Materialprufung (BAM) (1982).
78. Taylor, H.P., Sharp, J.V., Fatigue in offshore concrete structures, The structural Engineer. Vol. 564, pp. (1978).
79. Tilly, G.P., Fatigue of steel reinforcement bars in concrete: A review, Fatigue of Engineering Materials and Structures. V. 2, pp. 251–268 (1979).
80. Tilly, G.P., Moss, D.S., Long endurance fatigue of steel reinforcement, IABSE Colloquium. Fatigue of steel and concrete structures. Lausanne. 1982.
81. Trotter, H.G., High strength steel reinforcement, The Metallurgist and Materials Technologist, pp. 73–76 (1977).
82. Valiente, A., Criterios de fractura para alambres, Universidad Politécnica de Madrid. Escuela de Ingenieros de Caminos (1980).
83. Valiente, A., Elices, M., Morton, J., Determinación de factores de intensidad de tensiones mediante la utilización de tecnicas fotoelasticas con modelos entallados, Anales de Física. Vol. 77, pp. 122–129 (1981).
84. Valiente, A., Elices, M., Fracture of steel wires under lateral load (unpublished results).
85. Verpoest, I, Deruyttere, A., Aernoudt, E. Neyrinck M., The fatigue limit of steel wires as determined by the fatigue threshold and the surface characteristics, Proceedings of the 4th. ECF Conference (1982).
86. Yamane, S., Kasami, H., Okuno, T., Properties of concrete of very low temperatures, ACI-SP 55 pp. 207–221 (1978).

Subject index *

* In both Subject and Author Index, italicized numbers refer to chapters, roman numbers to pages within those chapters.

Author index